"*Anxious Creativity* provides an important new perspective on the broader debate over the status of creativity in American society. The disciplinary frames for this debate are extremely diverse, including academic perspectives from educational theory, sociology and cultural studies, as well as a range of important public policy debates. In addressing this complex array of issues and political concerns Trend deploys an impressive range of references, including empirical studies, scientific research, journalistic reports and a myriad of sources in contemporary critical theory. The result is a unique set of commentaries on the question of creativity that will be of interest to scholars in an equally broad range of disciplines, as well as general readers. There are many books that focus specifically on new forms of participatory culture or media, crowd sourcing, etc. but no one has yet drawn this material into dialogue with broader debates around creativity and the 'creative class,' theories of intersubjectivity, the arts and public policy. This book is destined to make an important contribution to policy debates over creativity and higher education."

—**Grant Kester**, University of California, San Diego, USA

"From the advent of paint-by-number artist's kits to the publishing phenomenon of the adult coloring book to alleviate stress, from Apple's 'Think Different' campaign to Google's DeepMind group, David Trend's *Anxious Creativity* maps the vast social domains where creativity is promised as a means to soothe the anxieties of Americans and solve the economic and political crises of capitalism's most recent twists and turns. Trend's encyclopedic knowledge of art, cultural, social and political theory, as well as of the ever-multiplying discourses surrounding art and creativity, make *Anxious Creativity* a necessary book for anyone working at the intersections of these fields. But perhaps even more important, for anyone who has ever picked up a creativity self-help book (*Big Magic,* anyone? *The Artist's Way* perhaps?) *Anxious Creativity* is a must read. Read this book and you'll never be able to think of creativity in quite the same way again. That makes *Anxious Creativity* a formidable act of creativity in itself."

—**Micki McGee**, Associate Professor, Sociology and American Studies, Fordham University, USA

"David Trend's *Anxious Creativity: When Imagination Fails* deals with several topics that are crucially important at the moment. He explores the varieties of creativity discourses and the forces that foster and impede them, making a significant contribution to the sociological literature on these topics. The book will be very attractive to scholars in critical sociology, social and cultural studies in education, cultural theory, media studies, psychology, and technology studies."

—**Kenneth Saltman**, University of Massachusetts, Dartmouth, USA; Author of *Scripted Bodies: Corporate Power, Smart Technologies, and the Undoing of Public Education*

Anxious Creativity

Creativity is getting new attention in today's America—along the way revealing fault lines in U.S. culture. Surveys show people overwhelmingly seeing creativity as both a desirable trait and a work enhancement, yet most say they just aren't creative. Like beauty and wealth, creativity seems universally desired but insufficiently possessed. Businesses likewise see innovation as essential to productivity and growth, but can't bring themselves to risk new ideas. Even as one's "inner artist" is hyped by a booming self-help industry, creative education dwindles in U.S. schools.

Anxious Creativity: When Imagination Fails examines this conceptual mess, while focusing on how America's current edginess dampens creativity in everyone. Written in an engaging and accessible style, *Anxious Creativity* draws on current ideas in the social sciences, economics, and the arts. Discussion centers on the knotty problem of reconciling the expressive potential in all people with the nation's tendency to reward only a few. Fortunately, there is some good news, as scientists, economists, and creative professionals have begun advocating new ways of sharing and collaboration. Building on these prospects, the book argues that America's innovation crisis demands a rethinking of individualism, competition, and the ways creativity is rewarded.

David Trend is Professor at the University of California, Irvine. His books include *Elsewhere in America: The Crisis of Belonging in Contemporary Culture* (2016), *Worlding* (2012), and *The End of Reading* (2010). Honored as a Getty Scholar, he is a former editor of the journals *Afterimage* and *Socialist Review*.

Critical Interventions

Politics, Culture, and the Promise of Democracy

Edited by Henry A. Giroux, Susan Searls Giroux, and Kenneth J. Saltman

The Great Inequality
Michael D. Yates

Elsewhere in America: The Crisis of Belonging in Contemporary Culture
David Trend

Scripted Bodies: Corporate Power, Smart Technology, and the Undoing of Public Education
Kenneth J. Saltman

Gender for the Warfare State: Literature of Women in Combat
Robin Truth Goodman

Disposable Americans: Extreme Capitalism and the Case for a Guaranteed Income
Paul Buchheit

The Public in Peril: Trump and the Menace of American Authoritarianism
Henry A. Giroux

Learning to Save the Future: Education After Economism and Digital Determinism
Alexander J. Means

Ecology and Revolution: Herbert Marcuse and the Challenge of a New World System Today
Charles Reitz

Anxious Creativity: When Imagination Fails
David Trend

Anxious Creativity

When Imagination Fails

David Trend

Routledge
Taylor & Francis Group
NEW YORK AND LONDON

First published 2020
by Routledge
52 Vanderbilt Avenue, New York, NY 10017

and by Routledge
2 Park Square, Milton Park, Abingdon, Oxon OX14 4RN

Routledge is an imprint of the Taylor & Francis Group, an informa business

Library of Congress Cataloging-in-Publication Data
A catalog record for this title has been requested

ISBN: 978-0-367-27506-8 (hbk)
ISBN: 978-0-367-27509-9 (pbk)
ISBN: 978-0-429-29643-7 (ebk)

Typeset in Bembo
by Taylor & Francis Books

Contents

Introduction

Imagination in Crisis

Most Americans will tell you they live in the world's most creative country. But when pushed, they'll admit they don't feel creative themselves. These findings come from the largest survey of creativity to date, conducted by the Adobe Corporation and backed by previous research.[1] It's no big surprise Americans perceive creative abundance, given the nation's expansive media and consumer culture. More shocking is the disconnect with people's actual lives. Many see a country brimming with expressive opportunity, much like the American Dream itself. But when reality sets in, they find themselves disregarded, uninspired, and unable to imagine a way out. Often, they turn on each other, their leaders, or even themselves—never realizing that larger factors may be in play.

You really can't blame Americans—immersed as they are in creative reminders, with everyone posting selfies, streaming movies, or shopping online. On one hand, the creative impulse is something everybody loves and wants to support. People see artistry as a public good and a road to personal growth, while increasingly recognizing the economic utility of the nation's "creative industries." Buoyed by slogans like "America First," they hold the *highest views of their nation's creativity*—ahead of countries like Britain, France, Germany, and Japan. Americans envision a land where anything is possible with enough energy and drive. But as most soon discover, not everyone wins in the creativity game. Among national populations, Americans report the *lowest views of personal creativity*, as an astonishing 84 percent say they can't reach their potentials. Most blame job pressures or ratcheting competition, with 72 percent saying workplace "risk aversion stifles creativity" and that "creativity is discouraged by the education system."[2]

Over a decade ago, innovation guru Richard Florida began warning of "America's Looming Creativity Crisis"—writing in the pages of the *Harvard Business Review* that "the land of opportunity and innovation—is on the verge of losing its competitive edge."[3] The problem didn't come from external coercion or any natural advantage of other countries, according to Florida. Instead, the U.S. was losing a far more important capacity. "America's growth miracle turns on one key factor: its openness to new ideas, which has allowed it to mobilize and harness the creative energies of its people." Not only was business losing its taste for novelty, but the country was closing its doors to talented newcomers. Within a few years, *Newsweek* joined the fray with a cover about "The Creativity Crisis."[4] This time the blame lay in too much media watching and failing schools. "All around us are matters of national and international importance calling out for creative solutions," *Newsweek* said. But rather than acquiring such problem-solving skills, "kids spend hours in front of the TV and playing videogames," with schools "overwhelmed by curriculum standards leaving no room in the day for a creativity class."[5]

Anxious Creativity: When Imagination Fails examines America's desperate hunger for inspiration, as millions fret about a nation in decline. Once a beacon of prosperity and freedom, the U.S. today seems lacking—while other nations seem to be gaining economic and moral high ground. Insecurity now looms as both cause and effect of a creeping neoliberalism, in which wealth is the measure of all things and failure a matter of personal shame. Many experts speak of failing ideas, both the loss of old ones and the lack of anything new, with a sense of apprehension now haunting American society. Of course, there's nothing wrong about a little anxiety to get a person going. What psychologists call "positive stress" can be a motivator in being productive or facing a challenge—sometimes giving a person just the right nudge. And philosophers long have argued that existential anxiety is what drives creativity.

But as most people know, *too much* anxiety can have the opposite effect—paralyzing the fretful creator with writer's block or stage fright. Something similar is happening in America today. News reports abound about rampant "economic anxiety," "racial anxiety," "election anxiety," and, of course, "Trump anxiety." This is setting aside the fact that one in five Americans suffer diagnosable anxiety disorders and the country spends over $2 billion on anti-anxiety drugs.[6] The *New York Times* has run a series of articles about "Anxious Americans" and "America's New Anxiety Disorder," with the *New York Post* dubbing Millennials the "Anxious Generation."[7] Many people have that nagging feeling that something might go wrong, often unable to name a clear reason or cause. The feeling blossoms when anticipation meets uncertainty.

In all of this, one often hears that with a bit more "creativity" things might turn around, whether this means novel thinking or more open attitudes. And who can argue with that? As both an "old" and "new" idea, creativity long has taken many faces—from imagination and the spirit of invention to enchantment and the genesis of life itself—as societies have projected creativity onto their deepest desires. But what happens when this magical substance becomes an object of public policy, corporate agenda, or consumer desire? This book argues that creativity is a mixed bag at best, now both symptom and cause of the anxiety coursing through America. To examine these matters, *Anxious Creativity* divides into four sections, examining individuals, groups, institutions, and societies through creativity's lens.

Part I, "Creative Subjects," looks at how people engage creativity and why they get frustrated. America's fierce individualism certainly doesn't help. Premised as the cornerstone of liberty itself, individualism has gotten out of hand in the new millennium. Creativity plays right into this through the myth of the "solitary artist"—an image so ingrained that seven in ten Americans today say they must be alone to find inspiration.[8] Cultural obsessions with Silicon Valley "creative entrepreneurs" as lone geniuses only underscore such beliefs. In more general terms, marketing also reinforces the "self" with laser-like appeals to personal likes and preferences. Advertising always has done this in promising to make "you" more appealing, interesting, or happy. But matters have intensified in a "Think Different" era when no one can be creative enough. This consumer atomization has occurred as more people live alone than ever, and distrust of collective entities has reached all-time highs—whether this means a corporation, news outlet, or government office.

Part II, "Creative Differences," asks whether all people can be creative and what can be done to nurture and support this—as is done with intelligence and other attributes. Much as one might think otherwise, creativity rarely flourishes without encouragement. Famous prodigies such as Mozart and Picasso were heavily tutored and drilled as youngsters, and both had accomplished artistic parents. And if creativity is unevenly distributed, what are the best ways of proceeding in a society that places a value on human difference and equity? Even in an internet age suggesting "We're All Artists Now," creative fields remain deeply divided along socio-economic lines.[9] Factors like age, ethnicity, and geography also play huge roles in this, not to mention inequalities in education itself. Along with myths of unaided natural ability, the stereotype of the "crazy artist" perpetuates further misunderstandings. It romanticizes mental illness as a prerequisite for prodigious accomplishment, while casting creative people as inherently unstable and willingly poor. This marginalizes artists while minimizing an often deadly disease.

Part III, "Creative Industries," begins by examining why companies are reluctant to try out new ideas and how anxiety impedes innovation. Experiment requires risk, after all—and this is the opposite of what shareholders want these days. As a result, customers see more of the same, or modest updates on what they already buy. This tendency is nowhere more evident than in "creative industries" like movies and TV, where risk-avoidant studios pump out endless sequels, remakes, or adaptations of already successful books or comics. Government and foundation funding for basic research has been dropping as well. At the same time, creative education in America's schools and universities has been decimated in recent decades, both though direct funding cuts and the privileging of science, engineering, and business. Even though 90 percent of families say they want art education in K–12 schools, such offerings are at their lowest levels in 50 years.[10] Earlier and earlier, anxieties start about a youngster's career prospects or college aspirations, as teachers themselves get evaluated by standardized metrics. This puts pressure on everyone involved to concentrate on what can be reduced to data and statistics—usually to the detriment of experiment and creativity.

Part IV, "Creative Societies," addresses the role of policy and government in furthering creativity. Even as the U.S. languishes in anxious creativity, so-called "creative economies" are blossoming throughout the globe in nations such as Australia, Canada, Denmark, Finland, New Zealand, Singapore, and Sweden. In one way or another, creativity serves as an inspiration for nationalist ambitions everywhere, whether in the form of "new" enterprises (media, culture, and tourism) or ever-expanding markets (enabled via globalization, networks, and emerging technologies). In all of this there is a glimmer of good news, however. Owing partly to national agendas, a quiet revolution is beginning to take form—upending longstanding beliefs about how innovation and creativity come about. Individualism is giving way to a new spirit of collaboration and teamwork. Competition is softening in a new atmosphere of shared research and mutual support. This revolution is not coming from political ideologues or philosophical discontents. Leading economists, scientists, and creative professionals are making the case based on simple pragmatics—as well as an eternal truism: that throughout the ages societies have moved forward when people work together.[11] While individuals always will have their own ideas and should be rewarded suitably, it takes collective effort to get meaningful things done.

Growing from creativity's anxious status, a series of questions will be explored throughout this book. Central among these is whether creativity has an inherent virtue, as commonly is assumed, or whether creative value is always a matter of perspective. This throws into relief the question of whether creativity is a singular quality, an aspect of social interaction, or a multivariate "complex adaptive system."[12] Another question asks whether creativity is a ubiquitous quality possessed by everyone, or is something distributed variously by different means and measures. In either case, the overriding role of neoliberal capitalism comes into play in promoting creativity as both rationale for downward mobility and a possible way out of it. This begs the question of whether it is possible to maintain creativity as a social good, while also embracing its disruptive (and sometimes destructive) aspects. Put in straightforward political and ethical terms, *Anxious Creativity* poses several recurring questions. How is society made, more or less, by virtue of how we conceptualize creativity and the creative process? What critical relationship does "creativity" have to the market or possessive individualism? How might creativity contribute to social change? What are the key impediments to a progressive orientation to creativity and what can be done to challenge them?

Answering these questions will mean rethinking some common assumptions. Most people see creativity as a quintessential expression of a person's "self." While neuroscientists now say that certain brain structures indeed do make creativity more likely in some people, comparable talent often goes undeveloped in many others. This has led to a view of creativity more as an "effect" of the surrounding culture than a substance magically springing from an individual. Recent research into human cognition also shows how much people rely on others for their ideas. Not just opinions or beliefs, but also such simple things as driving directions or remembering to buy groceries. As social creatures, people draw heavily upon "distributed intelligence" stored in each other's minds, not to mention their extensions in books, maps, and, especially, these online sources.[13] Keep in mind as well that artists and inventors emerge from histories of previous efforts and cultures in which others often worked on the same topics.

Obviously, these conclusions won't sit well with everyone. On many levels, anxious creativity links to contradictions deeply embedded in American culture: tensions between individual and community, competition and cooperation, and the relationships between the private and public sectors. As this book will explore, creativity debates now resonate in education, economics, and public policy, as well as such arcane matters as technology distribution and arts funding. The stakes get even higher if one considers creativity a key ingredient in scientific research and successful entrepreneurship. Particularly in an American context, anxious creativity is rooted in foundational conflicts over growth, progress, and change as they play out in the often volatile atmosphere of democratic capitalism. All of which is to say that the question of *Anxious Creativity* has never been more urgent—as both philosophical ideal and a point of practicality.

Simply put, *Anxious Creativity* argues that the innovation crisis derives from systemic fractures in American society—subject to all manner of amplification and manipulation—especially in these stressful times. This matters because creativity and anxiety are so directly tied to how people *imagine* the future, whether this means hoping for the best or expecting the worst. With some estimating that as many as 30 percent of Americans now work in creative industries or related fields, it certainly can be said that everyone consumes or practices some form of creative something every day.[14] So how did America get it so wrong? Ultimately, the answers lie in facing up to the nation's anxious demeanor, recognizing that help is needed, and being open to something different. Maybe something a little creative? Clearly, digging into business-as-usual is worsening matters. As a book, *Anxious Creativity* will walk readers through widely accepted explanations for the nation's innovation woes, while pointing out that solutions are not really that hard to find. They lie in resources Americans can find in each other.

Notes

1 *State of Create Study: Global Benchmark Study on Attitudes and Beliefs about Creativity at Work, School, and Home*, Adobe Corp. (2016) www.adobe.com/aboutadobe/pressroom/pdfs/Adobe_State_of_Create_Global_Benchmark_Study.pdf (accessed Jan. 26, 2019).

2 All statistics in this section from *State of Create.*
3 Richard Florida, "America's Looming Creativity Crisis," *Harvard Business Review* (Oct. 2004) https://hbr.org/2004/10/americas-looming-creativity-crisis (accessed Aug. 18, 2018).
4 Po Bronson and Ashley Merryman, "The Creativity Crisis," *Newsweek* (Jul. 10, 2010) www.newsweek.com/creativity-crisis-74665 (accessed Jan. 26, 2019).
5 K.H. Kim, "The Creativity Crisis: It's Getting Worse," *Idea to Value* (Apr. 2017) www.ideatovalue.com/crea/khkim/2017/04/creativity-crisis-getting-worse (accessed Jan. 26, 2019).
6 Nitsuh Abebe, "America's New 'Anxiety' Disorder," *New York Times* (Apr. 18, 2017) www.nytimes.com/2017/04/18/magazine/americas-new-anxiety-disorder.html?mcubz=0 (accessed Aug. 18, 2018).
7 T.M. Luhrmann, "The Anxious Americans," *New York Times* (Jul. 18, 2015) www.nytimes.com/2015/07/19/opinion/sunday/the-anxious-americans.html?mcubz=0 (accessed Aug. 18, 2018); Karol Markowitz, "They Can't Even: Why Millennials are the 'Anxious Generation,'" *New York Post* (Mar. 20, 2016) http://nypost.com/2016/03/20/they-cant-even-why-millennials-are-the-anxious-generation (accessed Feb. 18, 2019).
8 *State of Create Study: Global Benchmark Study on Attitudes and Beliefs about Creativity at Work, School, and Home*, Adobe Corp. (2012) www.adobe.com/aboutadobe/pressroom/pdfs/Adobe_State_of_Create_Global_Benchmark_Study.pdf (accessed Jan. 26, 2019).
9 Laura M. Holson, "We're All Artists Now," *New York Times* (Sep. 4, 2015) www.nytimes.com/2015/09/06/opinion/were-all-artists-now.html (accessed Jan. 1, 2019).
10 Randy I. Cohen, "American Public Says YES to Arts Education," Americans for the Arts (Mar. 5, 2016) http://blog.americansforthearts.org/2016/03/05/the-american-public-says-yes-to-arts-education (accessed Aug. 21, 2018).
11 John Hartley, Wen Wen, and Henry Siling Li, *Creative Economy and Culture: Challenges, Changes and Futures for the Creative Industries* (New York: Sage, 2015); Giuseppe Cocco and Barbara Szaniecki, eds., *Creative Capitalism, Multitudinous Creativity: Radicalities and Alterities* (New York: Lexington, 2015).
12 Murray Gell-Mann, "What is Complexity?" *Complexity* 1, no. 1 (1995) http://complexity.martinsewell.com/Gell95.pdf (accessed Aug. 7, 2018).
13 Steven Sloman and Philip Fernbach, *The Knowledge Illusion: Why We Never Think Alone* (New York: Riverhead, 2017).
14 Richard Florida, *The Rise of the Creative Class: And How It's Transforming Work, Leisure, Community, and Everyday Life* (New York: Basic Books, 2002) p. 74.

Part I

Creative Subjects

This section introduces *Anxious Creativity* by looking at how individuals perceive their own creativity (or lack thereof)—as well as ways advertising and self-help encourage consumers to "create" better versions of themselves. To many, creativity means making something from nothing—as in sagas of the "creation of the universe" or the birth of the species. With roots in Greek mythology, this *ex nihilo* ("out of nothing") principle deeply influenced the Western mind, especially as seen in Christian faith and American-style capitalism.[1] In cultures worldwide, creation stories provide explanations for civilization's most vexing questions: the origins and purpose of life, the meaning of individual existence, mysteries of the cosmos and the unknown. And, of course, in today's world the concept implies a host of enviable abilities. Saying someone exhibits a "creative personality" or finds "creative solutions" imbues the person with a knack for invention or helpfulness, but of a sort that can't quite be identified. The ineffability of creativity is part of what gives it its celebrated "magic."

But anxious times bring changes in temperament. In an America proud of its inventiveness and "can-do" spirit, more and more people worry they can't measure up. Creativity has joined qualities like beauty and fitness as things everybody wants but nobody has in sufficient measure. The resulting insecurity feeds broader anxieties, as economic worries make people more cautious in their thinking and less generous to others. Creativity suffers as companies spend less on research, people give less to arts institutions, government funding gets cut, and creative education dwindles in schools. The crisis isn't just in creative fields. Economists now speak of a sweeping innovation crisis in science and technology, as the U.S. shows signs of falling behind other nations. Heightened competition and social isolation only seem to make things worse.

The book begins by looking at an America plagued by writer's block, along with other anxieties over money, politics, and cultural controversies. Chapter 1, "Anxious Moments: Anticipation Meets Uncertainty," examines anxious creativity through the writings of Brené Brown, Martin Heidegger, Søren Kierkegaard, Jacques Lacan, Joseph LeDoux, and Rollo May. Among other questions, discussion examines how and why manageable anxiety can become a destructive force. While creativity can accompany mild forms of worry and sometimes alleviate stress, its advocates overstate their case in pushing it as a universal cure-all. In his canonical 1964 work *The Anxious Object*, art critic Harold Rosenberg wrote of the difficulties that result when definitions of art lose coherence and societies become confused about aesthetic meaning.[2] Might today's advocacy of the "creative industries" be doing the same thing? Drawing on recent studies of working artists, this chapter points out that not everyone in the "creative class" is faring well in today's ebullient embrace of artistry.

Worry over America's declining innovation is bringing creativity into the public spotlight as never before. Chapter 2, "Creative You: Self-Help to the Rescue," looks at how the resulting "crisis" talk (and its reality) makes creative qualities all the more desired, even as they grow more elusive and rare. In personal terms, most people feel creativity is missing in their lives—evidenced in a rising self-help industry catering to one's "inner artist" or forgotten childhood. Amazon.com currently lists over 57,000 books devoted to creativity, representing a 30 percent increase in the past year alone.[3] Analyzing this in her book *Self-Help, Inc.*, sociologist Micki McGee explained the growing demand for self-improvement as a symptom of widespread worry over money and jobs.[4] Such insecurities underlie the anxious "self" obsession infecting the U.S. today, much as Christopher Lasch described the malady decades ago in *The Culture of Narcissism.* [5] Further symptoms now appear in new evidence-based programs in wellness and arts therapy from entities such as the National Endowment for the Arts. These issues are examined through the thinking of Wendy Hui Kyong Chun, Hillary Davidson, Melanie Klein, Christian Smith, and Slavoj Žižek.

Donald Trump's infamous antics as "Performance Artist in Chief" open Chapter 3, "The Neoliberal Imagination: When More Is Not Enough." While initially startling, the President's slash-and-burn agenda of upending Washington soon was revealed as a corporatist scam. Economist Joseph Schumpeter coined the expression "Creative Destruction" in 1942 to describe the aggressive upending of liberal orthodoxies in favor of market-friendly agendas. Indeed, critics of neoliberalism now note the doctrine's frequent use of crisis and confusion to get its way, not unlike the tactics of fascist regimes. Schumpeter said that "the creative impulse incessantly revolutionizes the economic structure from within, incessantly destroying the old one, incessantly creating a new one," concluding that "the process of Creative Destruction is the essential fact about capitalism."[6] Drawing on thinking by Michel Foucault, David Harvey, Naomi Klein, and C. Wright Mills, this chapter links creative destruction to the heightened emphasis of the creative industries on privatization and individual competition, as well as the structural precarity the industries generate in workers' lives.

Notes

1 Alfred North Whitehead, *Adventures of Ideas* (New York: Free Press, 1933) p. 179.
2 Harold Rosenberg, *The Anxious Object: Art Today and Its Audience* (New York: Horizon, 1964).
3 Laura M. Holson, "We're All Artists Now," *New York Times* (Sept. 4, 2015) www.nytimes.com/2015/09/06/opinion/were-all-artists-now.html?_r=0 (accessed Mar. 26, 2019).
4 Micki McGee, *Self Help, Inc.: Makeover Culture in American Life* (New York: Oxford University Press, 2007).
5 Christopher Lasch, *The Culture of Narcissism: American Life in an Age of Diminishing Expectations* (New York: Norton, 1979).
6 Joseph A. Schumpeter, *Capitalism, Socialism, and Democracy* (London: Routledge, 1994) pp. 82–3.

Chapter 1

Anxious Moments

Anticipation Meets Uncertainty

Imagination may be one of humanity's greatest gifts. But the toll it takes is anxiety. Philosophers long have argued that the ability to see beyond the present is the engine of artistry, innovation, and even freedom itself. Yet this capacity for abstraction also is a window to what might go wrong, especially when the future seems uncertain. Anxiety is one of those feelings that people accept in the right doses, but know is toxic when out of hand—as when healthy caution devolves into paranoia. Stress can prompt an author to start typing, but also paralyze with writer's block. This chapter is about the widespread jitteriness now palpable in American culture, the damage it inflicts on creativity, and what can be done to fix things.

It's become cliché to speak of an "anxious America," overcome with worries about a sluggish economy, terrorist threats, and, in many people's minds, a generalized loss of hope. Nervous about the future, individuals retreat into worlds of familiarity and self-interest. Business seems to follow similar patterns, with short-term thinking now inducing a mood of risk-avoidance. None of this is good for innovation, as firms lean toward predictability and certainty. "Politicians like to say the U.S. is the most innovative country in the world. But our economy may be too risky for many entrepreneurs," a recent report stated.[1] Worse still, economists now say novelty is suffering on the "demand" side of the equation, as consumers seemingly prefer more of the same over anything new. Experts have been warning of a U.S. "innovation crisis" for over a decade—as other nations seem to be pulling ahead. American companies won't gamble on new ideas, government research is dwindling, and school arts programs continue to decline.

"Creativity" has become a new buzzword in this panic over innovation—manifest in policy summits, think-tank meetings, TED talks, and news accounts. At research universities, disciplines like science, engineering, and medicine are clamoring for creativity to spur fresh thinking. Following the success of Richard Florida's bestselling book *The Rise of the Creative Class*, so-called "creative industries" also have garnered attention as their own economic force, prompting a projection of creativity into many non-artistic fields—along with quasi-creative ones like entertainment and advertising.[2]

Advocates for the arts have joined a movement claiming that America's expanding "creative economy" accounts for over $800 billion and exceeds many conventional sectors. Along the way, the hype boosts a self-help industry asserting that one's inner creativity can calm a fretful mind.

Keep in mind that artists never have been especially well paid. Aside from a handful of gallery superstars, most artists can't make a living from their professional work. They piece together part-time jobs or compete for adjunct gigs at universities. The creative industries have perpetuated such fractional or temporary hiring, while promoting the benefits of "flexible" employment. And indeed, surveys show that many young people seem willing to sacrifice good pay and benefits for the personalized rewards of "meaningful" work. This takes a toll on creatives, and not just in monetary terms. One of the little-discussed consequences of poverty is stress—and the worries of late bills or looking for work. This makes artists vulnerable to clinical strains of anxiety and depression, which together affect one in four Americans.[3]

Is it possible to reconcile these anxious conditions with the promises of the creative economy? In many ways this question is as old as American capitalism itself—and the inherent tensions it generates. The challenge of the new creativity lies in finding answers without succumbing to extremism. Dualisms tend to generate oppositions, which easily fall prey to ideological suasion. Certainly no political party can own an idea as large as creativity—a premise attaching over time to comforting pleasures, radical disturbances, and everything in between. Trying times breed anxiety, suspicion, and, as recent history has shown, often new forms of contention and inequity. In such a moment it remains all the more important to remain wary of familiar-sounding solutions to complicated problems.

This chapter highlights the contradictions of anxious creativity. While America's economic worries seem to call for new ideas and better products, these impulses often are pushed aside by the certainties of tried-and-true formulas and goods. The much-publicized creative industries get promoted as an economic panacea, but they tend to see artistry only in commercial terms. Meanwhile, citizens are encouraged to "think creatively" or develop the resilience of artists, even as real-life working artists and other creatives remain poorly paid and often marginally employed. Rather than emphasizing the nurturing values that give artistry its emotional appeal, the "new" creativity seems more driven by individual competition and profit than humanistic impulse.

The Anxious Moment

Let's talk about the pervasive jitteriness in America—and how it affects creativity. Common wisdom holds that anxiety helps to motivate people and that artists in particular are driven (sometimes "tortured") by such negative feelings. The truth is that anxiety is a mixed bag, helpful in certain amounts but damaging when excessive. Contrary to popular stereotypes, studies of artists show they produce little when in the throes of clinical mood disorders such as anxiety and depression, often unable to conjure new ideas or do much of anything. Famously "mad" artists like Byron or Van Gogh only were productive when their conditions were under control. The same can be said about American business these days. Panicked by earnings worries, many companies are losing their abilities to innovate. They run to the safety of familiarity and predictability rather than investing in novelty and experiment. Clearly a sense of balance needs to return.

Despite a rising economy and the lowest crime rates in decades, polling shows most people believing matters are worsening. Gallup reports, "Pessimism has increased despite a strong stock market, rising consumer confidence, and a persistent low unemployment rate."[4] The same is true with crime, which has declined at a constant rate for 25 years, according to FBI statistics.[5] Episodically evident in electoral volatility, this jitteriness now pervades a population brimming with worry, frustration, and even anger. "Occasional anxiety is a normal part of life," says the National Institute of Mental Health (NIMH), as when one takes a school test or faces a challenge at work.[6] Defined as a "Feeling of worry, nervousness, or unease, typically about an imminent event or something with an uncertain outcome," anxiety can change how people and groups behave.[7] It can make them cautious, apprehensive, and mistrustful—especially when well-being might be at stake. This becomes a disorder when the anxiety persists or intensifies.[8] As the NIMH put it, "Anxiety *disorders* involve more than temporary worry or fear. For a person with an anxiety disorder, the anxiety does not go away and can get worse over time."[9]

Innovation once was America's national brand. Long credited with the country's global preeminence, the term "Yankee ingenuity" was coined shortly after independence, referring to colonial abilities to improvise with few resources. The same "can-do" spirit attached to later waves of immigration, as newcomers brought fresh ideas and gritty determination in pursuing the American Dream.[10] In the 19th and early 20th centuries, independent Edison-type inventor-entrepreneurs brought new energy to the industrial revolution, later spurred further by scientists and engineers working in federally sponsored labs. Herbert Hoover's Depression Era "Up-By-Your-Bootstraps" slogan typified the *ex nihilo* (out of nothing) approach to innovation later mythologized in Silicon Valley's geeky garages. But eventually things would change, as small companies grew into huge corporations—and shareholder returns began driving priorities. In an age of multinational giants, an emphasis on quality products slowly gave way to the abstraction of the "financialized" balance sheet. Finally, the recessionary climate of the 2000s made people reluctant to launch small businesses in the face of giants like Amazon and Walmart. Innovative risk-taking began to shrink at exactly the time consumers were getting more cautious with their spending.

All of this has made anxiety famous in the business world, where pressures to innovate affect everyone. Cutting-edge companies such as Apple and Google have introduced exercise and mindfulness to reduce worker stress and stimulate creativity. But among most CEOs and boards of directors, it seems worry is blocking invention. Recent financial press headlines tell the story, with the *Economist* asking, "Has the Ideas Machine Broken Down?" and the *Wall Street Journal* adding, "Is the Engine of Innovation in Danger of Stalling?" *Forbes Magazine* replied with "Why U.S. Firms Are Dying: Failure to Innovate," stating that the "failure" came from declining investment in new ideas.[11] *Forbes* cited reports showing that only 5 percent of workers felt motivated to innovate and less than 19 percent said they lacked resources to even try. The "why" came from shifting management priorities "that focused attention on stock prices and short-term performance" rather than basic research.[12] To economist Robert J. Gordon, this has led to a fixation on balance sheets over the quality of products or their benefits to consumers. In his book *The Rise and Fall of American Growth*, Gordon argued that "some inventions are more important than others," but the distinction gets lost when reduced to mere sales figures.[13] A final reason for the innovation decline lies in the nature of companies themselves. In "America's Innovation Crisis," Daniel Vinik wrote that "[f]ewer startups are opening their doors" as older firms increasingly dominate the landscape: "The U.S. economy is filled with aging behemoths—less creative destruction and more old stagnation."[14]

On one level there is nothing wrong with a little anxiety to spur one's imagination—as philosophers and artists have long asserted. The problem occurs when worry gets out of hand, and is allowed to paralyze constructive thinking. Clinical research on creative anxiety has yielded contradictory findings, with studies showing that inspiration can be driven by negative as well as positive emotions, and in turn artistic endeavors can induce similarly disparate feelings.[15] Sometimes creative worry has a rational basis deriving from competition or a known type of risk, giving anxious creativity an "adaptive" utility. But all of this vanishes when worry gets out of control. As with any kind of malady, identifying (or admitting) the problem is the first step toward getting better.

So, what worries Americans most these days? There are obvious and not-so-obvious answers, according to national polling. But as recent headlines have made clear, such surveys don't always give a clear picture of what people think or feel. According to recent Gallup surveys, the most pressing concerns fall into familiar categories—with 55 percent of Americans anxious about a sluggish economy and 53 percent fearful about crime and terrorism.[16] But in announcing these findings, analysts noted that public perception seemed strangely disconnected from empirical evidence. One explanation may lie in widening income disparities in the U.S., which have tended to shift the benefits of the economic recovery to the upper end of the income scale.[17] Worries about violent harm also seem persistent. Despite lower crime rates and the absence of foreign terrorism on U.S. soil since

2001, Americans report growing safety concerns and that "they are 'very' or 'somewhat' worried about personally becoming a victim" of an attack.[18]

This raises questions about what people believe and why. Louis Althusser wrote about public worry in theorizing ideology as an imaginary rendering of society.[19] To Althusser, people often distrust each other for unwarranted reasons, but such feelings are so deeply ensconced that many are unaware of their origins. Drawing on psychoanalysis, Althusser said that fears of loss and harm reside within everyone, deriving from childhood memory, lived experience, or sometimes intergenerational trauma. Because such anxieties are unconscious, they can be very difficult to grasp or unlearn. Moments of security or comfort remain fleeting in the presence of lingering anxieties, which can be activated for cause or simply by happenstance.

Building on this view, more recent media theory has warned about public disorientation due to information overload. Today's constant barrage of tweets and sounds bites undermines people's ability to make sense of the world, leaving many feeling overwhelmed, disoriented, or isolated from those around them. Along the way, perceptions of consistency and certainty inevitably give way to insecurity and anxiety.[20] Time seems in short supply and moving too quickly. Continually pulled into an ever-changing present moment, it becomes difficult to maintain historical perspective or coherently contemplate the future. Not helping matters is the ongoing erosion of face-to-face interaction in an age of virtual friendship and online shopping. Feelings of alienation and disempowerment increase as distinctions disappear between "public" and "private" worlds. Unsurprisingly, then, resentment grows among citizens who hate the very government they have put in place, even as they become equally wary of corporate enterprise.

Sorting out public anxiety is no simple matter in an age when facts and fiction are not easily discernible. Witness the famous 2016 American Presidential campaign, and the astonishing rise of Donald Trump and Bernie Sanders, both of whom were marginal figures in the years prior to the election. Though ideologically as opposite as night and day, Trump and Sanders found traction by appealing to the 70 percent of the U.S. population either "very angry" or "somewhat angry" about "the way things are going" in America.[21] At the time, Trump proclaimed that he was "very, very, angry" and "gladly accepting the mantle of anger," with Sanders chiming in, "I am angry and millions of Americans are angry."[22] These sentiments show no signs of diminishing, as they have resounded in elections ever since. Trying to make sense of this from outside the U.S., a *BBC News* story entitled "Why Are Americans so Angry?" said the U.S. system "seems to only be working for the insiders with money and power, like those on Wall Street or in Washington," while drawing parallels with a Britain torn apart by its infamous Brexit referendum.[23]

Rather than settling matters, the electoral process clearly intensifies citizen anger—dividing an anxious America against itself in ways unthinkable only a few years ago. Past and continuing money worries certainly play a big part in rising anti-immigrant sentiment, racial hatred, and a virulent rejection of government on any terms. Despite upbeat news from Washington, America's low- and middle-income populations simply haven't seen much from the rising economy. While rarely described as such, the recession of the 2000s had a redistributive effect—initially taking from everyone, but then giving back only the wealthy. Data-aggregating firm FiveThirtyEight reports that "wages may have rebounded from the recession but they have largely been flat for a decade when adjusted for inflation."[24] Meanwhile, college degrees are no longer the pathway to the middle class they once were. And the growing consensus among leading economists is that the U.S. has entered a prolonged period of slow growth. It is hardly surprising that many Americans are looking for something—anything— in these troubled times.

Might America's anxious mood present a creative opportunity? Some have argued that nothing brings out new ideas like a crisis. Pressured situations have a way of taking one out of the status quo—often involuntarily—and thrusting an unforeseen set of exigencies in one's way. "Necessity is the mother of invention," some will say. Unlike the anxious creativity that emerges from generalized worry or depression, emergency situations trigger rapid responses, sometimes even instinctual ones. Sociologists and philosophers have written about this phenomenon, given its pervasiveness in the twenty-first century. Semiologist Paulo Virno describes anxious creativity as an episodic event, rather than an ongoing process. One can't be in creative mode all the time, according to Virno. It takes a stimulus or need to set creativity into motion. And to Virno, such needs are nearly always a form of crisis. As he writes, "I propose a very limited, almost narrow, acceptation of 'creativity': the forms of verbal thought that allow for a change of one's behavior in an emergency situation."[25] In this sense, innovation doesn't simply emerge from a desire for novelty as an end in itself. Some purpose or problem always drives creativity, with "emergency" constituting the essence of the process. Notably, Virno also points out that emergencies often demand "new ideas" and actions, especially when the crisis comes from a failure of the status quo. What remains an open question is the extent to which individuals and groups can exercise reasoned judgment in emergency situations, especially as external interests apply pressure.

Industrial Creativity

What if the "new idea" America needed turned out to be novelty itself—rebranded as creativity? This not-so-original premise started percolating in U.S. business schools and think tanks in the early 2000s—as creativity began getting attention through books such as Richard E. Caves' *Creative Industries: Contracts Between Art and Commerce*, John Howkins' *The Creative Economy: How People Make Money from Ideas*, and, perhaps most famously, Florida's *The Rise of the Creative Class: And How It's Transforming Work, Leisure, Community, and*

Everyday Life.[26] Part of what made Florida's book a runaway bestseller was its assertion that the "creative class" had grown in recent decades to 30 percent of American workers—including those in such stereotypically creative fields as architecture and music, but also in jobs where creativity plays a role, like science and engineering. This paralleled another key transition and possible explanation for the creativity buzz—as old-school U.S. "industrial goods" (manufacturing and agricultural products) continued losing ground to so-called "service goods" (information and banking services, media and entertainment, education and training, travel and tourism).

The honeymoon for the creative class was short-lived, however. Within a few years, anxieties began rising over a "crisis of innovation" in U.S. corporate and academic sectors. America needed fresh thinking and new things to sell, financial analysts and economists announced, as they began documenting a slowdown of product development and decline in U.S. trade exports.[27] The nation's once-robust investments in information technology had leveled off by 2003, as web and e-commerce reached saturation point. In international terms, America's foreign competitors in Europe and Asia simply were coming up with better ideas—extending a pattern that had been decades in the making. By 2004, the federal government had fallen into a record-breaking $360 billion budget deficit, exacerbated by massive expenditures on homeland security and the Iraq war.[28]

Soon corporations and the government began cutting back on research funding, foreshadowing a pattern that would become all too familiar in the years to come—with financial declines driving a panic mentality that put new ideas on hold in the name of economic pragmatism. Quick profits became the order of the day. Given his visibility in the field of innovation, Florida was among the loudest voices protesting this move, writing in a 2004 issue of the *Harvard Business Review* of "America's Looming Creativity Crisis," as dozens of his colleagues expressed similar worries. Not only was the U.S. forgoing the enormous economic potential of its creative class, Florida argued, but America's foreign competitors—such as Ireland, Finland, Canada, Australia, and New Zealand—were doing just the opposite. Framing the matter as an epic battle, Florida wrote, "The United States may well have been the Goliath of the twentieth-century global economy, but it will take just half a dozen twenty-first-century Davids to begin to wear it down."[29]

Panic over innovation soon drove the finance industry to seek its own kind of novelty. Around 2005, hedge fund managers began gambling that a little "creative investment" in the sub-prime mortgage market could pay off with big returns. This set the stage for the financial collapse that would trigger the Great Recession of 2007–2009. As wages fell and people lost their jobs, all manner of anxious emotions arose in American society. With the hindsight of history, it hardly seems surprising that a relatively unknown Presidential candidate named Barack Obama would gain traction with campaign slogans of "Change We Can Believe In" and "Yes We Can."

Through a savvy use of online media and other grassroots organizing tools, Obama electrified younger voters with support from artists and musicians like Shepard Fairey and the Black Eyed Peas.

As the campaign was heating up in 2008, *MTV News* opined that Obama had reached "a voting bloc that other candidates thus far have not: the so-called 'creative class.'"[30] After the election, no less a publication than *Forbes Magazine* would credit Obama's victory to a radical reorientation in American society. In "The Triumph of the Creative Class," Joel Kotkin effused that "Obama's triumph portended a shift in the economic center of gravity away from military contractors, manufacturers, agribusiness, pharmaceuticals, suburban real-estate developers, energy companies, old-line remnants on Wall Street and other traditional backers of the GOP," elaborating, "In their place, we can see the rise of a different set of players, predominately drawn from the so-called creative class of Silicon Valley, Hollywood and the younger, go-go set in the financial world."[31]

Of course, not everybody was happy about this. Already the creative class had been identified with a decidedly progressive ideology. What other agendas might it have? Soon questions arose about the social and demographic impact of this well-educated and typically white urban-dwelling *class*. What mindset led these people to creative work? Was elitist privilege part of the package? And would creative industries alter local economies for better or worse? Early on, conservatives cast the new creatives as inherently arrogant and often self-indulgent types, part and parcel of the notorious "left-wing media," as well as the rising "anti-family" LGBT movement. But suspicions weren't limited to conservatives.

Before long, leftist scholars would react to the business-friendly creative class, labeling it yet another fabrication of a weary capitalist machine. These critics said that free enterprise ultimately stumbles over itself, requiring ever more clever ways of selling things and manipulating labor. The creative industries exemplified such impulses—aided and abetted by a negatively defined neoliberalism.[32] The work of sociologists Luc Boltanski and Ève Chiapello often is cited for this view. In *The New Spirit of Capitalism*, Boltanski and Chiapello asserted a certain "creativity" within capitalism, which allowed it to co-opt oppositional efforts and otherwise "discover routes to its survival in critiques of it."[33] This occurs partially because artistic critiques and capitalism share individualist "ideals of liberation and/or of individual autonomy, singularity and authenticity."[34]

The creative class started faltering as the recession wore on—and things still haven't gotten much better since then. Contrary to upbeat portrayals of well-heeled software developers and movie producers, many creatives now work at low-paid, part-time jobs without benefits—often living in marginal bohemian districts. Common depictions of the creative industries describe this underemployment and geographic "clustering" as mutually beneficial to all concerned. To be fair, this claim is borne out in research showing that today's Millennials often willingly sacrifice conventional career benefits (good pay and job security) in exchange for less tangible rewards (freedom and a sense of purpose), with as many as 65 percent reporting satisfaction with part-time jobs.[35] Complementing this from the corporate side, companies also often say they prefer a flexible workforce conveniently living in "creative clusters" nearby.

Throughout the American economy, fractional employment has become commonplace in post-recessionary times. The Creativity Group (TCG), with 31 offices across the U.S., bills itself as the leading staffing consultant for the creative industries. "Why hire full-time employees for tasks?" TCG quips in a blurb headlined "Why Flexible Staffing Strategy Is Crucial for Business Success." As TCG further explains, "Rigid staffing structures are expensive and inefficient. They don't allow firms to move quickly and strategically. In order for businesses to have the right people at the right place at the right time, they need flexibility."[36] While "flexibility" may drive the staffing services of TCG, it also generates anxiety in the lives of workers, not all of whom can survive on idealism alone. This hits some cohorts of workers harder than others.

In part, this is simply a recessionary matter. Tight money makes companies anxious about the bottom line, rationalizing temporary or part-time employment as a humane form of pragmatism. But make no mistake about it, the new flexibility privileges balance sheets over anything else, and this mindset now prevails throughout the economy. It takes an extra toll on artists and others in creative fields, where work always has been scarce and undervalued. Seen as "naturally" inventive and resourceful, artists are stereotyped as not needing (or wanting) the accommodations of "real" jobs.

This parsimonious employment market demands flexible "creativity" from all jobseekers, whether inside or outside the creative industries, per se. Working online, from home, or when opportunities arise—isolates the individual worker. Even as many firms tout the benefits of teamwork and collaboration, it's nearly impossible for employees who rarely see each other to collectively organize, much less bargain for a better deal. While firms like TCG may paint flexibility as a fair-minded way of dealing with business challenges, it clearly shifts the economic anxiety burden from bosses to workers. This new regime of fractional employment has become so naturalized that many workers, especially younger ones, simply accept its premises as givens, blaming their difficulties and failures on themselves rather than structural conditions. In such an atmosphere, apprehensions of bad outcomes become the rule rather than the exception.

Starving Artists

In an earlier moment of rampant public worry, art critic Harold Rosenberg issued his 1964 *The Anxious Object*. [37] While heavily laden with Cold War jargon about international tension and looming conflict, the book framed anxiety in aesthetic terms. Even as the U.S. was competing for global preeminence, Rosenberg saw a sinister danger coming from within. Like the Red Menace, threats were growing over the definition of art itself. Emerging strains of conceptualism made viewers "anxious" because they couldn't understand the new art, Rosenberg contended, even as the art marketplace rewarded fashionable styles and big names. All of this worked against the inherent freedom that certain American artists embodied. While some audiences were conditioned to accept anything "new" as a matter of faith, others reacted with scorn (as did some reviewers). To Rosenberg, these reactions were a sad commentary on people's expectations, a fundamental misunderstanding of the generative role of art in society. After all, "The nature of creativity is that it contains the unexpected," he wrote.[38]

Rosenberg wanted artwork to be clearly defined. While seeing freedom in abstract expressionism and "action painting" (a term he coined), the critic expressed anxiety over new strains of installation, performance, and other "impermanent art" emerging in the 1960s. And Rosenberg virulently condemned Andy Warhol's pop art, which suggested anyone could be an artist. "A profound crisis has overtaken the arts in our epoch," Rosenberg wrote. It was no longer possible to know with any certainty what was and what was not a legitimate artwork. "Today art exists, but it lacks a reason for existing," he lamented.[39] Painting should be a liberating act, Rosenberg asserted—an existential struggle against conventional "Value—political, aesthetic, or moral."[40] Though inspiring to many readers, Rosenberg helped perpetuate a view of artists as combative outsiders whose hardships went along with their temperaments.

Definitions of art are even murkier in today's anxious times, but indomitability remains an artistic stereotype. Imbued with adaptability and imagination, the eternally resilient creator is said to find beauty even in the most squalid conditions. Then add clichés about "starving artists," stereotypes framing poverty as a preferred way of life. Such antiquated views conform to broader prejudices toward low-income people, who purportedly put themselves in misery as a matter of choice or temperament. These myths date to well-meaning reform movements in the twentieth century, which sought to replace determinist beliefs of destitution passed from generation to generation with "culture of poverty" theories in which certain groups (such as welfare recipients and artists) remain poor voluntarily. These views continue to reinforce beliefs that artists work best when struggling to pay rent or buy food—or, worse still, that poverty is a prerequisite for creative inspiration.

Always the exemplars of the creative class, "working artists," in fact, reside near the very bottom of the wage scale—typically employed on an occasional basis as freelancers, adjuncts, or project-based workers. In the largest study to date, the Strategic National Arts Alumni Project (SNAAP) reported that over 50 percent of artists hold two or more jobs concurrently, with six in ten defining themselves as self-employed and earnings running 15 percent below comparable fields.[41] Other research paints an even darker picture. "Keeping Your Day Job: Identifying Artists Who Have Dual

Careers" from the National Endowment for the Arts (NEA) gave dismal statistics on those who identify as "professional" artists while needing other kinds of work to pay bills.[42] According to the NEA, less than one in 100 Americans earn enough to define themselves as "working" artists—a number that also translates to less than 10 percent of those who received BA in Art or BFA degrees.[43] Artists with such degrees in New York City make a median income of $25,000, which is less than half the national average.

The creative industries put the artistically minded in a double bind—valorized in the new economy and undercompensated at the same time. Meanwhile, these contradictions are internalized by creative workers themselves, as the SNAAP and NEA studies indicate a 50-50 divide between creatives expressing satisfaction with the intrinsic rewards of artistic work and those saying they can't bring themselves to commit to such low-paying careers.[44] All of this is occurring as definitions of the creative class keeps broadening into fields traditionally considered non-artistic, in effect diffusing artistry into broader categories of labor, which muddies any analysis. Add to this research showing that "meaningful" creative employment also makes people work harder, longer, and with more passion—making such jobs all-consuming and potentially unhealthy. Social psychologist Robert Vallerand is among a growing number of scholars urging creatives to ask themselves, "Are you able to stop working when you want to or do you feel driven to do just more thing?"[45] In other words, "meaningful" jobs sometimes create a certain workaholism. And when this happens, other parts of one's life can begin to suffer.

Keep in mind that these labor transformations also are occurring within a broader context of economic stratification. While artists may thrive or struggle in the new creative economy, monstrous amounts of money are being exchanged at Sotheby's and Christie's, where paintings have fetched prices above the $100-million mark for decades. While most visual artists scrape by on miniscule earnings, art speculation and profit-making only seem to accelerate in the commercial realm. According the *Art Market Monitor*, the increasing interaction of Asian and Western markets has resulted in a 212 percent upsurge in sales during the past decade: "Considering global economic and financial conditions, the Fine Art market has demonstrated its maturity as a genuine investment alternative investment channel, with Western art markets generating $11.2 billion" in the last year alone.[46]

In many ways, the income stratification in the art market parallels economic disparities long seen in the entertainment industry. Living in Los Angeles for the past two decades, I have witnessed job-related creative anxiety first-hand, as well as the toll it can take. When I arrived in the 1990s, the Hollywood studios were transitioning from independent management into subsidiary companies owned by larger multinationals, such as Columbia (owned by Sony), Universal (owned by Viacom), and Warner Brothers (owned by Time Warner). Gone were the days of risk-taking moguls like Samuel Goldin and Louis B. Mayer—as visionary entrepreneurship gave way to an era of shareholder reports and mandatory profit-making. What little permanent full-time work there was disappeared for all but a small number of executives and staff. In the new era, each film or television show became its own mini-company, with staffing beginning and ending with the duration of the project.

Half of the people I know in LA have some sort of movie or TV job, with many coming here as aspiring actors, and then finding work among thousands of "below-the-line" personnel. My neighbors and friends include cinematographers, make-up artists, lighting engineers, dialect coaches, and screenwriters. None of them have ongoing work—not even the highly successful ones. And all of these creatives are relentlessly anxious—fretting constantly about getting the next job, where it will be, and how much they may rewarded or compromised. But even more poignantly, they fear the consequences of making mistakes, offending someone higher up, or otherwise damaging their reputations in a business run on word-of-mouth. Then add the worries for anyone over 30 about being seen as "old" in an industry famous for its youth obsession. Women feel this pressure more than men, of course, in a field also notorious for racial biases and inequities. But everyone feels the pressure of energetic newcomers and film-school graduates perennially glutting the job market with competition.

Nervous Disorders

Rarely does one hear about this psychological downside in rhetoric of America's growing creative economy. Zealotry has little patience for second opinions, and the upbeat boosterism of the new creativity is no exception. Part of this has to do with the way creativity's inherent *ex nihilo* premise maps onto American values of optimism, growth, and success—purportedly available in equal measure to every citizen. But as any artist will tell you, the choice of pursing a creative career is hardly a revenue-neutral decision in a mainstream America which—despite the upbeat rhetoric of the creative industries—still sees aesthetic work as frivolous, non-utilitarian, or simply too much fun. Creative people often find themselves on the economic margins, even before other kinds of bias and prejudice might come into play. Within this climate of disadvantage, the new creative capitalism adds a heavy dose of competition as yet another component of individualized anxiety. "In today's economy, creativity and competitiveness go hand in hand," Florida writes, for example, asserting the necessity of horserace mentality as a driver for novelty and innovation. Stressing this point, Florida joins others in asserting that America risks falling behind if it doesn't pick up its competitive pace.[47] Such thinking is deeply ingrained in American-style social Darwinism, which for two centuries has rationalized one person's gain as another person's loss in the dog-eat-dog world of business. In some accounts, it is precisely a worry over potential decline that underlies much of the ideology of competition, not to mention its free-market manifestations.

It hardly needs mentioning that competitive reminders are everywhere in today's America, where success and achievement are scrutinized in social contexts and reinforced by a media culture in which no one can ever have enough wealth, beauty, or fitness. Educators have become especially concerned with the effect of competition on kids, especially younger ones, in school environments that subject students and teachers alike to increasing measurement, comparison, and "accountability" on all levels. In his book *No Contest: The Case Against Competition (How We Lose in Our Race to Win)*, Alfie Kohn wrote of the way adults project their own fears of scarcity onto children, both through parenting styles and in the pedagogies that dominate K–12 education.[48] A hidden message of inadequacy often lies behind imperatives for personal growth and achievement. Fearful of being surpassed by others in the quest for approval, grades, college admissions, or a career later in life—many students become guarded in their studies, less willing to collaborate, and even hostile toward others. Regrettably, this is often reinforced in athletics and other extracurricular activities.

As adulthood approaches, the realities of anticipated scarcity become impossible to avoid. Most college-age young people growing up in the 2000s have memories of parents fretting over money, adding yet another dimension to their anxious outlook. One-third of these kids still live at home with their families—the highest number since the Great Depression.[49] And if you think about it, this also is the first generation coming of age knowing little of life *before* 9/11 and the War on Terror. In this context, is it any wonder that research shows Millennials with exceptional levels of worry about the future? Today's college-age young people also carry the highest level of student debt in history: more than $1.2 trillion in outstanding loans among 40 million borrowers, with an average balance due of $30,000.[50] Worse still, the steepest debt is accruing among the poorest students, thus adding yet another dimension to the growing wealth gap in the U.S.

"You wind up disadvantaged just as you begin," remarked social welfare professor Melinda Lewis, grimly adding that higher education no longer functions as "a force for upward mobility, or for an equitable chance at attaining it."[51] All of this takes a toll on student well-being and outlook on life. The American Psychological Association (APA) started sounding the alarm bells in 2011, with a report entitled "The Crisis on Campus," announcing a 16 percent spike in anxiety and depression among America's 15- to 25-year-olds.[52] The study also cited findings from the American College Health Association revealing that 46.5 percent of students reported feeling hopeless, so much so that 30.7 percent said they were having difficulties functioning in school. The same year, the National Survey of Counseling Center Directors announced rising levels of anxiety- and depression-related problems such as eating disorders (24.3%), substance abuse (45.7%), and self-injury (39.4%).[53] Five years later, the APA was ready to go to Congress in what it saw as a "concerning trend"—as 48.7 percent of college students sought mental-health-related counseling on their campuses.[54] Nowadays this is taking place as college counseling services are being reduced in a continuing tight economy.

This raises the question of mental illness, and whether America's jittery culture is having more serious health effects. Most clinicians are quick to draw distinctions between normal worries and more severe or persistent strains of anxiety constituting a psychiatric condition. That said, nearly one in five adults in the U.S.—18 percent—now qualifies for some form of a clinical anxiety disorder, with the World Mental Health Survey rating America as the most anxious among countries studied.[55] With 50 million people affected by such conditions, anxiety is second only to depression among mental health diagnoses in the U.S.—with Americans spending over $2 billion each year on anti-anxiety medications.[56] Keep in mind that worry is only partly driven by current circumstance. For as many as 30 percent of people, a propensity for anxiety lies within the mind.[57] Some anthropologists attribute this to animalistic survival mechanisms, although psychologists are quick to point out that most people can regulate instincts and keep them in check.

Some of the newest research on anxiety suggests looking backward—to critically examine the sources of one's fears. In his book *Anxious: Using the Brain to Treat Fear and Anxiety*, neuroscientist Joseph LeDoux says that anxiety is neither a hardwired instinct nor an existential given, as commonly thought.[58] His big insight was that unconscious "feelings" are separate from what people recognize as "emotions." Most fears are generated without one's awareness in one part of the brain (the amygdala emotion center). But it's how these feelings affect one's consciousness center (the neocortex region) that really matters. One can let a jittery feeling color everything one sees. Or a person can learn to recognize "fear triggers" and find ways of managing them. Going beyond the mere revisitation of fear-inducing events via "exposure therapy," LeDoux's work already is influencing treatment for trauma and post-traumatic stress disorder—suggesting both cognitive and behavior remedies. While this new research has not yet been applied to art-making, LeDoux believes that it helps explain how people perceive creative works and derive emotional responses from them.

Certain cohorts of the American population are more at risk for anxiety than others. People who have lost their jobs are obvious candidates, as are emergency responders and military personnel. In one startling statistic, the U.S. armed forces recently reported a 327 percent increase of anxiety disorders among those serving in the past decade.[59] According to *Psychology Today*, people who feel isolated or lack social support networks also are more likely to feel anxious. So are those who fall outside what it terms the "normalcy bias" in the U.S. This includes people who feel "different" for any variety of reasons, as well as the growing numbers of citizens who see themselves as not achieving normative standards of success. As L. Kevin Chapman puts it:

> The common thread among those of us who succumb to the normalcy bias is attempting to live above our means, trying to keep up with others to *appear* more successful, believing that money is the key to happiness, and that I will be accepted if I have more *things*. [60]

It is hardly surprising that anxiety has been called "the disease of the twenty-first century," although many Americans are loath to admit it. For a nation that prides itself on its optimism and the pursuit of happiness, admissions of inadequacy or failure are rarely talked about. The country's prideful attitude itself might be part of the problem, in that Americans tend to harbor a certain level of denial about mental health conditions. A relentless emphasis on high achievement and perfect health sends the message that anything less is unacceptable and unmentionable. This drives problems like anxiety underground, where they can linger and get worse. Aside from avoiding treatment, those suffering from anxiety (or other mental health problems) frequently try hiding the conditions or healing themselves. In a culture that valorizes individuality and self-reliance, this can foster internalized shame and self-blame. As blogger Joe Brewster put it, "I've been speaking with my friends in private about a dirty little secret no one is supposed to talk about. The shame people feel when they can't get a job."[61]And shame can be insidious. Unlike simple regret or guilt—which generally attach to a specific acts or events—shame affects one's personal assessment of one's core being. Guilt may arise from the bad thing a person has done, but shame is about the bad person one really is.[62] Anxiety works much the same way, as the edgy person anticipates the jitters and then feels badly for having them.

America seems to be of two minds about the role of creativity in all of this—linking it with the torment of artistic "mad genius," while expecting creative comfort from its own "inner artists." Negative portrayals of creative temperament seem to dissipate as people consider any artistry of their own, with creativity seen as a generative or reparative agent. Some of this has come from the rise of self-help therapy and mindfulness in popular culture. One of the most visible figures in this field is Dr. Brené Brown, who says that distress and anxiety are creativity killers, especially when attached to self-blame. Worry about one's inadequacies can lower any tolerance for risk and vulnerability, which make up "the birthplace of innovation, creativity, and change." Brown asserts.[63]

Other self-helpers make even stronger claims. As psychotherapists Carrie Barron and Alton Barron suggest in their popular book *The Creativity Cure*, feelings of "worry, fatigue, irritability, tension, and dread" can be fixed by a "combination of meaningful creativity (especially using your hands) and a deepened understanding of yourself."[64] The Barrons aren't especially picky about what works best, recommending everything from tulip gardening and grilling tuna to writing poems or playing a musical instrument. What matters is finding something that is self-affirming, pulls one into the present moment, and drives away those apprehensive impulses. Publishing in this area has been astonishingly robust, seen in books like *Art Heals, Awakening to Inner Creativity, Color Me Calm, Coloring for Tranquility, Crafting Calm, Creative Healing, Creative Living Beyond Fear, Mindful Creativity, Peaceful Coloring*, and *Stop Worrying, Start Writing.* [65] The issue isn't so much that these books exist or even what they say. After all, a little tulip gardening can't hurt, and mindfulness is certainly in short supply these days. But the recent flurry of such books underscores the epidemic of worry and apprehension now pervading American culture. More sobering still is the message of self-treatment via disconnection from anticipatory thinking, which also implies a foreclosure of any *different future* or alternative imagination.

Philosophically on Edge

But wait a minute. Is everyone whining too much about anxiety and a search for "inner" peace? Haven't philosophers been talking about this stuff for ages? The answer is yes and no. Certainly, everyone worries at some point about how they are doing or the way others see them. In Western philosophy, Plato, René Descartes, Georg W.F. Hegel, and Karl Marx would speak of the individual's "estrangement" or "alienation" from what surrounds it.[66] Eventually, this would be described as a dualistic relationship between self and other (or subject and object) in what many say is the central problem of Enlightenment thinking. Hegel spoke extensively about the inherent anxiety in this, arguing that the resultant "dialectical" tension drove learning itself through the experience of difference.

Søren Kierkegaard saw anxiety as an existential by-product of freedom. In arguably the first book ever to address the topic directly, Kierkegaard saw anxiety emerging as one grapples with the boundless possibilities of one's own existence. In his 1844 work, *The Concept of Anxiety*, Kierkegaard wrote that "anxiety is the dizziness of freedom, which emerges when the spirit wants to posit the synthesis and freedom looks down into its own possibility."[67] Or put another way, without possibility there would be no anxiety. Not unlike Hegel, Kierkegaard also offered a caution in his analysis, noting that "whoever is educated by possibility is exposed to danger" via a misunderstanding of anxiety that "does not lead to faith but away from faith."[68] While some of this undoubtedly came from Kierkegaard's religious orientation, his open-ended view of anxiety's wildness is worthy of some thought. As anyone who has known anxiety can certainly attest, the emotion has the ability to either mobilize or paralyze depending on its severity.

Questions of existence and the self would continue to occupy philosophers for the next century, especially in the context of a changing and chaotic Western world. To Martin Heidegger, the very fact of "being-in-the-world" (what he termed *Dasein*) casts the self into a kind of worrisome confusion. Writing in his 1927 book *Being and Time*, Heidegger highlighted the anticipatory character of anxiety in relationship to an unknown future in which death lingers in the background. The unspecified character of anxiety distinguishes it from fear, which nearly always has a clearly defined object. Even worse, these vague worries about an unknown future can color one's experience of the present moment. As Heidegger put it, "Anxiety takes away from Dasein the possibility of understanding itself," by throwing the self "back upon that which it is anxious about—its authentic potentiality-for-Being-in-the-World."[69] And what does the self then do with this anxious potentiality? To Heidegger the only choice is to push back against the world—to make things, invent technologies, or otherwise devise methods to impede nature and forestall human mortality: in other words, to be creative.

As the Cold War was breaking out in the 1950s, psychology began taking creativity seriously, with J.P. Guilford, President of the American Psychological Association, calling for research on the topic. This launched a wave of studies and theories on the traits of creative personalities, their cognitive capacities, and neurological make-up—largely in the interest of keeping a worried nation ahead of its enemies and economic competitors. During this fertile period, psychologist Rollo May added yet another chapter to the anxious creativity story. Responding to Kierkegaard's earlier work, May wrote a doctoral dissertation titled *The Meaning of Anxiety*, which further explored the existential quandary of freedom, positing a response in the form of inventive thought and action. In his analysis, May argued that anxiety and creativity are always tied together: "Because it is possible to create—creating one's self, willing to be oneself, as well as creating in all the innumerable daily activities (and these are two phases of the same process)—one has anxiety."[70] In detailing his ideas, May dwelled on matters of temporality and change, which would support popular views linking creativity to progress. He wrote that "creativity always involves destroying the status quo, destroying old patterns within oneself, progressively destroying what one has clung to from childhood on, creating new and original ways of living." And bringing the role of anxiety back into focus, May would add, "to put the matter figuratively, in every experience of creativity something in the past is killed that something new in the present may be born."[71]

Psychoanalyst Jacques Lacan also had spoken of anxious creativity in relationship to existential doubt. Pondering the feelings of emptiness people often experience, Lacan was able to reference both the uncertainty of being-in-the-world and the struggle of expressing this feeling. Lacan was quite specific about the role of creativity in this. Whether using artistic materials, dramatic gestures, or symbolic imagery, he would write that "man is the artisan of his support system."[72] Put in linguistic terms, Lacan located this search for support or insight in the gap between thought and object, referent and sign, signified and signifier—suggesting that within the gap lies a yearning that animates most artistic endeavors.

Lacan described artistic products and insights emerging "out of nothing." In his essay "Creation *Ex Nihilo*," Lacan wrote of the way ceramicists fashion vases around voids and how architects design structures to enclose spaces, concluding that out of emptiness "emerges a work of art."[73] Over and over in philosophy and psychoanalytic theory, these themes of absence and yearning return, as they also do in recent strains of postmodern discourse. While one has to be careful about dragging ideas from the past into the current moment, a quasi-consensus has been circulating in millennial critical theory of a crisis in consciousness brought about by a series of disappointments—a distrust of leadership and authority, a faltering of economy, the threat of terrorism, a vacuous consumerism, a loss of future hope—all of which have contributed to a generalized cultural malaise and mood of nihilism. America and other nations have gone through periods like this before, often coming to an end by some distraction, cataclysm, period of innovation, or cyclical improvement. The question is what can be done to improve the current climate *now*. Is there anything a single person can do? In different ways, each of the philosophers discussed above would answer "yes," but only if a positive and humanistic purpose motivates the creative response to anxiety.

Think Different

What exactly is gained and lost in America's embrace of its "inner artist"? And what does this say about the famous eccentricities of the painter or composer? These questions have a history paralleling those about artistic poverty. The mythic troubled artist emerged in part from American ideals of normalcy—and common suspicions about those considered different or unusual. The nation's normalcy fixation dates to nineteenth-century anxieties about public well-being and proper health. Amid fears of epidemic and disease, anyone appearing to be ill or behaving oddly became an object of distrust. As time progressed, human beings would be categorized as either "normal" or "deviant," with such distinctions becoming more rigid with America's drive to social conformity. In this mindset, the image of the "Average American" excluded racial minorities, new immigrants, the disabled, homosexuals, and a raft of eccentrics including artists and anyone with a behavioral difference. Also informing the crazy-artist stereotype were beliefs linking creativity to psychic torment. In this line of thinking, a disturbed mind was what drove artists to novel or unusual ideas, as popularized in accounts of figures like Salvador Dalí, Paul Gaugin, and Vincent Van Gogh. Even today, the self-destructive proclivities of creative types often are naturalized as the inevitable trade-off for a successful career.

Opinions vary on the demographics of creativity, owing to the contradictory ways people see the topic. Is creativity something everyone has in varying amounts? If so, how does one account for these differences? And if there is such a thing as genius, where does it come from? Most experts in the area say that creativity is something akin to intelligence, as a trait that one inherits to varying degrees, which can be cultivated with encouragement, educational opportunity, and other forms of support. Still, some children and adults seem to possess uncanny aptitudes that are hard to explain, especially in families with no hint of such abilities in prior generations. Complicating matters further are the enormously varied forms creativity takes. Technical virtuosity with the piano or paintbrush? A wild imagination? The ability to conjure new ideas?

The new creative economy provides a partial resolution to these definitional problems by translating creative value into the metrics of deliverables. In this context, it is no coincidence that artists and educators are deeply divided over the effects of creative industries—on one hand pleased that recent enthusiasm gives credibility to arts education and careers, but also worried that the move reduces creativity to its most instrumental aspects. Put another way, the new creativity simultaneously broadens the artistic enterprise in practical terms, while narrowing it philosophically. In some ways, this parallels changes in American society at a time when "diversity" has become a rising orthodoxy, even as the ideologies of individualism, especially at the consumer level, are more aggressively promoted. In both instances, an incremental objectification of personhood is taking place, manifest, for example, in "self"-oriented advertising slogans such as "Think Different" (Apple), "I Am What I Am," (Reebok), and the recently updated "Be Your Way" (Burger King).

Despite its narrowed goals, the new creativity also embraces cultural diversity on the employment side. For nearly a decade, leading technology companies such as Microsoft, Oracle, Infosys, and Intel actively have been looking outside the U.S. for skilled engineers, programmers, and financial personnel. Meanwhile, competition has continued to grow within other sectors, notably the creative industries. A recent study by the Conference Board (CPB) entitled "Ready to Innovate" found huge numbers of American corporations jumping on the creativity bandwagon, with over 90 percent reporting that creativity — and creative employees — are especially needed for their business, Unfortunately, CPB says that "85 percent of employers looking to hire creative people say they are unable to find the applicants they seek."[74] Hence, anxiety over finding creative talent has become epidemic in American corporations. The type of creativity these companies are seeking is not what one might expect, with procedural intelligence seeming to trump artistic sensibilities. CPB says that skills like "problem-solving" and "identifying patterns" landed people jobs twice as often as traits such as "risk-taking" and "communicating new ideas." [75] Once again, this is bad news for innovation.

"Cognitive flexibility" is the way the *Wall Street Journal* describes this preferred way of thinking, as outlined in a recent feature article, "The Secret of Immigrant Genius."[76] This is something employers believe newcomers to the U.S. have in abundance, according to the *WJS*'s Eric Weiner. It's not so much inventiveness that drives immigrant creativity, but rather the flexibility of mind that comes from adapting to a new place. To its credit, the *WSJ* doesn't jump on old stereotypes of the struggling American newcomer as an explanation. Instead, the answer comes from new research in social psychology, advancing the importance of so-called "schema violations" in cognitive development: "A schema violation occurs when our world is turned upside-down, when temporal cues are off-kilter."[77] Hence, it isn't merely determination or "grit" that makes the difference, as frequently assumed, but "the ability to see the world from a different perspective."[78]

This raises an interesting question about the relationship between innovation and diversity. I work at the University of California at Irvine (UCI), where the term "Inclusive Excellence" recently has entered the employment lexicon. As the term implies, the premise is that "excellence" (high quality or distinction) and "inclusion" (difference or multiplicity) are two sides of the same coin, rather than distinct and separate matters. As UCI puts the matter, "Our mission is simple and audacious: to establish UCI as a national leader and global model of inclusive excellence. This mission emphatically recognizes that excellence through diversity defines our campus community today and tomorrow."[79] Such high-minded idealism is not shared uniformly, however. Implementation of the Inclusive Excellence has been dogged by faculty complaints that intellectual disciplines develop objective and historically informed "standards" of quality, which get derailed when diversity is pursued for its own sake. Of course, this is exactly the kind of dichotomous reasoning the campus is trying to avoid. Obviously, mere difference in and of itself holds no ontological superiority. But by a similar token, sameness would seem to preclude novelty or improvement. Moreover, "objective" standards tend to sidestep questions of insularity within the group establishing the measures. Social groups are bound together by common interests, norms, and rules—often so naturalized that they fail to recognize their own blind spots. How easily can any social group evaluate the excellence of an idea about which it was previously unaware or perhaps was developed via alternative means and reasoning?

This matter of objective standards has broader implications for this chapter's discussion of innovation and anxious creativity. As the American population continues to worry and stress about its future, the nation's citizens gravitate toward anything offering certainty and a sense of security. In such an atmosphere, humanistic versions of creativity are easily overcome by the material standards of value. This narrowing of creativity for commercial ends may be bad enough in its own terms. But an anxious America is doing something even worse, as the market's invisible hand is failing to provide better goods. Producers and consumers show a growing avoidance of anything new in a culture increasingly craving familiarity and sameness. Unlike previous "cultures of conformity," this new mood is not attributable to industrialization, automatization, or mass media. Rather than the robotic group-think so vilified in the 1950s, this new age of anxiety has more to do with the very values of selfhood and autonomy embodied in the American individual—a figure now more alone and threatened than ever. In such an anxious moment, it hardly seems surprising that creativity would stage a return.

This chapter has introduced anxious creativity, focusing on its effects on students, workers, and anyone seeking more artistic fulfillment in their own life. After all, what is more individual than one's own creativity? Like money and beauty, creativity always is coveted. No one can ever have enough of it. In some accounts, anxious creativity is a symptom of a broader reorientation of American society along utilitarian lines, with more and more of everyday life defined in terms of specific purposes and goals. This gives the new creativity a certain hollowness and, on the surface at least, a degree of neutrality. But as discussed earlier, not everyone sees the new creativity as an innocent development. In an effort to draw a distinction between new and old, one recent anthology contrasts what it terms the "morbid" creativity of the recent creative upsurge with the "vital" creativity of traditional aesthetic idealism.[80] While a very old debate in the field of art history, the distinction is hardly something most people know about. This is why the narrowed field of creative capitalism seems to be constricting more open-ended strains of "multitudinous creativity"—yet again, to the detriment of new knowledge. This constriction hardly is incidental to the premises of this book in the questions it raises about the motivations and rewards that people derive from creative activities—while also pointing to a key dilemma in the definition of creativity itself. If creativity's ex *nihilo* ontology promises something new or beautiful in some way, does it not follow that creativity's value must be recognized by an already existing sensibility (the famous "eye of the beholder")? Does relativizing creativity in this way enhance or diminish it? Or does it further romanticize the ineffability for which creativity has for so long been known? Finally, what does it say about innovation and free expression in these anxious times?

Notes

1 Danny Vinik, "America's Innovation Crisis," *Politico* (Oct. 14, 2016) www.politico.com/agenda/story/2016/10/americas-innovation-crisis-000222 (accessed Jan. 26, 2019).

2 Richard Florida, *The Rise of the Creative Class: And How It's Transforming Work, Leisure, Community and Everyday Life* (New York: Basic Books, 2002); Richard Florida, *The Rise of the Creative Class—Revisited: 10th Anniversary Edition* (New York: Basic Books, 2012).

3 "Facts and Statistics," Anxiety and Depression Association of America (2017) https://adaa.org/about-adaa/press-room/facts-statistics (accessed Jan. 28, 2019).

4 Justin McCarthy, "Worry About Terror Attacks in U.S. High, but Not Top Concern," Gallup (Mar. 23, 2016) www.gallup.com/poll/190253/worry-terror-attacks-high-not-top-concern.aspx (accessed Mar. 19, 2019).

5 Matt Ford, "What Caused the Great Crime Decline in the U.S?" *The Atlantic* (Apr. 15, 2015) www.theatlantic.com/politics/archive/2016/04/what-caused-the-crime-decline/477408 (accessed May 19, 2018).

6 "Anxiety Disorders," National Institute for Mental Health (2017) www.nimh.nih.gov/health/topics/anxiety-disorders/index.shtml (accessed Jan. 29, 2019).
7 "Anxiety," Oxford Dictionaries (2017) https://en.oxforddictionaries.com/definition/anxiety (accessed Jan. 29, 2019).
8 Andrew Brink, *The Creative Matrix: Anxiety and the Origin of Creativity* (New York: Peter Lang, 2000).
9 "Anxiety Disorders."
10 Alana Samuels, "How Immigrants Have Contributed to the American Dream," *The Atlantic* (Feb. 7, 2017) www.theatlantic.com/business/archive/2017/02/immigrants-american-inventiveness/515928 (accessed Jan. 29, 2019).
11 Christopher Mims, "Is Engine of Innovation in Danger of Stalling?" *Wall Street Journal* (Nov. 20, 2016) www.wsj.com/articles/is-engine-of-innovation-in-danger-of-stalling-1479668795 (accessed May 28, 2018); "Has the Ideas Machine Broken Down?" *Economist* (Jan. 12, 2012) www.economist.com/news/briefing/21569381-idea-innovation-and-new-technology-have-stopped-driving-growth-getting-increasing (accessed Dec. 29, 2018); Steve Denning, "Why U.S. Firms Are Dying: Failure to Innovate," *Forbes* (Feb. 27, 2015) www.forbes.com/sites/stevedenning/2015/02/27/is-there-an-innovation-crisis-at-us-firms/#2589455864cb (accessed Jan. 28, 2019).
12 "Why U.S. Firms Are Dying."
13 Robert Gordon, *The Rise and Fall of American Growth: The U.S. Standard of Living Since the Civil War* (Princeton: Princeton University Press, 2016) p. 2.
14 "America's Innovation Crisis."
15 Christine Charyton et al., "Creativity as an Attribute of Positive Psychology: The Impact of Positive and Negative Affect on the Creative Personality," *Journal of Creativity in Mental Health* 4, no. 1 (2009): 58.
16 "Worry About Terror Attacks in U.S. High."
17 Sean Williams, "The Stunningly Large Obamacare Enrollment Number That's Flying Under the Radar," *The Motley Fool* (Jan. 10, 2016) www.fool.com/investing/general/2016/01/10/the-stunningly-large-obamacare-enrollment-number-t.aspx (accessed June 2, 2018); *Medicaid & CHIP: January 2016 Monthly Applications, Eligibility Determinations and Enrollment Report*, Centers for Medicare & Medicaid Services (Apr. 13, 2016) www.medicaid.gov/medicaid-chip-program-information/program-information/downloads/january-2016-enrollment-report.pdf (accessed Oct. 5, 2018).
18 "Worry About Terror Attacks in U.S. High."
19 Louis Althusser, "Ideology and Ideological State Apparatuses (Notes Toward and Investigation)," *Lenin and Philosophy and Other Essays*, trans. Ben Brewster (New York: Free Press, 1983) p. 284.
20 Slavoj Žižek, *Welcome to the Desert of the Real: Five Essays on September 11 and Related Dates* (London and New York: Verso, 2002).
21 Vanessa Barford, "Why Are Americans so Angry?" *BBC News Magazine* (Feb. 4, 2016) www.bbc.com/news/magazine-35406324 (accessed May 20, 2018).
22 Ibid.
23 Ibid.
24 Ben Castleman, "Americans Are Still Really Worried About the Economy," *FiveThirtyEight* (Jan. 29, 2016) http://fivethirtyeight.com/features/americans-are-still-really-worried-about-the-economy (accessed Apr. 5, 2018).
25 Paolo Virno, "Wit and Innovation," trans. Arianna Bove, *Transversal Texts* (June 2004) http://eipcp.net/transversal/0207/virno/en (accessed Jun. 15, 2018).
26 Richard E. Caves, *Creative Industries: Contracts Between Art and* Commerce (Cambridge, MA: Harvard University Press, 2000); John Howkins, *The Creative Economy: How People Make Money from Ideas* (New York: Penguin, 2001);

Richard Florida, *The Rise of the Creative Class: And How It's Transforming Work, Leisure, Community, and Everyday Life* (New York: Basic Books, 2002).

27 Richard Florida, "America's Looming Creativity Crisis," *Harvard Business Review* (Oct. 2004) https://hbr.org/2004/10/americas-looming-creativity-crisis (accessed Jun. 1, 2018).

28 "The U.S. Economy 2004," *The Pulse of Capitalism* 5, no. 1 (April 2005) http://www.comw.org/poc/0504.html (accessed Jul. 6, 2018).

29 "America's Looming Creativity Crisis."

30 Connor Bezane, "Barak Obama and the 'Creative Class': Has He Tapped into a New Voting Bloc?" MTV News (Feb. 21, 2008) www.mtv.com/news/1582081/barack-obama-and-the-creative-class-has-he-tapped-into-a-new-voting-bloc (accessed Jul. 6, 2018).

31 Joel Kotkin, "The Triumph of the Creative Class," *Forbes Magazine* (Nov. 5, 2008) www.forbes.com/2008/11/04/creative-class-obama-oped-cx_jk_1105kotkin.html (accessed Jun. 6, 2018).

32 Terry Flew, *The Creative Industries: Culture and Policy* (London and Thousand Oaks: Sage, 2012) p. 187.

33 Luc Boltanski and Eve Chiapello, *The New Spirit of Capitalism*, trans. Gregory Elliot (London and New York: Verso, 2007) p. 8.

34 *The New Spirit of Capitalism*, p. 176.

35 Leha Arnold-Smeets, "Here's Why Millennials Want to Work Part-Time," PayScale.com (Aug. 15, 2015) www.payscale.com/career-news/2015/08/millennials-want-to-work-part-time (accessed Jun. 10, 2018).

36 *Secret Sauce: 2016 Salary Guide*, The Creative Group/Robert Half, Inc. (2016) www.roberthalf.com/sites/default/files/Media_Root/images/tcg-pdfs/the_creative_group_2016_salary_guide.pdf (accessed Oct. 7, 2018).

37 Harold Rosenberg, *The Anxious Object: Art Today and Its Audience* (New York: Horizon, 1964).

38 *The Anxious Object*, p. 257.

39 Harold Rosenberg, as quoted in Barry Gwen, "State of the Art," *New York Times* (Dec. 11, 2005) www.nytimes.com/2005/12/11/books/review/state-of-the-art.html (accessed Jul. 31, 2018).

40 Blair Asbury Brooks, "What Did Harold Rosenberg Do?" *Artspace* (Jul. 31, 2014) www.artspace.com/magazine/art_101/know-your-critics/what-did-harold-rosenberg-do-52448 (accessed Dec. 5, 2017).

41 "Snaapshot," Strategic National Arts Alumni Project (2014) http://snaap.indiana.edu/snaapshot (accessed Jun. 11, 2018).

42 "The National Endowment for the Arts Announces New Research on Arts Employment," NEA (Apr. 28, 2014) www.arts.gov/news/2014/national-endowment-arts-announces-new-research-arts-employment (accessed Jun. 7, 2018).

43 Alexis Clements, "Indicting Higher Education in the Arts and Beyond," Hyperallergic.com (Oct.20, 2014) http://hyperallergic.com/156068/indicting-higher-education-in-the-arts-and-beyond (accessed Jun. 7, 2018).

44 Dan Berett, "The Myth of the Starving Artist," *Inside Higher Ed* (May 3, 2011) http://snaap.indiana.edu/uploads/THE_Article_on_SNAAP.pdf (accessed Jun. 18, 2019).

45 Claude Fernet, Stephanie Austin, and Robert J. Vallerand, "The Effects of Work Motivation on Employee Exhaustion and Commitment," *Work & Stress* 26, no. 3 (Sep. 2012) http://selfdeterminationtheory.org/SDT/documents/2012_FernetAustinVallerand_WS.pdf (accessed June 10, 2018).

46 Thierry Ehermann, "The Art Market 2015," artprice.com (2016) http://imgpublic.artprice.com/pdf/rama2016_en.pdf (accessed Jul. 2, 2018).
47 "America's Looming Creativity Crisis."
48 Alfie Kohn, *No Contest: The Case Against Competition (How We Lose in Our Race to Win)* (New York: Houghton Mifflin, 1986, 1992).
49 Catherine Rampell, "Want to Blame Someone for Basement-dwelling Millennials? Try the Boomers," *Washington Post* (May 26, 2016) www.washingtonpost.com/opinions/mom-and-dads-basement-is-looking-pretty-good-right-now/2016/05/26/d23c7f6c-237c-11e6-aa84-42391ba52c91_story.html?hpid=hp_no-name_opinion-card-f%3Ahomepage%2Fstory (accessed Mar. 27, 2019).
50 Kelley Holland, "The High Economic and Social Costs of Student Loan Debt," CNBC.com (June 15, 2015) www.cnbc.com/2015/06/15/the-high-economic-and-social-costs-of-student-loan-debt.html (accessed Jun. 13, 2018).
51 Ibid.
52 Arielle Eiser, "The Crisis on Campus," *American Psychological Association Monitor* 42, no. 8 (Sept. 2011) www.apa.org/monitor/2011/09/crisis-campus.aspx (accessed Jun. 21, 2018).
53 Ibid.
54 Amy Novotney, "Students under Pressure," *American Psychological Association Monitor* 45, no. 8 (Sept. 2014) www.apa.org/monitor/2014/09/cover-pressure.aspx (accessed Mar. 21, 2019).
55 T.M. Luhrmann, "The Anxious Americans," *New York Times* (Jul. 18, 2015) www.nytimes.com/2015/07/19/opinion/sunday/the-anxious-americans.html?_r=0 (accessed Mar. 20, 2019).
56 Ibid.
57 Ibid.
58 Joseph LeDoux, *Anxious: Using the Brain to Treat Fear and Anxiety* (New York: Penguin, 2015).
59 Marc Thompson, "Anxiety Disorders on the Rise in the Ranks," *Time Magazine* (Nov. 5, 2015) http://swampland.time.com/2013/11/05/anxiety-disorders-on-the-rise-in-the-ranks (accessed Mar. 21, 2019).
60 L. Kevin Chapman, "Why American Culture Is Plagued by Anxiety—Two Good Reasons," *Psychology Today* (Jan 12, 2012) www.psychologytoday.com/blog/the-color-anxiety/201201/why-american-culture-is-plagued-anxiety-two-good-reasons (accessed Mar. 21, 2019).
61 Joe Brewster, "The Mental Disease of Late-Stage Capitalism," medium.com (Apr. 4, 2016) (accessed Jun. 15, 2018).
62 "Dr. Brené Brown on Shame, Guilt, and Addiction," *Huffington Post* (Apr. 18, 2013) www.huffingtonpost.com/2013/04/18/brene-brown-shame-guilt-addiction-oprah_n_2966351.html (accessed Jan. 15, 2019).
63 Helen Walters, "Vulnerability Is the Birthplace of Innovation, Creativity, and Change: Brené Brown at TED 2012," TEDblog.com (Mar. 2, 2012) http://blog.ted.com/vulnerability-is-the-birthplace-of-innovation-creativity-and-change-brene-brown-at-ted2012 (accessed Mar. 14, 2019).
64 Carrie Barron and Alton Barron: *The Creativity Cure: How to Build Happiness with Your Own Two Hands* (New York: Simon & Schuster, 2017) p. 12.
65 Shaun McNiff, *Art Heals: How Creativity Cures the Soul* (New York: Shambala, 2004); Babatunde Kolawole, *Awakening to Inner Creativity: Following Your Inner Guide* (Lagos: National Library of Nigeria, 2014); Lacy Mucklow, *Color Me Calm: 10 Coloring Templates for Meditation and Relaxation* (New York: Race Point, 2014); Individuality Books, *Coloring for Tranquility* (New York: Create Space, 2016); Maggie Shannon, *Crafting Calm: Projects and Practices for Creativity*

and Contemplation (New York: viva, 2013); Michael Samuels, *Creative Healing: How to Heal Yourself by Tapping Your Hidden Creativity* (New York: Wiley, 1998); Elizabeth Gilbert, *Big Magic: Creative Living Beyond Fear* (New York: Riverhead, 2016); Karen Daniels, *Mindful Creativity: Easy Tips and Meditations to Unleash Your Creativity and Purpose* (New York: CreateSpace, 2011); Maria Montclair, *Peaceful Coloring* (New York: CreateSpace, 2015); Sarah Painter, *Stop Worrying, Start Writing: How to Overcome Fear, Self-Doubt and Procrastination* (New York: Siskin, 2017).

66 David E. Cooper, *World Philosophies: An Historical Introduction* (Oxford and Cambridge: Blackwell, 1996) p. 5.

67 Søren Kierkegaard, *The Concept of Anxiety: A Simple Psychologically Oriented Deliberation in View of the Dogmatic Problem of Hereditary Sin* (1844) trans. Alastair Hannay (New York: Liveright, 2015) p. 152.

68 Ibid.

69 Martin Heidegger, *Being and Time* (1927) (New York: Harper, 2008) p. 187.

70 Rollo May, *The Meaning of Anxiety* (New York: Ronald Press, 1950) p. 38.

71 *The Meaning of Anxiety*, p. 39.

72 Jacques Lacan, *The Ethics of Psychoanalysis, 1959-1960: The Seminar of Jacques Lacan, Book VII,* trans. Jacques-Alain Miller (London and New York: Routledge, 2008) p. 119.

73 *The Ethics of Psychoanalysis*, p. 117.

74 "Ready to Innovate," The Conference Board (2008) www.americansforthearts.org/sites/default/files/ReadytoInnovateFull.pdf (accessed Jun. 15, 2019).

75 Ibid.

76 Eric Weiner, "The Secret of Immigrant Genius," *Wall Street Journal* (Jan. 15, 2016) www.wsj.com/articles/the-secret-of-immigrant-genius-1452875951 (accessed Mar. 14, 2019).

77 Ibid.

78 Ibid.

79 "Message from Vice Provost Douglas M. Haynes, PhD," UCI Office of Inclusive Excellence (n.d.) http://inclusion.uci.edu/about (accessed Mar. 16, 2019).

80 Clarissa Moreira in Guiseppe Cocco and Barbara Szaniecki, eds., *Creative Capitalism, Multitudinous Creativity: Radicalities and Alerities* (Boulder and New York: Lexington Books, 2015) p. 96.

Chapter 2

Creative You

Self-Help to the Rescue

"Yes You Can," (Sprint), "Be All You Can Be" (U.S. Army), "Because You're Worth It," (L'Oréal) in "Your World, Delivered" (AT&T). You've seen these new ads: pitches for products or services to let you "be yourself" or "take control" of some aspect of your life. It's a new strategy called "empowerment marketing," based on the premise that in media-savvy age people are smarter about advertising and need to be approached in a way that flatters their evolved sensibilities. As a recent feature in *Your Business* put it, "Traditional marketing depends on creating anxiety in the customer in convincing her that she has a need that only the product or service sold can help her fill." In contrast, "Empowerment marketing subverts traditional marketing techniques by recasting the consumer as the hero who has the power to effect change and use the product or service being sold to achieve success."[1]

Nice as this sounds, it is really a case of putting old wine in new bottles. The example *Your Business* uses is the familiar Nike "Just Do It" campaign, which doesn't so much promote a certain shoe as much as "the message that anyone can be an athlete if they're willing to work hard."[2] Indeed, this is exactly the message that appears on the first page of Nike's current website: "Your daily motivation with the latest gear, most effective workouts and the inspiration you need to test your limits—and unleash your potential" with a fashion item lower on the page captioned "Dress like a champion."[3] In other words, the new empowerment advertising doesn't really forgo conventional appeals to consumer anxiety. It simply personalizes the pitch with the lure of enhanced autonomy. The Nike ad itself sums up this contradiction perfectly in stating: "Life isn't about finding your limits. It's about realizing you have none."[4]

Just think about the potency of this message in an America where people feel controlled and managed at every turn. Where is the one place to exercise absolute creativity? What is the canvas that is yours alone? It's the body, the face, the external image of the self. Nike pitches heavily to younger customers pressured by school or job seeking and still in the vulnerable stage of identity formation that psychologists term "differentiation." And, of course, a lot of empowerment advertising has targeted women, ever since the technique was first introduced decades ago with the Virginia Slims

"Have it Your Way, Baby" campaign for cigarettes. One doesn't have to look far to find other examples, such as the recent "Be Who You Are" ad campaign by Bobbi Brown cosmetics, offering "supremely flattering makeup for all skin tones … to make any woman feel like herself."[5] Ads by Covergirl ("Because We Rule") and Famous Footwear ("Victory Is Yours") work the same way.

The concept of creativity drives brands like "Creative Make-up" (Sephora), "Creative Cosmetics" (Thames & Cosmos), and the entire "BE Creative Make Up" line, the latter of which urges customers to "Be Beautiful, Be Yourself" because "You Are Wonderfully Unique."[6] No less a publication than the *New York Times Style Magazine* recently ran a cover story along the same lines entitled "To Thine Own Self Be True," proclaiming its "praise of idiosyncratic beauty." Summarizing all of this in *Business Life Magazine*, columnist David Mattin argued in a piece called "The New New Thing" that a novel truth has emerged in American marketing—"The endless search to become the people dream of being." To Mattin, the commodity of personal fulfillment—whether through appearance, personal fitness, happiness, or knowledge—has become the new holy grail in advertising. As he writes: "Whatever you sell, you'd better be selling personal fulfillment. Increasingly, brands that fail to understand that powerful truth will find themselves sailing into oblivion."[7]

This may sound appealing, but it also has troubling implications—the notion of creativity and "personal fulfillment" as a sponsored aspect of the human psyche. Here again, this isn't so much a matter of anything "bad" being promoted or sold, but of "good" qualities transformed into products. It's a form of seductive conditioning that draws consumers into purchasing with the lure of self-expression. Despite its proponents' claims, this process works by making people feel badly about themselves. Why would people need beauty or fulfillment products if they already had these things? This is "traditional marketing" reaching into fresh territory, but still leaning on familiar social norms and ideals to convey inadequacy (unachieved fitness, skinniness, newness, for example), and along the way implying prejudice or outright distain toward anyone not fitting the image. And if you can't afford the product, you are in double trouble.

None of this should come as a surprise in a culture seemingly obsessed with the topic of identity. The rise of internet technology has made it possible to "create" an image of oneself in ways earlier generations could never imagine: online profiles, timelines, personal blogs, dating apps, game avatars, etc. More than ever, the "self" has become one's most treasured possession—something to be cultivated, stylized, beautified, and presented for public view. And, of course, identity has to be protected as well—not only in the sense of a combative politics of identity, but also as an aspect of self that can be diminished, tarnished, or even erased. "Someone you've never met can post your picture on the internet," wrote Daniel J. Solove in *The Future of Reputation*. "These transformations pose threats to people's control over their reputations and their ability to be who they want to be."[8] To Solove and many other privacy experts, the expansiveness and permanence of internet databases threaten to further compromise the boundaries between self and others. Now ordinary citizens must contend with the kind of public scrutiny once reserved for celebrities. But things can get strikingly personal too. When the *New York Times* recently asked, "Do you snoop on your partner?" one in four respondents said "yes."[9] The solution Solove proposes is a "new and broader notion of privacy and by reaching a better balance between privacy and free speech." And then there is the matter of "identity theft" in which the once-philosophical notions of the self-as-possession is literalized, often with criminal intent as when credit card fraud results.[10]

This chapter takes a close look at America's new focus on personal empowerment, fulfillment, and creativity by asking a series of questions, many of which reveal serious ethical and political stakes. What lies behind this new preoccupation with the "self"? The seemingly narcissistic (and sometimes paranoid) obsession with personal identity, self-image, and reputation? Is this simply American individualism in yet another form, maybe amped up by media and ubiquitous social networking? Or is there something else behind the rising inward direction of American people? Might the neo-protectionism seen in Trump-era global policy reflect a deeper fearfulness of a nation and a people alone in a menacing world? Is it mere coincidence that empowerment marketing is rising as many feel politically disempowered, financially insecure, and alienated from their neighbors? Are self-expression and inner creativity becoming the final psychic refuge? And if the latter is the case, who gains and who loses from such a situation? And do prospects for change improve or lessen?

Neo-Narcissism

Think about the self as a kind of ultimate commodity. When money gets tight and the world seems bleak, there always is one final refuge: *You*. During the heady internet explosion of the early 2000s, *Time Magazine* shocked many by naming "You" as "Person of the Year." *Time*'s effusive accompanying commentary on the empowering potentials of a networked "You" showed how easily a term received individually could apply to millions of readers. This is a defining characteristic of much advertising and online social networking—the illusion of a "self" expressed in what feels like a personalized appeal. Writing in her book, *Updating to Remain the Same*, Wendy Hui Kyong Chun asserts that "You" is deployed in these contexts as "both singular and plural. In its plural form, it still refers to individuals as individuals, rather than creating another communal subject, a 'we', from more than one 'me.'"[11] Multitudes may receive the same information, but each recipient interprets the message personally. This key tenet of poststructuralist linguistics undermines the utopianism *Time* had in mind when it spoke of You in terms of "community and collaboration on a scale never seen before … wresting power from the few and helping one another for nothing."[12]

It is worth mentioning that the "self" is deeply engrained in American history. Remember that the very concept of self-directed individualism emerged around the time of the nation's founding, when the agency and consciousness of ordinary people (as opposed to royals or deities) was recognized for the first time. Shortly before America's founding, John Locke had written of the individual as "the proprietor of his own person," giving a materialist inflection to self-mastery.[13] The Enlightenment ideal of self-ownership had a particular appeal to populations arriving in the Americas, where "subjects" in the new land could *own* themselves, as well as other goods. The appeal of individualism wasn't exactly rocket science, owing to emerging philosophical beliefs that a self set apart from others is a crucial building block of subjectivity. This lonely soul would get credit for the gifts of reason, literacy, scientific empiricism—and, ultimately, concepts of democracy and capitalism that emerged as a novel expression of human agency. Decision-making of many kinds became newly vested in the reasoning abilities of individual agents, presumably able to assess truth claims. In this context, concepts like self-consciousness, self-interest, self-worth, self-improvement, and other forms of work on the "self" are artifacts of the modern era that began in the eighteenth century.

This fundamental message seems to be getting more virulent in the individualistic and privatized culture of the 2000s, however. Identity has become a prize like none other—so fetishized that it demands endless improvement, investment, promotion, protection, and guarding. Or at least this is what many people believe. There was a time when family or friends were thought to provide this kind of reassurance and sense of well-being. And, of course, this radicalized notion of individualism ignored decades of thought and research into the ways human beings are historically constructed, genetically programmed, relationally informed, and otherwise shaped by external experience. Even in neoliberal terms, the multibillion-dollar advertising industry is founded on the mutability of the self and its preferences.

There is a familiar term for this kind of egotistic preoccupation. Dictionary definitions of *narcissist* describe a person who is "overly self-involved" and "craving admiration," while also often displaying "extreme selfishness" and a "grandiose view of one's own talents."[14] This conforms with what most people call to mind when talking about someone like the current President. Dig a little deeper into mental health discussions of narcissism, and the meaning shifts in an important way, especially as narcissism escalates into a disorder. The Mayo Clinic says that such narcissists feel entitled, and become impatient about not having "the best" car, athletic club, etc. At the same time, the narcissist has "trouble handling anything that might be perceived as criticism," while sometimes also harboring "secret feelings of insecurity, shame, vulnerability, or humiliation."[15] The American Psychiatric Association adds such symptoms in its *Diagnostic and Statistical Manual of Mental Disorders* (*DSM-5*) as "being envious" and "taking advantage of others," as well as preoccupations with wealth, success, and power and a generalized "failure to emphasize with others."[16]

Without jumping to conclusions, it's fair to say that many of these same attributes often apply to today's U.S. consumer as well as the "anxious Americans" described in Chapter 1, especially those clinging to the prideful notion of an "exceptional" national identity. This correlation famously was described by historian Christopher Lasch, whose book *The Culture of Narcissism* appeared at the dawn of the 1980s neoliberal era. Lasch saw an America not unlike that of today, populated by self-absorbed individuals, fixated on personal goals, anxious about the future, and easily manipulated by powerful elites.[17] This created a dangerous situation, as citizens were becoming lost in the chasm of self-interest, distracted from its underlying politics, and left only with a cynical nihilism. Lasch wrote:

> People perceive their social position as a reflection of their own abilities and blame themselves for the injustices inflicted on them. Politics degenerates into a struggle not for social change but for self-realization. When the boundaries between the self and the rest of the world collapse, the pursuit of enlightened self-interest, which once informed every phase of political activity, becomes impossible.[18]

Creativity played a central role in Lasch's *Culture of Narcissism*, which depicted consumers increasingly attracted to "performing" characters drawn from advertising and media. Not unlike what Instagram and Facebook do today, Lasch saw boundaries disappearing between spectator and creator. The danger in this was the illusion of empowerment. "The merging of actors and audience does not make the spectator into a communicant," Lasch wrote.[19] It simply offers a narcissistic means to "admire oneself in the new role of pseudo-performer."[20] He said that repeated experiences of such creative role-playing generate an "indifference to the distinction between art and life" and a "collapse of the very idea of reality."[21] Citing a beauty ad captioned "Your Masterpiece—Yourself," Lasch argued that aestheticizing the real didn't make people feel very good either.[22] Constantly in a state of self-conscious performance, the modern subject finds itself "unable to express emotion without calculating its effect on others," resulting in "new forms of uneasiness and anxiety" and a "longing for spontaneous feeling."[23]

It wasn't creativity itself that troubled Lasch, but rather the way aesthetic sensibilities had fallen prey to anxious consumerism. This was seen in the vacuous and narcissistic way people expressed themselves through products and received ideas, often desperate for the approval or admiration of others. In language that might well describe the present time, Lasch wrote:

> Experiences of inner emptiness, loneliness, and inauthenticity are by no means unreal or, for that matter, and devoid of social content. They arise from the warlike conditions that pervade American society, from the dangers and uncertainty that surround us, and from a loss of confidence in the future.[24]

Like many post-war intellectuals, Lasch lay some of the blame for this on a weakening of widely held values that once bound American society together—with religion, the family, and even patriotism coming unraveled, along with people's attitudes toward self-sacrifice or helping each other. Worse still was a detachment from politics, with broad-based anger toward government and voter apathy occurring simultaneously. Lasch was careful to note the limitations of generalizing a diagnostic "narcissism" to an entire society. "Psychoanalysis deals with individuals, not with groups," Lasch pointed out, adding that "equating narcissism with everything selfish and disagreeable militates against historical specificity. Men have always been selfish, groups have always been ethnocentric; nothing is gained by giving these qualities a psychiatric label."[25] But as other intellectuals have done with terms like schizophrenia and mania, Lasch found in narcissism both a powerful metaphor and an explanatory narrative for American culture.

A healthier creativity might turn things around, Lasch argued, if people could channel their self-involvement in more positive directions. Freud famously wrote about this in his 1914 essay "On Narcissism," in which he introduced his concepts of "primary" and "secondary" narcissism.[26] Primary narcissism emerges as the infant initially imagines itself as omnipotent and all-powerful because it has not yet recognized that its caregiver is attending to its needs. As the child later formulates the external figure of the provider (typically a mother), a key template of want and desiring relationships is established, which Freud saw as inherently fraught with anxiety and later manifest in certain drives. Psychologically healthy people eventually resolve these tensions by directing their drives toward socially productive ends such as work or creative activities. Secondary narcissism was the pathological kind that concerned Lasch. In secondary narcissism, people turn affection back onto themselves, often worrying about self-preservation, becoming disinterested in society, and losing the ability to love others.

Secondary narcissism also can motivate public anger, as Lasch detailed in his discussion of psychoanalyst Melanie Klein. One of leading proponents of "object relations" theory, Klein worked more directly with children than Freud (who mostly analyzed adults) and reached more detailed conclusions. Notably, Klein found that infants were capable of internal rage, often directed toward the "object" of their mothers, and thus could develop both "good" and "bad" images of parents, authorities, and others. Klein saw narcissism as a response to the pain of separation from a key (typically maternal) object. Not every child would become a pathological narcissist, but the potential resided in everyone, nevertheless. The key variable lay in the child–caregiver relationship and the degree to which a comfortable bond is created. Klein and her colleagues saw adult narcissism in "greed, destructive emotions, and impulses that manifest themselves through envy, omnipotent denial of dependence, megalomaniac idealization of the self, and confusion."[27]

Later studies of children would look more closely at this issue of "attachment" as a determinant in the child's ability in building positive relationships to other people and the world in general.[28] A securely attached child would more willingly explore its environs and engage others, while an insecure one would remain anxious and distrustful. Although many adults can scarcely remember childhood experiences of comfort or insecurity, lasting patterns remain, nevertheless. Beginning at the time Lasch was writing *The Culture of Narcissism*, research began to show a range of attachment-like qualities in adult interpersonal relationships. Simply put, there seemed to be great variability in people's attributions of "good" or "bad" qualities to others, along with differentials in their capacities for trust and cooperation. Securely attached people seemed comfortable balancing intimacy with independence, for example, while anxiously attached individuals were more suspicious and attention-demanding.

Summing this all up, Lasch saw Western society torn between the isolating impulses of secondary narcissism and the interconnection of secure attachment: "Each society tries to solve the universal crises of childhood—the trauma of separating from the mother, the fear of abandonment, the pain of competing with others for the mothers love—in its own way," Lasch wrote, concluding that "loving memories constitute an indispensable psychological resource in maturity, and those who cannot fall back on the memory of loving relations in the past suffer terrible torments as a result."[29] The challenge for society lay in finding resources to counteract latent fears of abandonment rather than rekindling them in a narcissistic fashion. Following Freud, Lasch and many of his contemporaries found hope in the repurposing of selfish instincts into socially productive attitudes and behaviors—a view consistent with the founding principles of democratic capitalism through which individual and collective interest operated in reciprocal fashion.

Has the *Culture of Narcissism* returned today? Writing in a recent issue of *The Atlantic*, historian Rebecca Solnitt describes a recent shift in American culture, in which foundational values of connection and community have given way to what she terms an "ideology of isolation."[30] Throughout her essay, Solnit critiques the myth of the solitary cowboy "Marlboro Man" with factual instances of collective enterprise—whether manifest via business or government—in settling the American west, building railroads, defending the nation, and maintaining an economy. But in recent decades a new kind of radicalism has pushed this history aside, allowing not just more "freedom" to buy and sell things, but an ideological reconfiguration of "reality" itself. "Absolute freedom means you can have any truth you like," Solnitt wrote: "'Freedom' is just another word for nothing left to limit your options. And this is how the ideology of isolation becomes nihilism."[31]

Solnitt's main point is about irrationality—and ideology's ability to activate feelings at the expense of factuality. Appealing ideals such as freedom and individuality can spin out of control in a world defined by anxiety and fear—

feelings, as Lasch pointed out, which are certainly "real" in the minds of those experiencing them. What's important is how ideology attaches to what already exists in the mind, not something completely external or "false." As the above discussion of narcissism pointed out, all people carry the baggage of childhood memories, for better or worse, along with pressures of adult desires and responsibilities. These personal histories leave people with a range of vulnerabilities to different kinds of sales pitches, ideological claims, and decision-making options. But most of this boils down to the connection or lack thereof with others, first experienced early in life.

Channeling Creativity

Using creativity to break the grip of narcissism is no simple matter, especially in an America so enamored with success and personal interest. Freud saw what he termed "sublimation" as a means of moving the self in a positive direction. Goethe had used the word philosophically in the 1700s as a derivation of the Latin verb *sublimus*, meaning "eminent, exalted, or elevated."[32] A century later, sublimation acquired a psychological connotation when Friedrich Nietzsche drew on the term's scientific meaning of a solid turning directly into gas: "There is, strictly speaking, neither unselfish conduct, nor a wholly disinterested point of view. Both are simply sublimations in which the basic element seems almost evaporated."[33] What society needs, Nietzsche said, was a "chemistry of the moral, religious, aesthetic conceptions and feelings, as well as those of the emotions which we experience in the affairs, great and small, of society and civilization."[34]

Nietzsche prefigured Freud in many areas—including theories of the unconscious mind, psychological repression, and sublimation, among others. And in some ways, Nietzsche's thinking was more nuanced and specific. Freud saw in sublimation the direct conversion of a drive (usually an erotic one) into a socially acceptable activity, famously attributing Leonardo da Vinci's prodigious creativity to repressed homosexual desires.[35] Throughout his career, Freud would repeat a basic assertion that "sexual dispositions" could be "diverted in the direction of art" and that they provided "the energy for a greater number of our cultural achievements."[36] In contrast, Nietzsche never saw sublimation as a simple conversion of erotic impulses into productive activities, but rather a modulation between the two (as might be seen in da Vinci's longstanding interest in the male form).

Put another way, Nietzsche deferred from giving sublimation Freud's prescriptive meaning as a "healthy" channeling of libidinous impulses. This normative quality would get sublimation into trouble in the years after Freud's passing. While some like Lasch would embrace sublimation's seemingly useful aspects, Michel Foucault would cast sublimation as a way of policing the self in socially "acceptable" ways. Among other points, Foucault would assert that in constraining erotic desire, Freudian sublimation also encouraged some kinds of sexuality over others.[37] Things got more complicated when Herbert Marcuse coined a term for sublimation's *opposite*. Rather than simply speaking of sexual or capitalistic excess, Marcuse wrote of what he termed "desublimation," especially as propagated via mass media as a means of distracting people from their properly sublimated efforts for society. Specifically, Marcuse expressed concerns that the superficial (and often sexualized) pleasures of popular culture commodified genuine artistry while simultaneously masquerading as rebellion. Writing in his 1964 book *One-Dimensional Man*, Marcuse would say that this "repressive desublimation" removed a necessary "second" dimension of social critique from public discourse.[38]

Complicating sublimation yet further, Jacques Lacan applied linguistic theory to his own revision of Freud's psychoanalytic model of desire.[39] Lacan saw people as endlessly seeking an idealized object he simply termed *das ding* ("the thing") for which they would imagine various substitutes represented in words, images, or physical stand-ins. Derived from the infantile separation from the caregiver, *das ding* is never perfectly recovered and thus remains an empty feeling. Transitory pleasures may accrue in the search, but what most interested Lacan was the process itself, as the distance between the thing and its many instantiations is never closed. In his focus on process and the "void," Lacan departed from what he saw as a reductive Freudian sublimation-via-substitution and its implication of an obligatory "adaptation to the environment" encouraged by an all-powerful analyst.[40]

Various commentators on Lacan have noted how neatly *das ding* corresponds to consumer desire and capitalism's endless quest for new markets and greater profits.[41] But Lacan himself kept open a space for hope through the device of sublimation. Just as people can succumb to selfishness and superficial satisfaction, their wants also can be directed into worthy endeavors, gestures of loves, or generosity. To Lacan, this is where ethics enters the picture—as an element of both personal subjectivity and analytic practice. Ethics can be tricky, however. Not only can people have different ideas about what is "good" or "true," but as psychoanalysis pointed out, their ethics often can derive from sublimations they don't completely understand. Lacan cast an ethical dimension onto the search for insight, as one strives to understand one's underlying yearnings, motivations, and guilty feelings. With more presence of mind, a person is able to make less selfish decisions. Foucault similarly linked ethics to self-understanding, but extended the resources for insight to such matters as the study of history and the examination of one's conscience. Common to both Lacan and Foucault, as well as many other twentieth-century thinkers, was a rejection of religion as an external rule-making artifice.[42] In place of such templates of acquired belief systems, both saw ethics more as a process of questioning, facing the unknown, and continually searching for "truth" while knowing how elusive it can be.

In recent decades, philosophy has continued to wrestle with sublimation. One of the more intriguing lines of thought has explored further the gap between the self and the object of desire (caregiver, love interest, commodity, etc.), and recognized various spaces of openness and possibility. For example, Slavoj Žižek saw both psychoanalytic and material dialectics as eternally unresolvable, but not in a meaningless or pessimistic sense. Contradictions and frustration are exactly the point, although they also generate pressures to seize upon easy answers. Of course capitalism promises big money and cheap thrills, Žižek said. But because the market always fails to deliver lasting satisfaction, it can't help but leave itself open to exposure of the "less than nothing" inside of it.[43] More radical still, Žižek suggested that recognition of this openness offers its own forms of pleasure (*jouissance*). In this sense, sublimation can become less a matter of duty and rule-following than of purposeful intervention and exuberant disruption. This raises the provocative notion (in sharp contrast to Marcuse) of whether it is possible "enjoy" capitalism while also building alternative social spaces within it.[44]

Bernard Stiegler has been even more explicit in expanding the concept of sublimation. Using the tools of deconstruction, Stiegler wrote in his book *What Makes Life Worth Living* that sublimation is like a medicine: beneficial in the right dozes, but harmful in the wrong amounts. And also like drugs, history has shown the world good sublimations (as in contributing to an egalitarian society) as well as bad ones (as in conforming to a fascist regime). Turning to contemporary democratic capitalism, Stiegler described a special problem resulting from a culture of immediacy, as the kind of short-term thinking prevalent in contemporary society (instant messaging, one-click shopping, high-frequency trading, etc.) distracts people from considering the long-term consequences of their actions. This is not ordinary distraction, said Stiegler. In keeping with the metaphor of a medicine, he argued that the instant gratification of contemporary culture had a habit-forming effect: "[I]ts drive-based tendencies are systematically exploited while its sublimatory tendencies are systematically short-circuited," creating an "addictogenic society."[45] And as everyone knows, addicts are always thinking of the next fix rather than the "transcendental memory" or "transcendental imagination" that it takes to keep any group going long-term.[46]

Sublimation has one further dimension often ignored in psychoanalytic discussions: societies can't function without it. All too often, discussions of sublimation focus (as this one has) on its role in the psychic economy of individuals, their development, and maturation. All groups require participants willing to yield a bit of themselves in the interest of the collective project, often with exactly the type of long-term thinking outlined by Stiegler. The disciplines of cultural anthropology, psychology, and sociology all have examined how this works through kinship structures, group dynamics, and institutional forms. But what remains most important is that both individuals and groups *need* sublimation as the glue that holds them together in both mental and material terms. It might even be said that one cannot exist without the other—and that the familiar opposition of "individual" versus "society" is itself a meaningless concept.

At the same time, critiques of sublimation have made clear the limitations of one-size-fits-all applications. Psychoanalysis frequently has been taken to task for its unresponsiveness to cultural differences and changing historical circumstance. It's also unfair to reduce individuals or groups to diagnostic labels such as narcissism. Is the answer to "self" obsession and apparent narcissism simply a matter changing one's mind? Of acting like a grown-up and being "responsible"? Maybe there are legitimate reasons why so many people are turning inward, regressing, and responding to empowerment marketing.

Self-Help, Inc.

"Whether you can admit it to yourself or not, you are creative," read the ad copy for *Creative You*, one of the thousands of self-help books in circulation today about finding the "inner artist."[47] As authors Otto Kroeger and David B. Goldstein continued:

> In today's complex world, creativity is the key to finding and living your passion. Whatever that passion is—cooking, technology, writing or even plumbing—*Creative You* reveals your own personal style of creativity to help you build and environment of innovation at work and at home. [48]

This brief passage summarizes everything discussed in these pages thus far: the enhancement of the self, the commodification of creativity, the stresses of a complex world, and the imperative for success. Like much of the popular literature on creativity, *Creative You* says that an inner artist resides in everyone, but often gets lost in the pressures of everyday life. Drowning in feelings of personal inadequacy, seven in ten Americans are "actively disengaged" from their work and believe they not "living up to their creative potential," the authors assert.[49] "Being creative, on the other hand, turns this unhappiness around." Employing psychological inventories, personality assessments, and Jungian analytics, the book explicitly bills itself as a therapeutic work.

The "self-help" industry is booming in today's America, generating $9.9 billion in annual revenue and estimated to grow 5.6 percent each year through 2024.[50] In the past 12 months alone, *Publishers Weekly* reports a 14 percent increase in self-help sales.[51] At some point in their lives, one-third to a half of all Americans purchase such books. Traditionally targeting middle-aged Baby Boomers, the demographics of self-help customers have broadened to include Millennials. "Most Americans are unhappy," reported a recent feature in *Inc. Magazine,* owing to "living in a superficial society that values materialism, consumerism, and working way too many hours each week." As *Inc.*'s Matthew Jones put it, "The brutal truth is that several industries—including the self-industry—are profiting off of the emotional pain of people seeking quick fixes."[52] Happiness has become an object to be purchased, whether in a book, a work-out plan, or a creative pursuit. "The mechanism of this seeking for happiness is what reinforces your isolation from its presence," Jones explained.

There is a history behind today's rising self-help craze—and the prominence of creativity within it. Centuries ago, people found happiness in the people close to them and the reassurance of faith. But over time those traditional supports lost their potency, leaving many feeling isolated and under pressure to work and succeed. During the nineteenth century, a gradual secularization took place in American society, as people's notions of themselves and their "purposes" in life slowly moved in less morally conscious directions. Sociologist Max Weber famously attributed this to the dominant influence of Protestantism—a faith more accommodating of non-religious "callings" (vocations, businesses, etc.). Hard work and financial success acquired the ethical status once reserved for good deeds and Christian generosity. This business-friendly form of faith emerged in tandem with industry, of course—eventually causing several forms of backlash. One of these was the Romantic Era return to emotion and nature popularized in the writings and paintings of Americans such as Emily Dickinson, Nathaniel Hawthorne, and Thomas Cole. Promising a "new" American spirit, romanticism further pushed religiosity aside with promises of self-understanding. Hence, the therapeutic benefits of art and creativity preceded what we know today as psychology and psychiatry.

As psychotherapy emerged in the early twentieth century, it brought along a fascination with the inner artist. Freud had leaned on creativity in his formulation of the unconscious, based on the premise that artists could access their inner selves with exceptional directness. The work of artists like André Breton, Max Ernst, and René Magritte would draw heavily on such theory. Published in the tumultuous year of 1930, Freud's *Civilization and Its Discontents* would state that the "aesthetic attitude to the goal of life offers little protection against the threat of suffering, but it can compensate for a great deal … Civilization could not do without it."[53] Western art movements of the time (Primitivism, Dada, Surrealism, among others) were obsessed with finding (or appropriating) non-traditional means of operation. These impulses attached easily to Freud's assertions that artistic work could serve the dual purpose of working "with" culture (to define it) while alternately pushing "against" culture (to revolutionize it).

Following World War II, a wide variety of mental health treatments, medications, and therapies would emerge to calm an anxious America. Creativity, hobbies, and art-making quickly became part of this mix. Not unlike today, along with the combat-related ailments of the nation's 20 million veterans, the stress of the war had taken an enormous toll on the mental health of families and the population at large. Using the crude diagnostic categories of the era, "anxiety" became the most commonly identified public complaint. Severe problems were just as likely to be treated by general physicians as by psychiatrists, as lobotomies and electroshock treatments became commonplace ways of minimizing expensive long-term hospitalizations. America's psychiatric ills would lead to the establishment of the National Institutes of Mental Health—with a budget that rose from an initial $364,000 in 1949 to $14 million in 1955.[54] Medications like tranquilizer meprobamate (Miltown) would come to U.S. markets, with 37 million Americans using the drug by 1957 (one-third of all prescriptions written in the U.S.).[55] The sedative hypnotic drug thalidomide also would also be introduced to Americans in 1957—as a remedy for anxiety, insomnia, and morning sickness—with 2.5 million "experimental" doses given until birth defects began appearing in the early 1960s.[56]

With Americans everywhere hungering for ways to feel better psychologically, the emergence of "art therapy" during this time seems no big surprise. Building on readings of Freud, Austrian immigrant Edith Kramer would advance the "healing" qualities of creative practice, as emotionally disturbed children productively engaged in the process.[57] Kramer theorized that the art therapist could guide youngsters through the mixing of paint, making images, and thereby "working through" troublesome memories or psychological issues. Kramer's ensuing popularity grew from her expansive claims that such common problems as worry, ambivalence, and loneliness could be helped through creative activities children already liked. Later art therapy would take this supportive premise further, with the practitioner serving as a co-interpreter of artworks made by children and adults alike. In the decades following its introduction, enthusiasm for the healing properties of art therapy faded somewhat within the mental health community, although psychotherapists still would find children's drawings, for example, to be useful diagnostic tools—especially among youngsters unable or unwilling to verbally express their feelings.

Through all of this, art therapy retained a place within popular culture in the broader context of mindfulness, meditation, and other stress-reduction modalities. But things really took off for art therapy in the recessionary 2000s—following two separate trajectories. The first was an offshoot of the self-help industry discussed above. The second occurred within mainstream mental health treatment, where evidence-based art therapy gained legitimacy. According to sociologist Micki McGee, the self-help side of art therapy is more complicated than many critics think. Certainly it is true that some people may embrace creativity as a narcissistic gesture of "self" worship. But what drives so many in this direction today, McGee argues, is how self-help functions as a way to cope with economic insecurity and the absence of social supports.

"The tremendous growth in self-help publishing parallels an overall trend of stagnant wages and destabilized employment opportunities for American workers," McGee wrote in her book *Self-Help, Inc.* "Americans face what some social observers have called a 'new insecurity' in the wake of the end of the standard job and family."[58] With social welfare programs all but gone, and with lifelong marriage and lifelong employment increasingly anachronistic, it is no longer enough to be married or employed. Instead, it is imperative to remain "marriageable and employable," as McGee put it. This twin dilemma accounts for much of what is seen in everything from beauty products and fitness products to self-improvement books and career guides. McGee wrote:

> The less predictable and controllable the life course has become, the more individuals are urged to chart their own course and to "master" their destinies. In addition to actual hours spent on the job—which have increased dramatically—Americans are compelled to constantly work on themselves.[59]

Such additional toil includes retraining and course work for new types of work, eternally looking youthful and vigorous, and searching for one's "true calling."

It's easy to see how artistry plays into this both directly and indirectly, on one hand literalizing the Romantic ideal of a person's life as a creative, aesthetic enterprise. To McGee, the anthem of therapeutic makeover culture goes something like this: "Imagine a self and then invent that self. Picture a life, and then create that life."[60] Dig a little deeper and a darker side of creative selfhood emerges, as other features of artistic lifestyles implicitly attach themselves: solitude, self-sacrifice, and a willingness to defer material gain. As in the industrial 1800s, self-actualization through work has become a guiding premise, with the ideal of the artist a model for post-industrial labor. This is how the self-supporting individual drifts further and further from collective sensibilities and social supports—increasingly alone while praised as "independent," "special," and "different."

Of course, none of this diminishes the demand side of self-help, personal creativity, or therapeutic products. People across America feel badly in huge numbers, looking for relief in whatever form they can find. Many find solace in self-help, especially those without access to psychotherapy, the money to afford it, or the willingness to make what to many seems a stigmatizing admission. Right now, the mental health field is beginning to grapple with the problem of population-wide anxiety, depression, and other maladies—exploring the potentials of internet practice, iPhone apps, and other ways of making treatment more accessible given the practical limitations of one-on-one work. Group encounters of the kind seen in wellness workshops, meditation classes, and art therapy increasingly are being recognized for their usefulness. This is all the more important as the mental health field more aggressively addresses prevention in its efforts for the population's health.

Art therapy already is used in more acute situations: hospitals, psychiatric facilities, schools, senior centers, rehabilitation programs, and private practices. But perhaps the most dramatic upsurge in the field has occurred in military contexts, owing largely to rising recognition of post-traumatic stress disorder (PTSD) and related issues. Beginning a decade ago with a modest "Military Healing Arts Network," the National Endowment for the Arts later launched its "Creative Forces" partnership with the U.S. Department of Defense, which was soon joined by the Bureau of Veterans Affairs. Even as President Trump vowed to eliminate the NEA when he took office, Creative Forces has garnered *increased* funding for the beleaguered agency ever since. Part of the reason is that the military angle has for the first time brought real science to art therapy, showing measurable benefits to the 500,000 service members with PTSD and traumatic brain injuries. Art programs are seen as an efficient way to reduce healthcare costs for veterans, saving the federal government as much as $1.7 billion.[61] They also seem useful for families of returning vets, many of whom experienced emotional difficulties during and after deployments. The arts are popular, too. As many as 85 percent of participants say they benefit from the program, which has expanded to 11 sites such as California's Camp Pendleton Marine base and the Army's Fort Hood in Texas.[62]

Not everyone jumps on the idea of being a "Warrior Artist" right away. "I thought this was a joke," recalled Staff Sergeant Perry Hopman, who served as a flight medic in Iraq. "I wanted no part of it because, number one, I'm a man, and I don't like holding a dainty little paintbrush. Number two, I'm not an artist. And number three, I'm not in kindergarten."[63] Hopman was one of many service members at the National Intrepid Center of Excellence (NICoE) housed at Walter Reed Hospital in Maryland, where one of Creative Forces' signature efforts involves asking participants to paint or draw on blank masks. "I was ignorant, and I was wrong, because it's great," Hopman later said. "I think this is what started me kind of opening up and talking about stuff and actually trying to get better." Today, visitors to NICoE are greeted by a wall covered with the masks depicting themes like death, physical pain, and heroism. Using masks, participants are able to document their past experiences, reflect current feelings, or project possible futures. This can be especially effective with certain kinds of physical trauma. "Brain injuries caused by blast events change soldiers in ways many can't articulate," explained Caroline Alexander in a *National Geographic* cover story on Creative Forces. This occurs as regions of brain affecting cognition and speech are damaged. "Some use art therapy, creating painted masks to express how they feel," Alexander reported.[64]

As Sergeant Hopman's comments illustrate, the new legitimacy accruing to arts therapy is broadening its appeal as a destigmatized entry point to mental health care. While music and painting have been long seen as ways of reaching people with conditions like dementia and autism, the arts increasingly are used to treat psychological trauma resulting from illness, injury, or abuse. Typifying how art therapy is finding new applications, California's Arts in Corrections program now receives $8 million per year in serving all of the state's 35 facilities. The effort is described by the California Arts Council as a cost-effective way to "enhance rehabilitative goals, improve the safety and environment of state prisons, and combat recidivism" as participants "change how they interact with others, how they see themselves, and the overall trajectory of their lives."[65]

Happy Together

This chapter about the creative "self" began by focusing on today's relentless pursuit of personal satisfaction and empowerment—often delivered by a product or service of some kind. My overriding premise has been that widespread anxiety in American society has sent people looking for relief, often searching in vain for an elusive happiness. Taking a sardonic look at this issue, Showtime's TV series *Happyish* told the story of Thom Payne, a burned-out 44-year-old ad executive pitted against cocky Millennials in his workplace. In one dismal meeting, company researchers presented a dilemma facing Coca-Cola, explaining that while Coke had always attached itself to a positive message, "Now happiness isn't testing very well." The quandary facing Thom and his team was "how to sell happiness in a culture of disillusionment." Shortly after the meeting, Thom had lunch with a corporate job consultant, to whom he confessed that he hated his job. But rather than sympathy, the consultant snapped back that Thom should stop complaining. "You think you're not happy? You couldn't possibly be happier," she quipped. "You've reached your 'joy ceiling' without even realizing it."

The take-away message of *Happyish* was found in the show's title. Happiness is overrated and has become something people think they can buy *for themselves*. In reality, the best one can expect from this personal approach is partial fulfillment or feeling "happyish." Not surprisingly, *Happyish* went off the air after one season. Most critics simply didn't like the message, calling the show "a reactionary screed" (*Slant Magazine*), "filled with personal rage" (*Washington Post*), and "not a laugh-fest" (*YahooTV*).[66] This failure of a show like *Happyish* is no great surprise in the current "Make America Great Again" zeitgeist in which everyone (rich, poor, old, young) has been told that "life, liberty, and the pursuit of *happiness*" are narcissistic rights. In this context, I found *Happyish* a refreshing antidote to the endless barrage of neatly resolved storylines that typify most TV and a valuable object lesson about what people seek in today's America.

Like most states of mind, happiness carries lots of variability, suggestibility, and the ability to shift and change over time. But when most people think of happiness, they often resort to a personal view ("Am I happy?"). This is hardly surprising in a culture so predicated on personal desire, gratification of the "self," and the endless products aimed at enhancing one's health, appearance, confidence, or "creativity." Even Freud's sublimation has been transformed from the socially conscious process he envisioned—instead, promoted in individual "workaholism" or feelings of guilt when one isn't productive. Countless studies have shown that relationships with people are what truly bring people happiness, even though most don't realize it. Even the *Journal of Consumer Research* has said that "things won't make us happy when we need a boost."[67] Quite the contrary, in fact. The journal showed links between materialism and loneliness, with more materialistic people reporting increased feelings of loneliness, and more lonely people reporting increased feelings of materialism. These views square with the conclusions discussed above of self-help—and its tendencies to diminish people's abilities to find strength and support from each other.

For some time, the field of positive psychology has explored all of this, with one of its leading figures finding that happiness derives more from interconnection than solitude. Martin Seligman has said that happiness breaks down into both "gratification" (momentary pleasures, self-gratification) and "authentic" (meaningful engagement, relationships with others) forms, adding that "authentic happiness" comes from using "your highest strengths and talents to belong to and serve something you believe is larger than yourself."[68] Recent brain research has confirmed this as well, with neuroscientist Morten Kringelbach mapping two overlapping "pleasure centers" in the brain—one of experiential "liking," the other of anticipatory "wanting." To Kringelbach, these two processes are both always present and interacting—with important social implications. "Humans are intensely social, and data indicate that one of the most important factors for happiness is social relationships with other people," he wrote.[69] Eating a candy bar might give you a childlike feeling of enjoyment, but having a relaxed meal with friends is likely to be more memorable.

Then there is giving to others. While much of this chapter has been devoted to the "Me" and its tendencies toward selfish behavior, anthropologists have long recognized that people derive happiness from sharing things. Social science has confirmed this as well. In his classic 1923 work *The Gift: Forms and Functions of Exchange*, Marcel Mauss studied populations existing before or outside of capitalism in tribal regions in Asia, Australia, and the Americas. Mauss saw a "symbolic economy" of giving in which a psychological exchange undergirds a material one. His most striking finding had to do with the impermanence of possession. Mauss found that not all cultures got stuck on the idea of owning things forever. Some of this was because primitive societies had to share food and other goods for survival. But this necessity for sharing also led to certain value systems and rituals. For example, Malaysian and Trobriander peoples believed they had three obligations in life: to give, to receive, and to reciprocate. A person was not only expected to share with others, but also to agreeably accept things received. In this way, cycles of giving and receiving were ongoing and mutually reinforcing. In nearly every culture he examined, Mauss found that such giving often was sacralized in religious faith. Mauss limited his work to those he studied, and refrained from extrapolating his findings to contemporary Western cultures. But he identified what to many seems a fundamental insight: that when people get something, they often feel like giving back, and vice versa.

Also, an emotional payback often comes from giving. Psychologists say that being generous and doing for others simply makes a person feel better. In their book *The Paradox of Generosity*, sociologists Hilary Davison and

Christian Smith described a five-year study of over 5,000 individuals—in which they found multiple benefits from giving. They said:

> It involves everything from developing a sense of self as generous to being more socially networked to being more physically active. We argue that it involves neurochemical changes in the brain, that it gives people more pleasure chemistry in their brain, a sense of reward for having done something good.[70]

Another large study of people who volunteer time for charities found significant pleasure deriving from such generosity. In "How Giving Makes Us Happy," psychologist Therese Borchard described how "68 percent reported that it made them feel physically healthier; 89 percent that it 'has improved my sense of well-bring' (e.g., happiness) and 73 percent that it 'lowered my stress levels.'"[71]

Nowadays, savvy charities and foundations often use psychologists to consult on fundraising campaigns. "Altruism isn't exactly simple," explained consultant Krystall Dunaway in discussing the motivations behind philanthropy. To Dunaway, common views of "pure" altruism often miss the extent to which "performing the selfless act (i.e., donating to a cause) also benefits the self through positive feelings about doing a good thing."[72] Dunaway explained that the "good feeling" often associated with giving also can be attributed to the narcissistic reward or a complement to the contributor's social status or achievement, or authority. This helps explain why people giving to institutions often want to be recognized in reports, on wall plaques, or even in the names of buildings or programs. But it also puts into perspective how the self plays into what are often considered "selfless" acts. The point is that there are many ways of reconciling the perceived distance between self and others—and that this doesn't need to mean a sacrifice or loss. Unfortunately, many people just don't see this.

This chapter began by posing a series of questions about the ethical and political stakes in empowerment marketing, notions of "finding oneself," and the promotion of particular kinds of creativity. While nobody would deny the importance of self-improvement, the new push for radical individualism also isolates many people and can encourage antagonism among groups. The promotion of the lone artist-as-role-model tends to valorize attitudes of alienation and disconnection from others, potentially worsening feelings of anger and resentment common in the U.S. today. Contrary to its claims, America's booming "self-help" industry often encourages this as well. Fortunately, models of creativity long have existed for using personal expression in positive ways via sublimation. Another promising use of creativity is found in forms of art therapy fostering connection and reflection rather than the "self" alone. In an America so predicated on "Being All You Can Be" and individual personhood as one's most precious commodity, there need to be ways of finding happiness that don't alienate people from one another. Otherwise, any hopes for a creative society will continue to fragment and lose effectiveness.

Notes

1 Jean Marie Bauhaus "What Does Empowerment Mean in Marketing?" *Business & Entrepreneurship* (n.d.) http://yourbusiness.azcentral.com/empowerment-mean-marketing-28220.html (accessed Feb. 6, 2019).
2 Ibid.
3 "Start Your Journey," Nike.com (Aug. 1, 2016) www.nike.com/us/en_us/c/justdoit (accessed Feb. 6, 2019).
4 Ibid.
5 Bobbi Brown, "The Bobbie Brown Philosophy: Empowering Women to Look Like Their Best Selves," AOL.com (Mar. 21, 2015) www.aol.com/article/2014/03/21/bobbi-brown-makeup-philosophy/20849490/ (accessed Jul. 17, 2018).
6 "Be Beautiful. Be Yourself." BE Creative (n.d.) www.becreativemakeup.com (accessed Mar. 28, 2018).
7 David Mattin, "Trends: The New New Thing," *Business Life* (Jun. 2016): 10.
8 Daniel J. Solove, *The Future of Reputation: Gossip, Rumor, and Privacy on the Internet* (New Haven: Yale University Press, 2007) p. 3.
9 "Do You Snoop on Your Partner?" *New York Times Magazine* (Aug. 7, 2016) (accessed Feb. 7, 2019).
10 *The Future of Reputation*, p. 4.
11 Wendy Hui Kyong Chun, *Updating to Remain the Same: Habitual New Media* (Cambridge and London, MIT Press, 2016) p. 118.
12 Lev Grossman, "You—Yes, You—Are TIME's Person of the Year." *Time Magazine* (Dec. 25, 2006) http://content.time.com/time/magazine/article/0,9171,1570810,00.html (accessed Mar. 28, 2019).
13 John Locke, as cited in Étienne Balibar, "My *Self* and My *Own*" in Bill Maurer and Gabrielle Schwab, eds., *Accelerating Possession: Global Futures of Property and Personhood* (New York: Columbia University Press, 2006) p. 27.
14 "Narcissist," dictionary.com (2016) www.dictionary.com/browse/narcissist (accessed Feb. 6, 2019); oxforddictionaries.com (n.d.) www.oxforddictionaries.com/us/definition/american_english/narcissism (accessed Feb. 6, 2019).
15 "Narcissistic Personality Disorder: Symptoms," mayoclinic.com (2016) www.mayoclinic.org/diseases-conditions/narcissistic-personality-disorder/basics/symptoms/con-20025568 (accessed Feb. 8, 2019).
16 Ibid.
17 William Grimes, "Christopher Lasch is Dead at 61: Wrote About America's Malaise," *New York Times* (Feb. 15, 1994) www.nytimes.com/1994/02/15/obituaries/christopher-lasch-is-dead-at-61-wrote-about-america-s-malaise.html (accessed Feb. 2, 2019).
18 Christopher Lasch, *The Culture of Narcissism* (London and New York: W.W. Norton, 1979) p. 28.
19 *The Culture of Narcissism*, p. 89.
20 Ibid.
21 *The Culture of Narcissism*, p. 87.
22 *The Culture of Narcissism*, p. 92.
23 *The Culture of Narcissism*, p. 93.
24 *The Culture of Narcissism*, p. 24.

25 *The Culture of Narcissism*, p. 32.
26 Sigmund Freud, "On Narcissism" (1914), trans. C.M. Baines, *Standard Edition, 14* (1957) pp. 67–102.
27 R. Steiner, "On Narcissism: The Kleinian Approach," *Psychiatric Clinics of North America* 12, no. 3 (Sep. 1989): 741–70.
28 John Bowlby, *A Secure Base: Parent–Child Attachment and Healthy Human* Development (London: Routledge, 1988).
29 *The Culture of Narcissism*, p. xvii.
30 Rebecca Solnitt, "The Ideology of Isolation," *The Atlantic* (July 2016) http://harpers.org/archive/2016/07/the-ideology-of-isolation (accessed Feb. 24, 2019)
31 Ibid.
32 "Sublimus, Sublima," latindictionary.net (n.d.) www.latin-dictionary.net/definition/36017/sublimus-sublima (accessed Mar. 31, 2019).
33 Friedrich Nietzsche, *Writings of Nietzsche, Vol. III*, ed. Anthony Uyl (Ontario and Woodstock: Devoted Publishing, 2017) p. 207.
34 Friedrich Nietzsche, *Human, All Too Human* (1878) trans. R.J. Hollingdale (Cambridge: Cambridge University Press, 1986).
35 Sigmund Freud, *Leonardo da Vinci and a Memory of His Childhood*, (1910) ed. James Strachey (New York: Norton, 1990) pp. 57–137.
36 "On Narcissism," p. 124.
37 Michel Foucault, *The History of Sexuality: Volume 1: An Introduction*, trans. Robert Hurley (New York: Vintage, 1978).
38 Herbert Marcuse, *One-Dimensional Man* (London and New York: Beacon, 1964).
39 Jacques Lacan, *The Ethics of Psychoanalysis (Book VII) (The Seminar of Jacques Lacan)*, ed. Jacques Alain-Miller, trans. Dennis Porter (London and New York: W.W. Norton, 1997) p. 52.
40 Jacques Lacan, *Écrits: A Selection*, trans. Alan Sheridan (London: Tavistock, 1977). p. 230.
41 Fabio Vighi, *On Žižek's Dialectics: Surplus, Subtraction, Sublimation* (London and New York, Continuum, 2010).
42 Simon O'Sullivan, "Lacan's Ethics and Foucault's 'Care of the Self': Two Diagrams of the Production of Subjectivity (And the Subject's Relation to Truth)," *Parrhesia* 10 (2010): 51–73.
43 Slavoj Žižek, *Absolute Recoil: Towards a New Foundation of Dialectical Materialism* (London and New York: Verso, 2014) p. 4.
44 *On Žižek's Dialectics.*
45 Bernard Stiegler, *What Makes Life Worth Living: On Pharmacology*, trans. Daniel Ross (Malden, MA: Polity, 2010) p. 27.
46 *What Makes Life Worth Living*, p. 19.
47 Excerpt from Otto Kroeger and David B. Goldstein, *Creative You: Using Your Personality to Thrive* (New York: Atria, 2013). See "Creative You," Center for Applications of Creative Type (n.d.) www.capt.org/catalog/itmdtl.htm?pnum=60541&bhcp=1 (accessed Jun. 15, 2019).
48 Ibid.
49 *Creative You*, p. xxiii.
50 John LaRosa, "What's Next for the $9.9 Billion Personal Development Industry," Market Research.com (Jan. 17, 2018) https://blog.marketresearch.com/whats-next-for-the-9-9-billion-personal-development-industry (accessed Feb. 7, 2019).
51 Catherine LaSota, "All the Feels: Self-Help Books 2017–2018," *Publishers Weekly* (Oct. 17, 2017) www.publishersweekly.com/pw/by-topic/new-titles/a

dult-announcements/article/75073-all-the-feels-self-help-books-2017-2018.html (accessed Feb. 7, 2019).
52 Matthew Jones, "11 Billion Reasons the Self Help Industry Doesn't Want You to Know the Truth About Happiness." Inc.com (Oct. 17, 2017) www.inc.com/matthew-jones/11-billion-reasons-self-help-industry-doesnt-want-you-to-know-truth-about-happiness.html (accessed Feb. 7, 2019).
53 Sigmund Freud, *Civilization and Its Discontents* (1930) trans. James Strachey (New York: W.W. Norton, 1961) p. 29.
54 Roy W. Menninger and John C. Nemiah, *American Psychiatry After World War II (1944–1994)* (New York: APA, 2000) p. 216.
55 Tony Dokoupil, "America's Long Love Affair with Anti-Anxiety Drugs," *Newsweek* (Jan. 1, 2009) www.newsweek.com/americas-long-love-affair-anti-anxiety-drugs-77967 (accessed Feb. 12, 2019)
56 Michael Winerip, "The Death and Afterlife of Thalidomide," *New York Times* (Sep. 23, 2013) www.nytimes.com/2013/09/23/booming/the-death-and-afterlife-of-thalidomide.html?_r=0 (accessed Aug. 12, 2018).
57 Edith Kramer, *Art Therapy in a Children's Community: A Study of the Function of Art Therapy in the Treatment Program of Wiltwyck School for Boys* (1958) (London: Schocken, 1990).
58 Micki McGee, *Self-Help, Inc.: Makeover Culture in American Life* (New York and Oxford: Oxford University Press, 2005) p. 11.
59 *Self-Help, Inc.*, p. 12.
60 *Self-Help, Inc.*, p. 11.
61 Nolen Bivens, "From Combat to Community," National Endowment for the Arts (Oct. 25, 2016) www.arts.gov/art-works/2016/combat-community (accessed May 31, 2019).
62 Ibid.
63 Caroline Alexander, "Behind the Mask: Revealing the Trauma of War," *National Geographic* (Feb. 15, 2015) www.nationalgeographic.com/healing-soldiers/blast-force.html (accessed Feb. 8, 2019).
64 Ibid.
65 "California Arts Council Expands Rehabilitative Arts Programs in State Prisons," California Arts Council (Jul. 6, 2017) http://arts.ca.gov/news/prdetail.php?id=251 (accessed Feb. 8, 2019).
66 "Critic Reviews: Happyish, Season 1," metacritic.com (Apr. 25, 2015) https://www.metacritic.com/tv/happyish (accessed May 31, 2019).
67 Rebecca Adams, "Think Retail Therapy Will Cure Loneliness? Think Again, Says a New Study," *Huffington Post* (Aug. 8, 2013) www.huffingtonpost.com/2013/08/08/retail-therapy-study_n_3727087.html (accessed Feb. 9, 2019).
68 Martin Seligman, *Authentic Happiness: Using the New Positive Psychology to Realize Your Potential for Lasting Fulfilment* (New York: Free Press, 2002). See also, Emily Esfahani, "There's More to Life than Being Happy," *The Atlantic* (Jan. 9, 2013) www.theatlantic.com/health/archive/2013/01/theres-more-to-life-than-being-happy/266805 (accessed Feb. 15, 2019).
69 Morten Kringelbach and Kent C. Berridge, "The Neuroscience of Happiness and Pleasure," *Social Research* 77, no. 2 (Summer 2010) http://lsa.umich.edu/psych/research&labs/berridge/publications/Kringelbach%20&%20Berridge%202010%20Neurosci%20happiness%20&%20pleasure%20Social%20Research.pdf (accessed Feb. 15, 2019). See also, Morten Kringelbach, *The Pleasure Center: Trust Your Animal Instincts* (Oxford: Oxford University Press, 2008).
70 Jordan Michael Smith, "Want to Be Happy? Stop Being So Cheap," *New Republic* (Sep. 21, 2014) https://newrepublic.com/article/119477/science-gener

osity-why-giving-makes-you-happy (accessed Jul. 30, 2018). See also, Hilary Davison and Christian Smith, *The Paradox of Generosity* (London and New York: Oxford University Press, 2014).

71 Therese J. Borchard, "How Giving Makes Us Happy," PsychCentral (Dec. 22, 2013) https://psychcentral.com/blog/archives/2013/12/22/how-giving-makes-us-happy (accessed Jul. 30, 2018).

72 Darice Britt, "Philanthropic Psychology Sheds Light on Charitable Giving, *South Source* (Mar. 2013) http://source.southuniversity.edu/philanthropic-psychology-sheds-light-on-charitable-giving-128334.aspx (accessed Jan. 2, 2019).

Chapter 3

The Neoliberal Imagination

When More Is Not Enough

History will remember the unlikely presidency of Donald Trump as one of the strangest periods in American politics. It may never fully be understood how the former entrepreneur and TV personality wound up in the Oval Office, amid the tangle of media sensationalism, rabid populism, and outright illegality involved. Certainly, a key element in the Trump phenomenon was the President's "creative" relationship with truth and factuality—for which he was well known well before he sought elected office. This chapter examines the Trump phenomenon and the insurgent populism sweeping the globe in the contexts of both today's "post-truth" era and totalitarian movements of the past.

Trump's behavior consistently has baffled his critics. At first many thought he was joking or simply trying to shock people. Then some labeled him a pathological liar. Others concluded he was just plain crazy. No less a periodical than the mainstream *Time Magazine* devoted a cover to the President's "Truth and Falsehoods"; the *Los Angeles Times* ran multiple "Why Trump Lies" editorials; and the *New Yorker* ran 16 installments in its "Trump and the Truth" series.[1] And along the way, the *Washington Post* clocked over 8,000 untruths spoken by Trump.[2] Unsurprisingly, the President doubled down on his claims, and—in keeping with his fondness for conspiracy theories—lambasted the entire field of journalism as "the enemy of the American people."[3] Pundits and commentators seemed unable to discern a logic in the President's bizarre behavior—in which mischief and chaos seem the only constants.

Trump didn't invent this disruptive approach to politics. Reactionary leaders through history (and around the world) have used the same techniques for a least a century. While some critics would question the President's grasp of "reality," others began to see a calculated shrewdness in his behavior—an underlying strategy not unlike what Naomi Klein discussed in her review of twentieth-century global upheavals in *The Shock Doctrine*. [4] Commenting on Trump's efforts "to deregulate markets, wage all-out war on 'radical Islamic terrorism,' trash climate science and unleash a fossil-fuel frenzy," Klein observed that the essence of Trump's program lay in "generating a tsunami of crises and shocks."[5] She correctly predicted economic shocks (as market bubbles burst), security shocks (as blowback from foreign belligerence would come home), weather shocks (as the climate was further destroyed), and industrial shocks (as oil pipelines spill and markets were destabilized).

"All this has been dangerous enough," Klein explained, "What's even worse was the way the Trump administration could be counted on to exploit these shocks politically and economically."[6] Trump himself forecasted as much often in promising a "radical break" from the past—described by *Fox News* as a "shock and awe campaign against the Washington establishment."[7] This new agenda bore little resemblance to earlier "culture wars" between conventional liberal and conservative camps. Moral idealism had no place in Trump's program of disruption and dishonesty. But his ability to confuse and deceive might have been taken more seriously. Hence, the Trump phenomenon provides a cautionary tale about the role of knowledge in contemporary society—and the ways different worldviews are conceived, put into circulation, and frequently politicized.

Again, these matters are nothing new. Philosophers have long pondered the messy distinction between truth and fiction, especially when emotion and human intention enter the picture. In the 1950s, Lionel Trilling argued in *The Liberal Imagination* that Americans were especially vulnerable to such confusion, despite their beliefs in reasoned "progress." Trilling joined a growing list of intellectuals pointing to the influence of "culture" in political thought. "Unless we insist that politics is imagination and mind," Trilling wrote, "we will learn that imagination and mind are politics, and of a kind we will not like."[8] This notion later was embraced by postmodern theorists to explain how representations (like words and images) often stand in for genuine experiences in people's minds. Taking this idea further in the twentieth century, media scholars would describe how commercial and political messages "manufacture consent" through a purposeful manipulation of information. The inherent instability of meaning continues to make mischief in today's political discourse—playing with people's abilities (or inabilities) to make sense of messages in the "knowledge economy."

Beginning with a look at President Trump, this chapter examines how Trilling's liberal imagination has taken neoliberal form in recent decades. A now-familiar extremism has overwhelmed classical liberal principles of limited government in a move promising the creative destruction of a failing system. Disaffected voters increasingly endorse this—often swayed by promises of more freedom to pursue their own interests. Along the way, selfish individualism overtakes civic responsibility in people's minds. Lost is the "sociological imagination" through which people value their communities and care about their neighbors. Culture and creativity become both incentive and method in this rising neoliberal trend, as Trump's shrewd use of media clearly demonstrated. The question becomes what, if anything, can be done to turn the situation around?

The Performance Art of the Deal

But let's return to Trump for a moment—and ponder his apparent disdain of "reality." In what follows I'll discuss the President's political theatrics through the lens of what might be called the "neoliberal imagination." With the liberal consensus of Trilling's day now gone, a new form of consciousness governs the world in which clear distinctions of many kinds have disappeared: private/public, truth/fiction, and even good/evil. In an era when the self has become a commodity and a reality TV star can rise to world leadership, is it any wonder that public confidence is so easily rattled? That many will scramble so quickly to reactionary extremes? Much of this derives from received ideas, as Trilling observed. But not all people realize how much imagination itself is shaped by the culture around it.

So was America's huckster-in-chief just putting on an act? Aren't all politicians "performers" in some sense of the term? You would expect that in an age of media spectacle most Americans are accustomed to this idea. The problem is that many still yearn for authenticity in their leaders' words and deeds. Hungering for a vision of a better world, voters will cast aside doubts about a candidate's claims and promises—only to be disappointed when reality sets in. This partly explains why Americans have become cynical about democracy. One often hears laments for an idealized past when "truth" prevailed, that it could be recognized when it appeared, and that the concept informed America's behavior in the world. In actuality, Americans (and their politicians) have always had a "creative" relationship with the truth. Has The Donald simply taken this tradition to its logical end?

"What you have to understand about Trump, first of all, is that he's a performance artist, not a politician," explained filmmaker Michael Moore.[9] As his presidency began to unfold, savvy critics began placing Trump's antics within the realm of artistic expression. And if you think about it, this makes a lot of sense. Before the presidency, Trump's most visible role had been that of a media showman in various incarnations of his well-known *Apprentice* TV shows. His books *The Art of the Deal* (1987), *The Art of Survival* (1991), and *The Art of the Comeback* (1997) are peppered with references to his "creative" financing and bookkeeping.[10] A quintessential proponent of creative capitalism, Trump wrote, "Deals are my art form. Other people paint beautifully on canvas or write wonderful poetry. I make deals, preferably big deals."[11] Art itself was a "con" to Trump, who once quipped that "successful painters are better salesmen and promoters than they are artists."[12] The Trump family made a fortune from the creative economy—from Donald Trump's own entertainment enterprises, to Melania Trump's career as a fashion model, to Ivanka Trump's clothing and accessory lines, to Donald Trump Jr's. real-estate investments in New York's artistic East Village. Recently, famed performance artist Karen Finley put the matter bluntly in stating that "Donald Trump *owes all of his wealth* to arts and culture."[13]

The Trump-as-performance-artist premise isn't as far-fetched as it might sound. Entrepreneurial success often hinges on creative traits like imagination, novelty-seeking, risk-taking, and "thinking outside the box." Also, the President's iconoclastic slash-and-burn assault on liberal sensibilities fit within certain well-established aesthetic traditions. Dada and Surrealist artists of the early twentieth century delighted in upsetting bourgeois sensibilities. Often linked to anarchist movements of the time, Dadaist "anti-art" upended conventional logic, aesthetics, and morality to shake up a complacent society and open the way toward a better world. André Breton's Surrealist manifesto of 1924 advocated "the absence of any control exercised by reason, exempt from any aesthetic or moral concern."[14] One outraged reviewer in *American Art News* called the movement "the most paralyzing and most destructive thing that has ever originated from the brain of man."[15]

Now, obviously, America's fearless leader may have balked at these comparisons. And keep in mind that early Dada and Surrealism (as well as their 1960s revivals) patently opposed both capitalism and militarism. Yet while Trump's brand of disruptive aesthetics lacked the utopianism of those movements, his antics indeed had another precedent—in the proto-fascist Italian Futurist movement that emerged from Dada. In 1908, Filippo Tommaso Marinetti's, "Futurist Manifesto" celebrated industry, speed, and technology with a rabid nationalism and a determined view of humanity's triumph over nature. Marinetti wrote:

> They will crowd around us, panting with anguish and disappointment, and exasperated by our proud indefatigable courage, will hurl themselves forward to kill us, with all the more hatred as their hearts will be drunk with love and admiration for us.[16]

Rebelling against harmony and the niceties of "good taste," the Futurists also valued intuition, not unlike President Trump. And, of course, the Futurist movement propelled the rise of Benito Mussolini—an authoritarian outsider whose populist appeal lay in his assault on establishment values.

Propaganda was a key tool of the Italian Fascist regime, as it also was in pre–World War II Germany. Mussolini reveled in pageantry, expansive rhetoric, and an image of infallibility. The corporatist state was valorized over liberal democracy, with a heavy emphasis on military power. Campaigning on the slogan, "Drain the Swamp," Mussolini repudiated immigrants and anything "foreign." And in yet another eerie parallel to Trump, "Il Duce" was portrayed as a workaholic insomniac who was infallible in his judgments.[17] Journalistic accounts that contradicted the regime were condemned, with newspapers often confiscated. News outlets approved by Il Duce largely ran stories about rampant crime and "monsters" loose on the streets. As a former news editor himself, Mussolini eventually established his own Ministry of Popular Culture to manage public opinion, much like the Third Reich's Ministry of Public Enlightenment and Propaganda. Early

Trump White House Strategist Steve Bannon cited the Reich Ministry as his source of technique and inspiration.[18]

While it is tempting to draw parallels between past authoritarian leaders and the contemporary Donald Trump, such comparisons aren't entirely accurate and they risk trivializing the horrors of the World War II era. Trump always has seemed less a creature of fascist ideology than of business pragmatism. He never was that much of a "conservative" either, as many republicans would grumble. For Trump, the deal has been everything. And like any good dealmaker, Trump has recognized the value of putting the other side at a disadvantage. Posturing, intimidation, and deceit are the main tactics Trump used against such hospitality giants as Hilton, Hyatt, and Holiday Inn. One key to Trump's success has been his skill in throwing opponents off balance while ignoring their arguments. According to Harvard University's Program on Negotiating (PON), this can be a shrewd technique. In "Secrets of Successful Dealmaking," PON experts wrote that strategic use of chaos and disruption allows "those who can adapt quickly to changing circumstances to gain an important edge on their competition."[19] In contrast, managers clinging to "rigid strategies risk being blind to unexpected perils—and to unforeseen opportunities, especially when engaging in integrative negotiation strategies."[20] The bigger and more complex the deal, the more useful chaos can be.

Creative Destruction

The term "creative destruction" describes capitalism's expansion into fresh territories, often "destroying" outdated forms of production and exchange along the way. Like the shock doctrine, creative destruction is a concept familiar to President Trump. As explained by Ilana Mercer in *The Trump Revolution*:

> Donald J. Trump was smashing an enmeshed political spoils system to bits: the media complex, the political and party system, the conservative power complex. In an age of unconstitutional government—Democratic and Republican—the process of creative destruction can only increase the freedom quotient.[21]

While he behaved at times like an "benevolent authoritarian," Mercer concluded that "the process of creative destruction begun by Donald Trump is likely the best Americans could hope for."[22]

In conceptual terms, creative destruction dates to the shift from agrarian to industrial economies in the 1800s. "Previously created productive forces are periodically destroyed," wrote Karl Marx and Friedrich Engels in describing the often brutal "creation" of new markets and commodity forms.[23] The actual term "creative destruction" was first used in 1942 by economist Joseph Schumpeter in a work entitled *Capitalism, Socialism and*

Democracy—in which he attempted to merge the three philosophies. Contrasting traditional capitalists with a new generation of "entrepreneurs," Schumpeter suggested the latter could be the genesis of an equitable free market, using creative destruction to break the power of undemocratic monopolies. He wrote that the creative impulse "incessantly revolutionizes the economic structure from within, incessantly destroying the old one, incessantly creating a new one," concluding, "The process of Creative Destruction is the essential fact about capitalism."[24]

Creative destruction took a decidedly conservative turn in the 1980s, as the Reagan administration began dismantling government programs with a vengeance. Soon creative destruction was being used by both the Democrats and Republicans to describe the new agenda. In "Neoliberalism as Creative Destruction," social theorist David Harvey would observe how early "experiments in creative destruction carried out in the periphery" were appropriated and transformed into an aggressive "model for the formulation of policies in the center."[25] Henceforth, creative destruction became synonymous with "political economic practices proposing that human well-being can best be advanced by the maximization of entrepreneurial freedoms within an institutional framework characterized by private property rights, individual liberty, unencumbered markets, and free trade."[26]

Steadily gaining momentum in the twenty-first century, "The wave of creative destruction that neoliberalization has visited across the globe is unparalleled in the history of capitalism," Harvey observed.[27] The ascension of Donald Trump to the nation's highest office removed any shreds of restraint in the process. Scapegoating public programs as the cause the country's woes, the President slashed federal support for agriculture, education, energy assistance, environmental protection, health care, housing, minority businesses, and research in a wide swath of medical and scientific fields. In this context, his efforts to eliminate such agencies as the Corporation for Public Broadcasting and the National Endowments for the Arts and Humanities surprised no one.

Partisan politics notwithstanding, the steady reorientation of American life in a further neoliberal direction has made it easier to accept market principles as normal ways of looking at things. In a nation already valorizing competition and personal freedom, the apparent "rationality" of supply and demand is difficult for many to resist. In this scheme, success or failure become matters of individual will—as attention is diverted from any structural advantage or disadvantage a person brings to the table. Inequality is easily justified as a "natural" phenomenon, rather than a consequence of human intent. The "marketization" of formerly non-commodified areas of life is not always direct and obvious, as political theorist Wendy Brown has observed. Private high schools don't need to tell perspective families that their children will make more well-to-do friends. Upscale dating services never promote themselves as pricing out low-income people. "Widespread economization of heretofore noneconomic domains, activities, and subjects, but not necessarily marketization or monetization of them, then, is the distinctive feature of neoliberalism," Brown wrote.[28]

In other words, neoliberal creative destruction affects more than government programs and policies. It changes the way people think about themselves, gradually transforming them into "financialized subjects." Max Haiven has described such financialization as an "ideological process whereby a set of narrative, metaphoric, and procedural resources imported from the financial world come to explain and reproduce everyday life and the capitalist totality of which we are part. But, in doing so, they also transform that reality more broadly."[29] This view has a certain history. Marx described the commodification of labor in the industrial age, as workers became dehumanized by manufacturers into little more than machine parts. The nature of work has changed significantly in today's post-industrial times, of course. With manufacturing distributed across national borders and electronic labor taking non-material form, labor no longer is confined by factory walls. Employees come and go in part-time jobs, check work email in coffee shops, and often complete tasks at home—with jobs residing more in their minds than any physical location. Direct supervision becomes unnecessary as the atomized field of workers monitors itself. This gives many workers the illusion of "freedom," as their financialized subjectivity becomes internalized.

The flip side of earning is spending. As with work, market apologists gleefully expound the inherent democracy of the marketplace, where buyers are "free" to choose anything they like. Once again, the lure of apparent autonomy plays a powerful role. Shopping now entails much more than picking a car or new pair of jeans—extending into areas unheard of a few decades ago. People shop for hospital care, places of worship, relationships, and designer babies in an age where one's sense of self often is defined more by brand identity than anything else. In a widely read essay entitled "What Isn't for Sale?" Harvard political philosopher Michael Sandel described the marketing of prison cell "upgrades" in California, fast-track access to highway lanes, and the right to pollute the environment (carbon credits) in an American consumer society "where almost *everything* is up for sale."[30] According to Sandel, "It's a way of life, in which market thinking and market values begin to dominate every aspect of life: personal relations, family life, health, education, politics, law, civic life."[31]

Earning and consuming come together in the "work-and-spend cycle," an expression coined by sociologist Juliet Schor to explain why people willingly put in more hours so they can buy more goods. For decades, academics like Schor have been charting the shrinkage of personal time in a U.S. where people already work more than their counterparts in countries like Canada, France, Germany, Great Britain, and Japan. Much of the blame comes from competition engrained in nearly every aspect of American life. As Schor observed in *The Overworked American*, "Time is a measure of commitment" in the competitive workplace, often obliging workers to curb the hours they allocate for family, community, and leisure activities.[32] Consequently, U.S. workers get fewer days off per year (15 days) than those in Europe (28 days) and the Asia-Pacific region (19 days).[33] The same comparative horserace also affects people's purchasing decisions, making those without the latest iPhone or Toyota feel less worthy. None of this was helped by the recent recession—in which a tightened job market gave employers the upper hand and made buying things all the more difficult.

The concept of creative economy plays into these transformations of work and spending. Jobs in creative fields often are temporary, project-based, and part-time, with off-site work also commonplace. Evidence also shows that many creative workers willingly sacrifice high pay for the extrinsic value of meaningful employment—which, despite its subjective benefits, also lowers standards of living while creating stresses and health risks. Meanwhile, marketing increasingly emphasizes consumer spending as a form of self-expression and definition, proffering "creative solutions" to everything from home decorating and repair to the selection of one's clothing and entertainment choices. Clearly, neoliberal creative destruction is much more than an economic program in its incursions into people's choices, habits, and desires. It has become a way of organizing the human imagination itself within a limited range of "creative" options. Can anything be done to reverse this way of thinking?

Neoliberalism in Context

Skeptics link the new creativity to "neoliberalism," a concept that appears frequently these days in discussions of contemporary culture. Once an obscure term in economics, neoliberalism now has become a catch-all phrase for a wide range of social and political changes in the past three decades, including those in the creative realm. Despite what its name suggests, there is nothing completely "neo" or new about neoliberalism. It mostly refers to a more business-friendly and less-regulated form of capitalism of the kind practiced in America before the 1930s and periodically since then. What is "new" is the extent to which its proponents would apply neoliberalism to all aspects of contemporary life. Hence, neoliberals tend to oppose big government, progressive taxation, minimum wage and equal pay measures, consumer protection, immigrant rights, media regulation, and environmental laws, among many other things—just as critics of neoliberalism favor such measures.

The ideological debates over neoliberalism are deeply rooted in America's past, pertaining to matters of personal liberty, free enterprise, egalitarianism, governance, and community responsibility. Hence, neoliberalism activates familiar left–right divides on many issues. Central in neoliberalism is a reconceptualization of nearly everything in life in monetary terms. This makes a certain kind of sense in an America still shaky from the recession, with memories of dinner conversations about how to pay bills fresh in people's minds. It also pushes the nation in a politically "rationalist" direction—whereby certain forms of reason and truth are held above all others, often at the expense of diverse opinion and democratic thought. But let's face it, when things get "real" as they did in the recession, values can shift and change. And with a little bit of prodding, they can shift quite a bit.

Analyses of contemporary neoliberalism hinge on four basic concepts—*Rationalization, Crisis, Virtualization,* and *Destabilization*—which together help explain neoliberalism's ubiquity and apparent invisibility. Even as opinion polls show that millions remain unhappy, frustrated, and anxious about the future, most people are unable to explain what is wrong. They blame other groups, big business, government, foreign powers, and often themselves, without seeing that something larger and more systematic might be behind it all. The fact is that neoliberalism is so expansive that it even transcends common categories like left or right, Democrat or Republican. The abstraction of the following concepts illustrate why neoliberalism is so difficult to pin down.

Rationalization may not be especially new, but from the earliest days of neoliberalism its advocates would claim that a particular form of economic "reason" should be applied to everything. This premise was introduced in the 1940s by Austro-Hungarian-born philosopher Friedrich Hayek, who later became a luminary in the famed Economics Department of the University of Chicago. Hayek wrote, "Economic control is not merely control of a sector of human life which can be separated from the rest; it is the control of the means for all our ends."[34] Within such sweeping logic, it simply seemed "sensible" to reward hard work and ingenuity, manage public programs with frugality, and remain vigilant against rising socialism. Unfortunately, this alluring "new" rationalism would enable an incremental reorientation of public policy and consciousness in the interest of "efficiency" rather than human concern, a redefinition of virtue as that which enhanced wealth, safety, or security. Anyone or anything not directly contributing to this agenda would be cast as a distraction, and impediment, or, sometimes, as a threat.

Crisis can trigger systemic change like nothing else, as outlined by Michel Foucault in his "Birth of Biopolitics" lectures given at the Collège de France in 1978–79.[35] Without a doubt one of the most trenchant early critics of neoliberalism, Foucault was unsatisfied with traditional Marxist theories of political economy. Rather than a symptom of capitalism's inevitable demise, Foucault attributed neoliberalism's enabling "crisis" to a broader loss of faith in the organizations that held society together, especially the all-important institution of the democratic state. People needed a place to look for answers in times of fiscal calamity and enemy threat. To Foucault, a new kind of reassurance began to emerge in the decades following World War II in the form of neoliberal "reason." The shock of global conflict and despair followed by economic growth and anti-communist sentiment—all of this set the stage for what Foucault would term *homo oeconomicus* (the "economic man") to emerge as a social ideal and model for subjectivity. Key in Foucault's theory was that this new rationalism need not ever be literally named or imposed as such, since it became a part of a more generalized *episteme* or "common sense" that could influence people "without touching" them directly.

Virtualization has underlain historical accounts of ideology from Napoléon Bonaparte to Louis Althusser and beyond. As neoliberalism began to ascend in the early 1980s, philosophers and social theorists began voicing concern about the rising role of media and culture in shaping the imaginary relationship of citizen to society. Postmodern critics would speak of a world in which "images" played a bigger role than direct experience. Or, as Fredric Jameson would write in *Postmodernism, Or the Cultural Logic of Late Capitalism*, there came a recognition of the "cultural" and the "economic" collapsing into each other (perhaps as Hayek would have liked) in a way that made them indistinguishable, obligating one to "talk about cultural phenomena at least in business terms if not in those of political economy."[36] More recently, the continued computerization of everyday life has paralleled a decline in industry and a rise of "financialized capital," as more and more of the economy is rendered in abstract language of credit, debt, and investment instruments. Fewer people use cash itself in an age of online shopping, bank cards, and Apple Pay.

Destabilization rises when nations lose direction and become confused about their governing ideals. Such disorientation pervades today's media-saturated America, as political meaning and economic reality become increasingly fluid and difficult to grasp, eroding stability in the balance between democracy and capitalism. As money itself become less "real," many people find themselves overloaded by a constant bombardment of information on their phones, computers, TV, and environments. Pulled into a constantly changing present moment, many find themselves disoriented and confused. Fed a constant diet of worry and fear, isolated individuals search for answers or hope when hardship hits. This leaves many vulnerable to compulsive consumerism and political opportunism. Wendy Brown recently wrote in *Undoing the Demos: Neoliberalism's Stealth Revolution* that in saturating "democracy with market values, neoliberalism assaults the principles, practices, cultures, subjects, and institutions of democracy understood as rule by the people."[37]

Recent cultural and demographic changes in U.S. society haven't helped matters either, especially as they have been amplified in an increasingly polarized political atmosphere. In a country that valorizes individualism and competition as foundational ideals, social alienation is reaching epidemic proportions. According to U.S. Census statistics, more people now live alone than ever before—a number that has grown from 5 percent in the 1920s to 17 percent today.[38] The shifts are especially visible in urban areas and among young people choosing to defer marriage. "Americans prefer privacy in living arrangements," the Census Bureau reported, while also expressing concern about spiking numbers of seniors becoming vulnerable due to such isolation. And while some believe the internet has made people more "connected," leading experts contend this isn't the case. In her book *Alone Together*, MIT Professor Sherry Turkle argued that virtual living does little to resolve tensions between individual and community. "We are lonely but fearful of intimacy," she writes, adding that "networked life allows us to hide from each other, even as we are tethered together."[39]

It bears restating that none of this is helped by the constant barrage of news reports, talk radio, and opinion programs that sensationalize difference and feed ideological tension. Resentment of authority has become palpable in a nation in which corporate CEOs and government leaders are equally despised. As fewer and fewer Americans attend religious gatherings of any kind, values of wealth and opulence greet them daily in the media they consume—even as the economy has flatlined. It's no secret that the income of a typical U.S. family hasn't risen a penny for two decades, when adjusted for inflation. At the same time, income inequality has risen dramatically, with the earnings of top 1 percent have rising by 256 percent.[40] As mentioned in Chapter 1, this partly explains the anger expressed by so many Americans during the last election cycle. *CNN Money* reports that things are especially bad for adults without college degrees, who find work at only 30 percent the level of those with higher education.[41] Meanwhile, conditions of structural racism, sexism, ageism, and other forms of bias continue to disadvantage huge cohorts of the population. Is it any wonder that many Americans feel the nation's recovery has not reached them?

The Wealth of Nations

Why were things allowed to get this way? Some of the answers lie in traditions of American thought that extend to the nation's origins, as well as the peculiar twists that public consciousness has taken in the information age. Taking this broader view helps explain the "progressive" neoliberalism of Bill Clinton and Barack Obama, as well as the "reactionary" version of Donald Trump.[42] In different ways, both "Yes We Can" and "Make America Great Again" evoked populist ideals of "freedom" and "independence" grounded in Enlightenment philosophical thinking of the colonial era. In the decades before and after 1776, uprisings and revolutions took place in nations now known as Austria, Belgium, Bosnia, Ecuador, France, Ireland, Greece, Paraguay, the Philippines, Poland, Serbia, and Scotland. All of this had to do with the rising stature of the "self" and associated concepts of personal autonomy—in contrast to prior beliefs that human existence was governed by natural laws, birth status, or godly authority.

This context gives Adam Smith's "invisible hand" principle a certain poetic resonance, although a cynic might say it was an early mystification of capitalism's dark side. Attributing ineffable intelligence to the workings of the market imbues the very real exchange of goods and money with a magical quality. Of course, few people remember that Smith was a moral philosopher as well as an economic theorist, and so his evocation of a deified invisibility should not be a big surprise. Smith's *Wealth of Nations* portrayed capitalism and democracy as inexorably intertwined around the new figure of the reasoning individual—expressing "self-interest" in the marketplace and the voting booth. Also often forgotten today, Smith fully advocated a role for government in such provisions as the funding of schools, the building of bridges and roads, and, of course, the upkeep of a national military.

Set against the backdrop of oligarchical tyranny, Smith and others saw a collective good in the unleashing of individual agency—with the broader society benefiting from the dynamism of acquisitive self-interest. At the time, this was termed a "liberal" approach to social policy (governance and economy). Smith and others genuinely believed that personal ambition indirectly benefited everyone, when he wrote:

> [The individual] intends only his own gain, and he is in this, as in many other cases, led by an invisible hand to promote an end which was no part of his intention. Nor is it always the worse for the society that it was no part of it. By pursuing his own interest he frequently promotes that of the society more effectually than when he really intends to promote it.[43]

In political terms, Western secular democracy emerged as an empowering expression of human agency. Decision-making became vested in the reasoning abilities of each person, presumably able to assess truth claims when afforded the information to do so. The novel concept of the American "citizen" entailed two reciprocal strains of freedom. *Private citizenship* entailed the freedom to pursue one's own goals, acquire goods, and advocate one's beliefs without interference. *Public citizenship* included the freedom to vote, play a role in governance, and to join others in collective efforts. From the beginning, the nation's founders recognized a productive tension between public and private realms, with discord and disagreement seen as healthy antidotes to prior ways of life driven by uniform beliefs or authoritarian rule. Individual motivation was a key element of this vision, but it was only half of the plan. Public interest was needed to give the system added purpose and dynamism.

Put in the language of democracy, capitalism was seen as a system in which citizens could vote with their money. Someone might enter a store selling various makes of shoes, and decide a certain kind was best. The buying of one brand of shoes would discourage the making of others. And as this pattern continued in the market, an "invisible hand" would raise footwear quality. It's a quaint theory in historical context, but not a universal paradigm for contemporary times. Smith and those of his day couldn't foresee how modern corporations would become able to aggregate consumers or how a mechanized shoe factory could multiply profit-making in geometric terms. Put another way, the founders couldn't predict how capitalist materialism might overtake democratic idealism.

And, of course, the tensions between capitalism and democracy *would* continue. Before long G.W.F. Hegel would write about this in his 1821 work *The Philosophy of Right*—which described the tendencies of bourgeois societies to generate excessive wealth, classes of paupers, and methods of advancing their interests through colonial adventures.[44] As a student of Hegel, Karl Marx later would elaborate upon the "surplus" profits generated by industry and the ways this affected workers and consumers.[45] Hence, the nineteenth-century critique of corporate capitalism would begin. Then as now, business interests would counter that profits translated into new jobs, higher wages, and added spending—in an endless growth spiral. But time would show that factory owners rarely were satisfied with simply selling more goods or adding workers. With more and more job-seekers flocking into cities, industrial employers applied market competition to hiring and retention practices. Marx termed this the "commodification of labor"—in which the human activity of working became abstracted as simply one more material cost of industrial production.

Class consciousness eventually developed as workers came to understand and communicate about their conditions. Strikes took place, unions were organized, and collective bargaining emerged—but sometimes with violent consequences. Eventually, government had no choice but to step in. Over time, new laws and labor regulations offered partial relief to the labor market, while also reigning in the not-so-invisible hands of monopolies. "Progressive Era" reforms in early twentieth-century America applied pressure to the expansive "liberal economy" of the era, as unions were formed, child labor curtailed, worker safety measures initiated, and conservation programs begun. Amid this atmosphere of political and economic reform and turmoil (not to mention the disruption of World War I), the rising influence of media and culture on class consciousness took a back seat to more directly material concerns. By growing increments, new communications technologies (newspapers, magazines, radio, movies) would change democracy and capitalism, enabling previously unknown forces to magnify differentials between private and public interests. As the pace of media innovation accelerated, regulation remained minimal or non-existent, especially in countries like the U.S. where free speech was historically politicized. Private interest came to trump public concern nearly everywhere in the American mediascape.

Classical economic liberalism's invisible hand theory was disproven by the Great Depression of the 1930s—showing that in unattended markets extreme profits and losses can spin out of control. To prevent such a catastrophe from happening again, John Maynard Keynes led a revolution in national fiscal policy by proposing that governments could keep economies from melting down.[46] While not proposing absolute federal management of the economy (as seen in socialist and communist Europe at the time),

Keynes advocated correcting market problems via regulation, taxation, and various forms of incentive. The premise of Keynesian economic mediation has a certain logic, although to this day staunch free-market advocates still insist that any government interference with the invisible hand is both wrong-headed and short-sighted. A nation stunned by the shock of the Great Depression would accept the government jobs and investment promises of Franklin Roosevelt's "New Deal," and with it Keynes's ideas for government intervention in macroeconomic matters. And as international conflicts roiled and the globe was lumbering toward World War II, the role of federal spending, employment, and economic intervention remained largely unchallenged.

The World War II military build-up also enabled an unprecedented economic alliance between defense contractors and government. With the cessation of fighting, then President Dwight D. Eisenhower made a startling speech as he was leaving office. In his historic "Military-Industrial Complex" address, the former Army Commanding General gave a name to a phenomenon with far-reaching implications, especially as new mood of ideological extremism began sweeping the nation. "The conjunction of an immense military establishment and a large arms industry is new in the American experience," Eisenhower stated. "Our toil, resources and livelihood are all involved; so is the very structure of our society. In the councils of government, we must guard against the acquisition of unwarranted influence, whether sought or unsought, by the military-industrial complex."[47]

Before long, cooperation between business and government would expand with laissez-faire encouragement. Reacting against the perceived overreach of Keynesian policies, Chicago economist Milton Friedman adopted the term "neoliberalism" in proposing an expansive reassertion of free-market principles throughout American society. Friedman's arguments quickly found support in an era of Cold War anti-Communist anxiety. Disregarding the economic benefits of government war and subsequent infrastructure spending, Friedman and other "Chicago School" ideologues relentlessly pushed business as the nation's only fiscal salvation. This neoliberal offensive would prevail through Republican and Democratic presidential administrations well into the 1970s.

"Government is not the solution to our problem; government is the problem," announced President Ronald Reagan in his 1981 Inaugural Address.[48] In what would become known as "Reaganomics," the new chief executive launched what many believe to be the most dramatic embrace of neoliberalism in modern history, immediately implementing across-the-board tax cuts, industrial deregulation, accelerated capital depreciation rates, cutbacks in welfare programs, and reversals in equal-opportunity hiring initiatives, as well as a broad range of conservative reforms in education, media, the humanities, and the arts (later termed the "Culture Wars"). Despite its economic tinkering, the Reagan Revolution threw the nation into a recession, however—owing to huge increases in defense spending and a series of military interventions overseas.

The next chapter in neoliberalism's story came in the Bill Clinton presidency of the 1990s under the guise of "third-way" politics and economics. At the time, third-way governments also came into power in Australia, Great Britain, and Italy—although their orientations varied. This definitional ambiguity made third-way movements unpopular on both the left and the right, while also giving them plenty of wiggle room within whatever "middle" ground they chose to occupy. It's worth noting that neoliberalism had been framed as a third way from its inception by Chicago School economists, who cast their program as a "compromise" between communist and capitalist extremism. This notion of a middle group between "left" and "right" confuses the issue, and distracts attention from the more serious compromise between what most Americans believe is a separation of government and business, and, more seriously, a prohibition against wealthy interests buying political favors. But it does help explain how Democratic Party presidents like Clinton and Obama came to adopt what has been termed "progressive neoliberalism" in the spirit of compromise.[49]

Worries about money in politics derive in part from a widely held misconception about the relationship between business and government more generally—with many people believing that a clear separation exists between the two, like the famous church/state firewall. In fact, a cooperative relationship between private and public interests was built into the American system from the start, based on the constitutional premise that the role of the government was, at least in part, to enable and protect free enterprise, the acquisition of property, and so on. The famous Citizens United Supreme Court decision of 2007 greatly undermined these earlier laws by allowing groups to make large contributions as part of their "free speech" rights. The Citizens United case in particular has fed growing consternation over the role of money in politics, especially when it comes to election campaigns.

Neoliberalism's ascendency has afforded the private sector an unprecedented role in governance. Over the past two decades, the amounts spent on Presidential campaigns have increased tenfold, from the $400 million spent in the 1996 Bill Clinton/Robert Dole race to over $4 billion expended in more recent campaigns.[50] But money in politics isn't limited to elections by any means. Consider the estimated $9 billion expended each year in lobbying efforts by corporations, unions, and other interest groups.[51] And this is leaving aside the many thousands of experts, consultants, and interest group representatives who regularly sit on government panels, councils, and funding committees, not to mention the additional legions of university professors, academics, and researchers from privately funded think tanks who similarly have voices in public policy.

In all of this, it's important to keep in mind that neoliberalism isn't the only way of doing things. There are all sorts of other models, even if one accepts a capitalistic framework. Germany's sustainable growth system has made it the dominant economic power of the European Union. Thriving nations like Canada, Denmark, the Netherlands, Norway, and Sweden operate with social-democratic economies. And, by the way, a report from the Organization for Economic Co-operation and Development finds that citizens in Northern European nations are among the happiest (and healthiest) in the world.[52] So, the solutions to America's problems are not matters of abstract speculation. They exist in real-life examples around the world. What is missing in the U.S. is the willingness to embrace change. What is missing is the ability to imagine a different way.

The Neoliberal Imagination

Money in politics isn't the only problem that neoliberalism brings. Equally concerning is the "neoliberal imagination" that maintains a grip on American society—as market values exert an influence on how people think about themselves, their relationships to each other, and the kind of society they envision. Since the mid-twentieth century, some critics have argued that media completely drive the public mind, as Marshall McLuhan suggested in 1964 when he wrote that "the medium is the message." Sensible as this seemed in the age of television, McLuhan may have underestimated certain matters—such as the emotional appeals embedded in "messages," the intentionality of their authors, and the role of institutions in promoting them. Also, McLuhan was writing before the abilities of audiences to think critically and "push back" were widely recognized.

That said, perceptions of truth or reason are always conditioned by the imagination to some extent. This became dramatically clear in recent presidential elections, as the nation has divided over visions of a country on the ascent or on the decline. Clearly, the electorate sees two very different "truths" about America and imagines quite disparate futures. Both appear "reasonable" to their fans, backed by facts and evidence, and, perhaps most importantly, both were supported by "interpretive communities" agreeing with their imaginary views.[53] Unfortunately, many people see truth and reason in absolute terms, with little room for compromise or nuance. This black-and-white way of seeing dates to the Enlightenment Age of Reason itself, and the legacy of dualistic thinking Descartes famously set into motion. For all of its benefits, various forms of "reason" also brought to the world the principles behind colonialism and imperialism, the subjugation of women, the institution of slavery, Nazi genocide, and the scapegoating of immigrants, the poor, and other groups.

Early sociologists recognized the centrality of imagination in group dynamics—and that traditions, beliefs, and communication were central to the definition of any society. By the 1950s, this "cultural" orientation would lead C. Wright Mills to coin the expression "sociological imagination" in describing the ways individuals connect their own experience to the wider society around them.[54] Key in the sociological imagination was the way both individual and collective thought often assumed a "normative" view of human existence, which foreclosed the possibility of "imagining" better or more inclusive ways of doing things.[55] The shift in sociology paralleled similar post-positivist movements across many academic disciplines—all stemming from the recognition that no single form of reason is universal in any ontological sense. This is to say that no matter how one defines "imagination," it comes as much from the society around one as it does from inside an individual. As most people recognize, imagination largely is influenced by communication, peer opinion, and other cultural factors. None of this is especially new, although it is common today to blame television and online media for public division, confusion, and apathy. The fact is that *perceptions* of reality have long influenced what people think and believe.

Visuality doesn't get much attention in most accounts of revolutionary America. But it's important to keep vision and seeing in mind as one thinks about how the nation's history unfolded. Vision also has a bearing on contemporary neoliberalism, owing to its historic role in supporting principles of reason and subjectivity. If you think about it, vision is wholeheartedly democratic (almost everyone has it), although historically it often has served the opposite ends. Early American settlers may have been troubled by British laws and taxation, but they also were wary of continual monitoring by colonial overseers—much in the same way that populations always have resented the intrusive eyes of police, military occupiers, and government surveillance of other kinds. Dutch settlers in Boston and New York coined the term "watchmen" for those paid to look over the behavior of citizens. Keeping in mind that as Franklin and Jefferson were advocating political and economic change, vision also was part of the "freedom and equality" they advocated. This was a time when "the right to look" was not universally held.[56] The wrong "look" by an underclass person could be considered an actionable offense, with many states in the U.S. South establishing "reckless eyeballing" laws prohibiting African-Americans from gazing at whites, especially white women.

Philosophically speaking, vision operated in colonial America as part of the emerging concept of the modern "subject." As early American settlers were embarking for the new land, René Descartes had just published his *La Dioptrique*, which conceived the human eye as a lens through which an "external" image passes to the "internal" mind within. Before this, seeing was superstitiously thought to be connective and embodied like touch (someone could be "bewitched" by a glance, for example). All of this changed with Descartes and the subsequent science confirming his views. As explained by historian Robert S. Nelson, "In the early modern period, the eyes shifted from being a gateway to the soul to being an instrument like the camera obscura."[57] In the eighteenth and nineteenth centuries, this externalization of "reality" would become a philosophical premise of its own, enabling the world and "other" people to become regarded as objects to be studied, organized, differentiated, and, if necessary, to be mastered or controlled—as the gaze became synonymous with individual looking and Cartesian rationalism. As astronomy was plotting planetary bodies and imagining the earth as the center of the universe, optical science would perfect principles of perspective reinforcing the viewpoint of the single observer.

Early capitalism had particularly important visual dimensions, including, of course, published engravings or other images of goods for sale (of added salience for illiterate viewers). But on a more primary level, the display of goods in commercial settings functioned as a key demarcation between "property" owned and not-owned. Well before the mid-nineteenth century, when photography would make the visual commodification of desire its own fetish, the mere display of products started the process. This was part of a broader emphasis on "seeing" in nineteenth-century culture, reinforced by ongoing improvements in printing technologies and a rising public fascination with optical amusements like panoramic painting and lantern projections. Meanwhile, the visualization of private property already had been engrained in paintings of wealthy patrons, their possessions, and land—further reinforced on a grander scale by depictions of colonized territories and the peoples they contained. Here, as elsewhere, images became equated with a form of possession, whether conceived as documentation, proof, or a way of capturing in memory.

It's no secret that this linkage of "seeing" and "owning" really took off in the twentieth century, as mass media and advertising elevated commodified vision to unprecedented dimensions. With photography's arrival in the 1840s and 1850s, demand grew for portraiture and the ability to possess images of loved ones and to "keep" memories. At this time, advances in publishing technology also made photographic images commonplace in newspapers and magazines—their apparent realism and objective "truthfulness" adding new potency in both commercial and journalistic applications. Movies would follow, and then television as well. By the 1950s, America was immersed in a culture of consumption driven by the unfolding presence of visual information. Much of this attaches to possessive individualism inherent in what it means to be American and notions that "ownership of the self" constituted a person's ultimate possession.

Fast-forward to today's media-savvy culture. It goes without saying that manipulating public imagination is a tricky business. Everybody knows that media influence public opinion—whether through the direct form of commercial advertising or the "free media" politicians get from news coverage. And no one is naïve about the role of media in encouraging people to buy things either, even though most people believe they can outsmart advertisers. Decades ago, communications scholars began to abandon old beliefs that mass audiences were easily tricked into "false consciousness" by unscrupulous advertisers or political hucksters. For one thing, people were suspicious of obvious ploys. But also they could get conflicting views from the morass of ads, pitches, news reports, and stories they consumed.

Knowing that audiences were wary of media messages, clever advertisers and politicians began appealing to people's emotions. This is hardly a novel strategy. From gruesome fairy tales to apocalyptic bible parables, scary stories have functioned throughout human history to manipulate both young and old. Fascism and totalitarian regimes thrived in the twentieth century through the use of fearful propaganda campaigns, with wars and violence often resulting. Underlying most of these were warnings of attack, hardship, or social ruin unless changes were made or some group eliminated. The flipside of this dynamic works with things people desire—like happiness or friendship—in which an associated product can function as a stand-in for the feeling itself. Also potent can be the sense of identity-confirmation that comes with voting for a candidate or buying something associated with a peer group or people with similar beliefs. Such symbolic connection to others can assume a heightened potency when face-to-face encounters are limited. Add to this mix the spirit of American rugged individualism that neoliberalism encourages. In fact, neoliberalism *aggressively* demands such a traditional go-it-alone attitude.

Neo Self-Expression

Neoliberalism conforms with the "self-expression" inherent in anxious creativity—especially as consumer culture encourages individualism to be acted out or "performed" as a form of personal empowerment. This appeal to self-expression acquires enormous potency at a time when so many Americans feel ignored by government and left behind economically, while they also find themselves taunted by images of the latest iPhones, cars, and clothing. Consumer culture offers an easy way for people to express themselves through the "symbolic creativity" of everyday goods, to be more fully discussed in the next chapter.[58] Meanwhile, the new creative industries promise meaningful work and flexibility if employees are willing to make a few sacrifices. But what happens when self-expression gets linked so directly to buying decisions? And what about the compromises the creative industries seem to require? Is the neoliberal culture of self-expression really as good as it purports to be? If it isn't, are there ways to resist its lure? Or maybe turn it around?

Every purchase of a product, every vote for a candidate, is an act of "answering" a message received, a "yes" to the pitch and the values supporting it. Participation of this kind is hardly neutral, as voters or consumers come to perform in the narrative of neoliberalism. Such is the way ideology works: in the foreground and background, in the message and in its reception. Every time one answers such "interpellation," the self is changed a tiny bit, often without one noticing.[59] And if one thinks of Hayek's expansion of neoliberalism into the "means for all our ends," interpellation inheres in the entire apparatus of social life, communication, institutions, government, and the marketplace.[60] Ironically perhaps, this also conforms with contemporary values of diversity—in which each individual's "special" identity is cherished and reinforced in a culture encouraging (or maybe commodifying) everyone to "Think Different." That's the theory anyway: a society in which self-expression can help anyone to cut through the clutter of social obligation and conformity, while at the same time celebrating the uniqueness of every person.

Let's look more comprehensively at how this emphasis on self plays out in contemporary culture. In today's media-saturated world, American society is immersed in an all-encompassing communications environment that supersedes any particular political or commercial pitch, reaching into everyday life with a ubiquity matched only by the consistency of its messaging. The term "transmedia" was coined by theorist Henry Jenkins in his book *Convergence Culture* to describe the ways that imaginary stories or "worlds" are created in viewers' minds through integrated media experiences.[61] As an example of this, Jenkins cites the phenomenon of product franchises that often accompany movie releases in which "integral elements of a fiction get dispersed systematically across multiple delivery channels for the purpose of creating a unified and coordinated entertainment experience."[62] An *Avengers* or *Star Wars* movie is released, referencing prior comic books or media products, then introducing new products in toy stores, branded merchandize at McDonald's, video games, and interactive websites—all backed up by billboards, TV commercials, and print advertisements.

One's entertainment preferences and personal choices are reinforced by a culture in which reminders of the franchise are everywhere. Imagination and creativity are key elements of the transmedia experience, giving the process an aesthetic dimension along with any cognitive aspects. One both *comprehends* and *feels* the effects of a transmedia story, often imagining oneself as part of the narrative (as when watching a movie) or quite literally participating in an unfolding drama (as with a computer game). Maintaining a critical distance from this media envelope is extremely difficult, since messages reach viewers in unexpected ways and in unguarded moments. Communications scholar George Gerbner first discussed this in the television age, comparing the experience of ubiquitous media to that of "fish in water." Gerbner wasn't especially shy about describing the implications of all of this. As he put it 20 years ago, "You know, who tells the stories of a culture really governs human behavior. It used to be the parent, the school, the church, the community. Now it's a handful of global conglomerates that have nothing to tell, but a great deal to sell."[63] Gerbner would add that none of this derives from a "democracy" of information, but, rather, a relentless repetition of certain messages. And this repetition has an often-unrecognized effect. One might not be persuaded by a single ad for a Lexus, but the repeated appearance of flashy automobiles in the transmedia environment instills desire of a "new" car nevertheless.

Consumer advocates believe this creates an environment in which people confuse basic "needs" (things like food or transportation) with "wants" (ice cream or the Lexus, for example).[64] Numerous studies have shown that

Americans of just about every income bracket believe they can't afford their fundamental *needs* when, in fact, they are craving the excess of artificially induced *wants*. Advertising is partly to blame. But social aspects of consumption play a part as well—as people compare what they have with that of others. Regardless of the cause, the end result leaves many people unhappy and with a feeling of "lack" or disappointment. Worse still, because this is a generalized condition of consumer culture (some even would say the human condition itself), no single purchase or "object" ever erases the sense that something is missing. One can never have "enough." Add to this the reality of income inequality and the cruelty of the needs/wants phenomenon becomes even more harsh.

It doesn't help that U.S. culture relentlessly encourages individualized beliefs in continued upward mobility, progress toward goals, and wish fulfillment. The guiding premise is that with just a little more effort or ingenuity, anyone can achieve the "American Dream." Failure to succeed is equated with a paucity of will, usually cast in individual terms. All of this feeds the generalized feeling held by so many Americans today that somehow they are being left behind or ignored as the nation slowly emerges from the recent recession. Such emotions are especially strong among groups who once held a demographic advantage in American society (European-American men, in particular). Amid this atmosphere of widely felt disempowerment, there is one source of freedom that Americans can rely on. No, I am not speaking of the freedoms to vote, worship, or bear arms. The overarching form of freedom most people actually "feel" in their daily lives is always just one click away at Amazon.com.

The principle of free "choice" animates this discussion on multiple levels, resonating so strongly, as it does, in America's libertarian psyche. Operating as a synonym for democratic and capitalistic freedom, choice is seen as the opposite of coercion, regulation, and authoritarian control. But while choice functions as a powerful philosophical ideal (electoral choice, religious choice, a woman's "right to choose"), the way choice plays out in daily life is not always so simple. Often choices are prescribed within preselected sets of options. Voting choices are limited by political parties. Purchasing choices are determined by merchant inventories. But neoliberalism isn't big on pointing out systemic limitations. The overriding logic of personal freedom and self-determination insists that everything comes down to individualized decision-making.

When people have problems or fail in some way, choice often is assumed to be the reason. The ideology of "choosing" may bolster one's sense of agency and control of one's life, but it can also license a lack of generosity toward others. As psychologist Art Markham wrote in *The Dark Side of Choice in America*, "It is as if our ability to make choices leads us to think that bad outcomes people suffer are largely a result their own poor choices."[65] Money problems, worrying, and certain medical conditions (obesity, hypertension, and lung cancer, for example) often are attributed to bad choices, even though the true source of the problem may lie beyond the individual's control. For this reason, the notion of choice is yet one more place where a rigid neoliberal "common sense" can paper over deeper problems and obscure more important questions. Obesity may result from eating too much high-calorie food. But how easy is it for a low-income family to resist a 99-cent meal at McDonald's? In these and many other ways, the high-minded neoliberal mantra of free choice neatly sidesteps matters of structural advantage/disadvantage, just as it also (not-so-coincidentally) opposes efforts by government or consumer advocacy groups to intervene or "regulate" the menus of "choices" set before the public.

The self-affirming kick from buying things has become more pervasive and empowering than voting to many people. As democratic theorist Benjamin Barber has put it, "the consumer trumps the citizen" in an age when purchasing has become the prime form of individual expression.[66] Echoing a similar view about choice, educator Henry A. Giroux was even more blunt in writing of citizens "treated by the political and economic elite as restless children invited daily to convert the practice of citizenship into the art of shopping."[67] And from outside the U.S., scholars Elizabeth Dore and John Weeks have commented: "In the United States the interests of capital have successfully redefined the nature of political and social existence. In place of 'citizens', people are defined as 'consumers' and 'taxpayers.' While these categories may seem blandly descriptive, they are profoundly ideological."[68]

Not surprisingly, the creativity industries also make much of choice—for both employees and employers—in offering new forms of self-expression, intrinsic reward, workplace organization, and, most of all, flexibility for all concerned. The very fact of a job offer or contract is seen by many as a kind of choice, after all, even when the applicant's options are limited in a tight economy. Despite the anxieties of frenzied and fragmented work, many in the creative industries nevertheless feel they exercise high degrees of autonomy, answering only to themselves rather than a conventional boss. Some might argue that the new creatives simply have internalized self-management so seamlessly that they no longer realize how much they are controlled by the virtual "boss" in their heads. As the workplace has become less tangible, other shifts have occurred as well. By transforming ever larger segments of the economy into the virtual sphere of creative activity, distinctions between paid time and personal time become increasingly blurry, with employers inevitably enjoying more of the benefits. Neoliberals are fond of the Rapunzel-like logic of creativity's *ex nihilo* premise. But outside the realm of children's stories, no credible economist truly believes in the fantasy of "something from nothing."[69] When the bills for anxious creativity inevitably come due, CEOs and shareholders certainly aren't the ones who end up paying. Instead, the difference comes from creative types conditioned to believe that meaningful work is more important than a paycheck, or that artistic employment doesn't merit the same pay as "real work." Perhaps this issue of "reality" itself is at stake in the easily mystified realm of the creative imagination.

If it isn't already clear, the choice of creative work can put employees on a slippery slope—on one hand offering new opportunities, fulfilling employment, and autonomy. But these often come at the expense of job security, benefits, and remuneration, not to mention the very freedom that such "flexible" labor seems to offer, as discussed in Chapter 1. Common-sense beliefs that artists don't want or need money feeds a culture that devalues cultural producers and leaves many in poverty. Of course, it's not news that most people see little connection between creativity and compensation, given longstanding views of art-making as an "amateur" or frivolous pursuit. *Webster's Dictionary* defines the amateur as a person who "does something poorly" or "for pleasure and not as a job," although the original meaning of the term *amator* ("one who loves") had more generous implications.[70] Is it possible to radicalize the artistic *amator* premise—by empowering artists and other creative workers to take up the cause of resistance and alternative "voice" as they sometimes have done in the past? Or perhaps instilling a more "critical" understanding of media and culture in the way educators and social activists have so long advocated? Or counting on youth and the disenfranchised to stand up to authoritarianism and unfairness?

One place to start is by looking closely at why neoliberalism is working, what kinds of impulses it feeds upon, and where all of this has come from. Maybe it's time to reclaim the very ideals and rewards that neoliberalism seems to offer, generalizing them as humanistic values rather than the economic rationalism foisted on the creative economy. When you think about it, neoliberalism shrinks and distorts elemental American ideals such as freedom, choice, and self-actualization—values that transcend ideology or partisan politics. Concepts like consumer choice and worker flexibility don't need to conform to the narrow logic of maximum profit. What if these premises could be reworked to actually deliver the liberating and fulfilling promises that give them their popular appeal? This isn't some faraway fantasy or radical vision. For more than a decade now, consumers have been turning the tables on entertainment conglomerates, the music industry, and online retailers by taking the infamous "means of production" into their own hands through online sharing, peer-to-peer merchandizing apps, and a burgeoning DIY culture that rejects big business and its crass commodification. Just witness the way iTunes and Amazon are giving way to YouTube and BitTorrent. And beyond this, reports are now surfacing of a rising "minimalist" trend in American society (partly inspired by artistic minimalism), which actively *resists* cultural predilections for consumption and over-accumulation.[71]

As this chapter has illustrated, America's Performance Artist in Chief may have helped this effort. Not as the positive role model he often claimed, but rather a dystopian poster child for everything wrong with selfish individualism. Artists long have used satire to make their points, of course. And in retrospect it's hard to see The Donald as anything other than a cartoon caricature of himself. The guy probably did more for *Saturday Night Live* ratings than any figure in history, largely because his level of unselfconscious buffoonery has been so astonishing. Halfway through one angry speech to the United Nations, Trump paused for a moment to find the General Assembly smirking at him. As reported in *Newsweek*, "He seemed to expect roaring approval from the audience. Instead, world leaders responded with laughter."[72] Not surprisingly, what brought on the mirth was Trump's over-the-top condemnation of socialism to a group of nations where public health programs, child and elder care, high-quality education, and social welfare efforts are highly valued aspects of government.

Just like dystopian movies, bad leaders like Trump dramatize what's wrong with the world. Moving society from a condition of antagonistic individualism and atomization is no easy matter. The answer would seem to lie in using the very strategies that have been available throughout American history—that is, in animating the tensions between capitalism and democracy using creativity as both a practical and a theoretical premise. After all, many ordinary forms of creativity derive more from manipulating/combining what already *is* than magically conjuring something from nothing. I would argue that many of the tools are already at hand, albeit sometimes in disorganized or underutilized forms. Impulses of personal agency, shareholding, and incorporation can be harnessed in the interests of capitalistic greed, but just as easily applied to values of fairness, egalitarianism, and collective benefit.

Making this shift may require little more than a reorientation of people's stakes in society itself—a "buy-in" to the principle that individual and collective actions of almost every kind make up the ecology of a society (and world). And, yes, artists do have a vital role in this, just as the recasting of creativity in universal terms can have positive effects—so long as neither of these are solely cast in the narrow definitions of cultural work or creative industries. "Imagination" itself is a better term for the psychic landscape of the personal, social, and governmental spheres conceived and represented as dreams or realities. Obviously, much of this hinges on the way people come to see themselves, understand their own mental processes, and project their ambitions within these worlds.

Notes

1 *Time* Staff, "President Trump's Interview with *TIME* on Truth and Falsehoods," *Time Magazine* (Mar. 22, 2017) http://time.com/4710456/donald-trump-time-interview-truth-falsehood (accessed Mar. 27, 2017); *Times* Editorial Board, "Why Trump Lies," *Los Angeles Times* (Apr. 3, 2017) www.latimes.com/projects/la-ed-why-trump-lies (accessed Apr. 7, 2018) David Remnick, "Introducing a New Series: Trump and the Truth," *The New Yorker* (Sep. 2, 2016) www.newyorker.com/news/news-desk/introducing-a-new-series-trump-and-the-truth (accessed Mar. 27, 2019).

2 Glenn Kessler, Salvador Rizzo, and Meg Kelly, "President Trump Made 8,558 False or Misleading Claims in His First Two Years," *Washington Post* (Mar. 3, 2019) www.washingtonpost.com/politics/2019/01/21/president-trump-made-false-or-misleading-claims-his-first-two-years/?utm_term=.1eb267272be6 (accessed Mar. 28, 2019).

3 David Jackson, "Trump Calls Media 'Enemy of the People,'" *USA Today* (Feb. 24, 2017) www.usatoday.com/story/news/politics/2017/02/24/donald-trump-cpac-media-enemy-of-the-people/98347970 (accessed Mar. 7, 2019).
4 Naomi Klein, *The Shock Doctrine: The Rise of Disaster Capitalism* (New York: Knopf, 2007).
5 Naomi Klein, "Get Ready for the First Shocks of Trump's Disaster Capitalism," *theintercept.com* (Jan. 24, 2017) https://theintercept.com/2017/01/24/get-ready-for-the-first-shocks-of-trumps-disaster-capitalism/ (accessed Mar. 7, 2019).
6 Ibid.
7 Sean Hannity, "Trump Mounts Shock and Awe Campaign on DC," Fox News (Jan. 24, 2017) www.foxnews.com/opinion/2017/01/24/sean-hannity-trump-mounts-shock-and-awe-campaign-on-dc.html (accessed Mar. 7, 2019).
8 Lionel Trilling, *The Liberal Imagination* (New York: Viking, 1950) p. 100.
9 Bradford Richardson, "Michael Moore: Trump a Performance Artist, Not a Politician," *The Hill* (Sep 15, 2015) http://thehill.com/blogs/blog-briefing-room/news/253477-michael-moore-trump-a-performance-artist-not-a-politician (accessed Mar. 5, 2019).
10 Donald Trump and Tony Schwartz, *The Art of the Deal* (New York: Random House, 1987); Donald Trump and Charles Leerhsen, *The Art of Survival* (New York: Warner, 1991); Donald Trump and Kate Bohner, *The Art of the Comeback* (New York: Times Books, 1997).
11 *Art of the Deal*, p. 1.
12 *Art of the Deal*, p. 34.
13 Karen Finley, "Karen Finley: Donald Trump Owes All of His Wealth to Arts and Culture," *Time* (Jan. 24, 2017) http://time.com/4644278/karen-finley-donald-trump-nea (accessed Mar. 2, 2019).
14 André Breton, *First Surrealist Manifesto* (1924) https://tcf.ua.edu/Classes/Jbutler/T340/F98/SurrealistManifesto.htm (accessed Mar. 4, 2019).
15 Fred S. Kleiner, *Gardner's Art Through the Ages* (12th ed.) (London and New York: Wadsworth Publishing, 2006) p. 754.
16 Filippo Tommaso Marinetti, *The Futurist Manifesto* (1908) www.societyforasianart.org/sites/default/files/manifesto_futurista.pdf (accessed Mar. 18, 2019).
17 "Mussolini, Doctrine of Fascism" (1932) www.historyguide.org/europe/duce.html (accessed Mar. 19, 2019).
18 Matt Pearce, "Steve Bannon Found Inspiration in Ancient Thinkers, Ronald Reagan and Nazi Propaganda," *Los Angeles Times* (Dec. 9, 2016) www.latimes.com/nation/la-na-bannon-influences-20161209-story.html (accessed Mar. 7, 2019).
19 Program on Negotiating Staff, "Use Chaos to Your Advantage at the Bargaining Table," *Dealmaking* (Dec. 16, 2016) www.pon.harvard.edu/daily/dealmaking-daily/turn-chaos-to-your-advantage-in-negotiation (accessed Mar. 1, 2019).
20 Ibid.
21 Ilana Mercer, *The Trump Revolution: The Donald's Creative Destruction Deconstructed* (New York: Politically Incorrect Press, 2016) p. 1.
22 *The Trump Revolution*, p. 2.
23 Friedrich Engels and Karl Marx, *The Communist Manifesto* (1848) (Harmondsworth, UK: Penguin 2002) p. 750.
24 Joseph A. Schumpeter, *Capitalism, Socialism, and Democracy* (London: Routledge, 1994) pp. 82–3.
25 David Harvey, "Neoliberalism as Creative Destruction," *The Annals of the American Academy*, 610 (Mar. 2007) http://cult320withallison.onmason.com/files/2014/12/Harvey_Creative-Destruction.pdf (accessed Mar. 8, 2019) p. 26.
26 "Neoliberalism as Creative Destruction," p. 30.

27 "Neoliberalism as Creative Destruction," p. 39.
28 Wendy Brown, *Undoing the Demos: Neoliberalism's Stealth Revolution* (New York: Zone Books, 2015) pp. 31–2.
29 Max Haiven, *Cultures of Financialization: Fictitious Capital in Popular Culture and Everyday Life* (London and New York: Palgrave Macmillan, 2014) p. 2.
30 Michael Sandel, "What Isn't for Sale?" *The Atlantic* (Apr. 2012) www.theatlantic.com/magazine/archive/2012/04/what-isnt-for-sale/308902 (accessed Mar. 2, 2019).
31 Ibid.
32 Juliet B. Schor, *The Overworked American: The Unexpected Decline in Leisure* (New York: Basic Books, 1992); quoted in Shelly Donald Coolidge, "Work and Spend Cycle Makes Company Slaves," *Christian Science Monitor* (Apr. 5, 1995) www.csmonitor.com/1995/0404/04092.html (accessed Mar. 5, 2019).
33 Chris Isidore and Tami Luhby, "Turns Out Americans Work Really Hard … But Some Want to Work Harder," CNN News (Jul 9, 2015) http://money.cnn.com/2015/07/09/news/economy/americans-work-bush (accessed Mar. 4, 2019).
34 Friedrich Hayek, *The Road to Serfdom: 50th Anniversary Edition* (1944) (Chicago: University of Chicago Press, 1994) p. 95.
35 Michel Foucault, *The Birth of Biopolitics: Lectures at the Collège de France, 1978–79*, ed. Michel Senellart, trans. Graham Burchell (New York: Picador, 2004) p. 144.
36 Fredric Jameson, *Postmodernism, Or the Cultural Logic of Late Capitalism* (Durham: Duke University Press, 1991) p. xxi.
37 Wendy Brown, *Undoing the Demos: Neoliberalism's Stealth Revolution* (New York: Zone Books, 2015) p. 9.
38 Tim Henderson, "More Americans Living Alone, Census Says," *Washington Post* (Sep. 28, 2014) www.washingtonpost.com/politics/more-americans-living-alone-census-says/2014/09/28/67e1d02e-473a-11e4-b72e-d60a9229cc10_story.html (accessed Jul. 16, 2018).
39 Sherry Turkle, *Alone Together: Why We Expect More from Technology and Less from Each Other* (New York: Basic Books, 2011) p 1.
40 "Income Inequality," Inequality.org (2015) http://inequality.org/income-inequality (accessed Jul. 17, 2018).
41 Heather Long and Patrick Gillespie, "Why Americans Are So Angry in 2016," CNN Money (Mar. 9, 2016) http://money.cnn.com/2016/03/09/news/economy/donald-trump-bernie-sanders-angry-america (accessed Jul. 17, 2018).
42 Nancy Fraser, "From Progressive Neoliberalism to Trump—and Beyond," *American Affairs*, 1, no. 4 (Winter 2017): 46–64.
43 Adam Smith, *The Wealth of Nations* (1776) (New York: Penguin Books, 1982) p. 456.
44 Georg Wilhelm Friedrich Hegel, *The Philosophy of Right* (1821) trans. Alan White (New York: Hackett, 2015).
45 Karl Marx, *Capital: A Critique of Political Economy, Vol. 1* (1867) trans. Samuel Moore and Edward Aveling (Moscow: Progress Publishers, 1887).
46 John Maynard Keynes, *A Treatise on Money* (New York: Harcourt, Brace & Co., 1930).
47 President Dwight D. Eisenhower, "A Farewell to the Nation" (Jan. 17, 1961) www.ourdocuments.gov/doc.php?flash=true&doc=90 (accessed Jul. 11, 2018).
48 "Reagan's First Inaugural: 'Government Is Not the Solution to Our Problem; Government Is the Problem,'" *First Principles Series*, Heritage Foundation (Jan. 20, 1981) http://thf_media.s3.amazonaws.com/CPP/FP_PS36.pdf (accessed Jun. 18, 2019).

49 "From Progressive Neoliberalism to Trump."
50 Danielle Kurtzleben, "2016 Campaigns Will Spend 44.4 Billion on TV Ads, But Why?" npr.org (Aug. 19, 2015) www.npr.org/sections/itsallpolitics/2015/08/19/432759311/2016-campaign-tv-ad-spending (accessed Jul. 25, 2018); Soo Rin Kim, "Midterms Spending Overview," abc NEWS (Nov. 6, 2018) https://abcnews.go.com/Politics/midterms-spending-overview-total-cost-expected-reach-52/story?id=58996037 (accessed Nov. 10, 2018).
51 Lee Fang, "Where Have All the Lobbyists Gone?" (Feb. 14, 2014) www.thenation.com/article/shadow-lobbying-complex (accessed Jul. 21, 2018).
52 Craig Brown, "World's Happiest Countries? Social Democracies," CommonDreams.com (May 11, 2009) www.commondreams.org/further/2009/05/11/worlds-happiest-countries-social-democracies (accessed Aug. 1, 2018).
53 Stanley Fish, *Is There a Text in this Class?* (Cambridge, MA: Harvard University Press, 1980) pp. 147–74.
54 C. Wright Mills, *The Sociological Imagination* (London: Oxford University Press, 1959).
55 Cornelius Castoriadis, *The Imaginary Institution of Society*, trans. Kathleen Blamey (Cambridge, MA: MIT Press, 1975).
56 Nicholas Mirzoeff, *The Right to Look: A Counterhistory of Visuality* (Durham: Duke University Press, 2011).
57 Robert S. Nelson, "Descartes' Cow and Other Domestications of the Visual," in Robert S. Nelson, ed., *Visuality Before and Beyond the Renaissance* (Cambridge: Cambridge University Press, 2000) p. 4.
58 Paul Willis, *Common Culture: Symbolic Work at Play in the Everyday Cultures of the Young* (Boulder: Westview, 1990).
59 Louis Althusser, "Ideology and Ideological State Apparatuses (Notes Towards an Investigation)," *Lenin and Philosophy and Other Essays*, trans. Ben Brewster (New York: Free Press, 1983).
60 *The Road to Serfdom.*
61 Henry Jenkins, *Convergence Culture: Where Old and New Media Collide* (New York: NYU Press, 2006).
62 Henry Jenkins, "Transmedia Storytelling 101," (2007) http://henryjenkins.org/2007/03/transmedia_storytelling_101.html (accessed Jul. 10, 2018).
63 George Gerbner, quoted in Myrna Oliver, "George Gerbner, 86: Educator Researched the Influence of TV Viewing on Perceptions," *Los Angeles Times* (Dec. 29, 2005) http://articles.latimes.com/2005/dec/29/local/me-gerbner29 (accessed Jul. 30, 2018).
64 Juliet Schor, *The Overspent American: Why We Want What We Don't Need* (New York: Harper, 1999).
65 Art Markham, "The Dark Side of Choice in America," huffpost.com (Jul. 4, 2011) www.huffpost.com/entry/dark-side-of-choice_n_888751 (accessed Jun. 3, 2019).
66 Bill Moyers, "Is Consumerism Killing Democracy? An Interview with Benjamin Barber" consumerist.com (Dec. 26, 2007) https://consumerist.com/2007/12/26/is-consumerism-killing-democracy (accessed Jul. 21, 2018).
67 Henry A. Giroux, "Why Americans Don't Care About Democracy at Home" truth-out.org (Oct. 12, 2012) www.truth-out.org/opinion/item/11872-henry-a-giroux-why-dont-americans-care-about-democracy-at-home (accessed Jul. 21, 2018).
68 Elizabeth Dore and John Weeks, "'America': Consumerism and the End of Citizenship," Social Europe (Apr. 11, 2011) www.socialeurope.eu/2011/04/america-consumerism-and-the-end-of-citizenship (accessed Jul. 21, 2018).

69 Lawrence W. Reed, "7 Fallacies of Economics," Foundation for Economic Education (Apr. 1, 1981) https://fee.org/articles/7-fallacies-of-economics (accessed Aug. 1, 2018).
70 "Amateur," Merriam-Webster (n.d.) www.merriam-webster.com/dictionary/amateur (accessed Jul. 6, 2016).
71 Kyle Chayka, "The Oppressive Gospel of 'Minimalism,'" *New York Times Magazine* (Jul. 21, 2016) www.nytimes.com/2016/07/31/magazine/the-oppressive-gospel-of-minimalism.html?_r=0 (accessed Aug. 4, 2018).
72 John Haltiwanger, "Trump Was Laughed at by World Leaders for Dissing Socialism," *Newsweek* (Sep. 19, 2017) www.newsweek.com/trump-was-laughed-world-leaders-dissing-socialism-667785 (accessed Mar. 17, 2019).

Part II

Creative Differences

Most Americans want to be creative but find themselves lacking. This section examines this anxious state of affairs, detailing why some people get support and recognition while others don't. Beginning with a chapter on the creative capacities in everyone, discussion next turns to the inequitable ways talent often is recognized, nurtured, and celebrated—sometimes extremely so. Some of this derives from how people are raised, educated, or otherwise socialized to see themselves having creative potential. But also factoring in are a plethora of artistic stereotypes, cultural biases, and institutional dynamics. Even as opportunities to express oneself abound in online culture, not everyone is noticed or recognized as artistic. In fact, many people don't see activities like posting or blogging as creative at all. Meanwhile, institutions like museums and movie studios promote narrow forms of artistry that admit only a few. The visibility of famous performers and artworld superstars often makes the average person feel insignificant by comparison.

Related to the special status of celebrated creatives also are a plethora of stereotypes. Artists and entertainers often are expected to be, among other things: childish, detached, irrational, secretive, vain, volatile, wealthy—and therefore prone to self-indulgence, sexual promiscuity, substance abuse, and suicide. Much of this is generated by media coverage sensationalizing aberrant behaviors and creating the impression they occur commonly. Audiences are drawn in via curiosity or shock in a phenomenon seen elsewhere in the over-reporting of crime and violence. This further distances creativity from the average person, while equating artistry itself with certain strains of "madness"—the label sometimes applied to artists as a class.

This section first considers forms of home-grown artistry from past decades, beginning with paint-by-number and model kits to the more recent phenomenon of grown-up coloring books. Chapter 4, "Everyday Creativity: Are We All Artists Now?" reviews ways these activities showed how much people sought to be artistic, even as the kits and books were mocked by art professionals. Things would change enormously in an internet age, of course, with every kid a YouTube producer and Instagram a ubiquitous reality. Going further back in history, the chapter next examines the role of art connoisseurship as a marker of elevated status and the oppositional rise of post-war "cultural studies" to revalue the preferences of ordinary people. Writings by Michel de Certeau, Donna Haraway, Richard Hofstadter, Thorstein Veblen, Max Weber, and Raymond Williams are explored. From this, discussion turns to insurgent movements for "cultural democracy" seen in works by artists Joseph Beuys and Allan Kaprow, and the more recent "social turn" in arts making and exhibiting practiced by Andrea Fraser and Komar and Melamid, among others, inspired by writings of Ernesto Laclau and Chantel Mouffe.

A look at the bizarre world of child prodigies opens Chapter 5, "Creative Differences: How to Raise a Genius." Most people have never heard of kids like Autumn de Forest and Aelita Andre, who raked in millions while still in grade school. Much of this is due to the stratospheric prices fetched in today's $66-billion-a-year art market, where top-selling artists make as much as movie stars.[1] Men still dominate museum exhibitions, much as they did in the 1970s when the activist Guerilla Girls propagated the slogan "Do Women Have to Be Naked to Get into the Met?" Similar disparities persist in the entertainment industry as well, despite well-publicized protests over the Oscar awards. Partly due to this demographic disconnect, audiences are walking away from conventional high and low culture, much as Pierre Bourdieu noted in his landmark *Distinction: A Social Critique of the Judgement of Taste.* [2] Polls and surveys documenting the continuing drop in museum and movie attendance are analyzed. Writers discussed range from Matthew Arnold, Immanuel Kant, and Friedrich Schiller to Douglas Crimp, Germane Greer, and Andreas Huyssen.

Widely held beliefs about widespread mental illness and substance use in the arts get attention in Chapter 6: "Divine Madness: The Crazy-Artist Myth." Such stereotypes have a history dating to the ancient Greeks and were reinforced in Enlightenment Era views of aesthetics as "unreason." With nineteenth-century concerns over public health, artists would find themselves lumped together with eccentrics, the mentally ill, and others seen as undesirable. Today, these attitudes persist in views of artists as moody or strange. Not helping matters have been retrospective "psychiatric autopsies" of figures such as Lord Byron and Van Gogh by well-known mental health professionals Nancy Andreasen and Kay Redfield Jamison.[3] While recent neuroscience confirms that certain brain activities sometimes are shared by artists and the mentally ill, researchers argue that the two are not connected directly. Others point out that artists (or anyone) in the throes of profound mental illness or drug addiction can't produce anything at all, much less important creative works. Informing this chapter are by writings by Anna Abraham, Teresa Amabile, Margaret A. Boden, Shelley Carson, Michel Foucault, Sigmund Freud, Sir Francis Galton, Havelock Ellis, Simon Kyaga, Jacques Lacan, Apara Ranjan, and Susan Sontag.

Notes

1 Karya Kazakina, "Art Market Nears Record with $66-Billion in Global Sales," *Bloomberg News* (Mar. 12, 2014) http://www.bloomberg.com/news/articles/2014-03-12/global-art-market-surged-to-66-billion-in-2013-report (accessed Jan. 5, 2019).
2 Pierre Bourdieu, *Distinction: A Social Critique of the Judgement of Taste* (New York and London: Routledge, 1979).
3 Nancy Andreasen, "Secrets of the Creative Brain," *Atlantic* (Jul./Aug. 2014) www.theatlantic.com/magazine/archive/2014/07/secrets-of-the-creative-brain/372299 (accessed Jan. 9. 2019); Kay Redfield Jamison, *Touched with Fire: Manic-Depressive Illness and the Artistic Temperament* (New York: Free Press, 1986).

Chapter 4

Everyday Creativity

Are We All Artists Now?

Most people don't see themselves as creative—they simply don't. And it's starting to make them anxious. This news comes from a most unlikely source, givens its dismal implications. Recent research from software giant Adobe (maker of Adobe "Creative Suite") found in the largest such study to date that the majority of people in the world believe they are not using their creative abilities, especially at work. In the U.S., half of all adults feel the same way. The reasons for this creative malaise follow much of what has been discussed in these pages, with respondents reporting feelings of being "pressured" to earn money and to appear "productive" in a competitive economy. Others said they worried about failure and avoided conceptual risk-taking. All of this directly contradicts what the new creative industries say they want to promote. And while you might expect that a study from a tech company would point to online life as a new creative outlet, it wasn't the case—with scarcely 15 percent saying the internet made things better.

What can be made of these findings? It seems that amid the recent buzz over the "creative economy," many people feel obligated to participate yet at the same time unable to do so. Adobe reported:

> The research showed that 8 in 10 people sense that unlocking creativity is critical to economic growth and nearly two-thirds of respondents feel creativity is valuable to society, yet a striking minority—only 1 in 4 people—believe they are living up to their own creative potential.[1]

Not helping matters is the high-profile visibility of innovators like Steve Jobs and Mark Zuckerberg, with one analyst observing that such hype makes creativity seem beyond people's reach, leaving them feeling insignificant and thinking, "I could never do that."[2]

What is causing this creative double bind? Definitional contradictions certainly play a big role, especially in the way creativity is seen as something "everyone" has, but is recognized in so few. Somewhat like beauty or cleverness, creativity has become a quality that is universally promoted and sought, yet no one ever seems to have enough of it. No wonder it makes people anxious. This chapter looks at the creative impulse as a native human trait, and then explores some of the curious pathways that "everyday culture" has taken in a society valuing egalitarianism yet enamored by artistic genius. Discussion opens with a look at the often-maligned "Paint-by-Number" craze of the 1950s, which elevated the creative aspirations of millions while also rankling critics who saw it as an affront to good taste. This raises questions about populist aesthetics, whether they can be developed, and how this can be done in a democratic society. As anthropologists and scientists have long asserted, the capacity for invention and self-expression is as old as the species itself. But societies have treated creativity differently, especially as they have differentiated themselves and stratified internally.

The ethical and political stakes of such social stratification have never been higher. This chapter explores how an America anxious about its intellectual standing often has denigrated the everyday creativity of ordinary people. Even in the world's leading democracy, critics bemoan the nation's favorite movies, television, and computer games—saying such entertainment distracts and otherwise "dumbs down" a population better served by literature and fine art. Contrasting these elitist sentiments, recent scholarship has looked more carefully at "popular" entertainment and found positive aspects. Key in this new thinking are principles of equity, pluralism, and the imperative for expanding the creative abilities of all of its citizens through education, policy, or philanthropy. And while many put their faith in the promises of new technology, little seems to change. Might it be that creativity itself is not the real issue, but merely a symptom of underlying contradictions within American society? And if that is the case, can creativity itself be used to fix the problem?

Paint-by-Number

The first "Paint-by-Number" kit was introduced by Craft Master in 1951 with packaging that read: "A beautiful oil painting first time you try." Within a few years, the kits became a national sensation, selling over 12 million units and later replicated as Color-by-Number (for children) and Needlecraft-by-Number versions. Artists and educators would mock the formulaic rigidity of these consumer products as both a symptom of 1950s conformity and an ongoing embarrassment for American society. But mostly they were just ignored. Is this all that can be said about the paint-by-number phenomenon? Put in more sympathetic terms, these assisted rendering kits created a fascinating bridge between art and craft. And they worked well for kids, too, especially in the more popular iteration of the coloring book. A resurgence in this form of craft has occurred in the past decade, with Amazon.com currently listing 34,410 titles in its "adult coloring book" category (none of them porn, in case you are wondering), with the new coloring craze accounting for five of Amazon's top ten bestselling titles.[3] Walmart made $100 million from such books in the last six months alone, with mobile phone apps like *Recolor* and *Colorify* drawing 2.3 million active users.[4]

No less a publication than *Forbes Magazine* said it was "nothing but bullish" about the booming adult coloring book market in a feature article entitled "The Adult Coloring Book Craze Continues and There Is No End in Sight." *Forbes* consulted behavioral economist Michal Ann Strahilevitz, who gave the familiar "inner child" explanation for the books. Every grown-up "wants to play" and be reminded of carefree childhood days, Strahilevitz said. "So many things in life are hard, but coloring in a coloring book is easy. It's a nice way to relax," she observed, adding: "These books let us be creative with choice of color and create something lovely and unique, and best of all, we don't have to be any good at drawing to make something pretty."[5] CNN jumped on the bandwagon with a story about "Why Adult Coloring Books Are Good for You," citing what it called "new research" about the mental health benefits of coloring, particularly in reducing anxiety. Citing the American Art Therapy Association, CNN said that adult coloring allows one to "explore feelings, foster self-awareness, manage behavior and addictions, develop social skills, improve reality orientation, reduce anxiety and increase self-esteem."[6]

As you might expect, not everyone is thrilled about the new adult return-to-childhood. *The New Yorker* pilloried the infantilizing tactics of coloring book maker Dover Creative Haven for marketing "an escape to a world of inspiration and artistic fulfillment," citing educational experts who contend that there is nothing "creative" about the "directed and restricted activity" of paint-by-number kits or coloring books, per se—and that such activity may, in fact, diminish genuine creativity. Author Susan Jacoby was even more strident in condemning grown-up coloring. A longstanding critic of American anti-intellectualism, Jacoby told the *New Yorker*, "There's a line from the Bible, 'When I was a child, I thought as a child, but as an adult, I put away childish things,'" adding, "The coloring book is an artifact of a broader cultural shift. And that cultural shift is a bad thing."[7] Jacoby lay much of the blame on a culture of easy gratification, a laziness manifest, for example, in the rising popularity of easy-to-read young adult literature. Jacoby saw Americans as increasingly prone to diversions from the world around them. Faced with hardship and insecurities, she said that many American adults simply don't want to grow up—a trend now seen in the vast numbers of jobless twenty-somethings moving back to their childhood homes.

Jacoby's comments echoed those of others lamenting an America awash in superficiality and popular diversion—the list of complainers including Charles Pierce (*Idiot America*), Mark Bauerlein (*The Dumbest Generation*), and Al Gore (*The Assault on Reason*), many of whom drew inspiration from Richard Hofstadter's 1966 Pulitzer Prize–winning book, *Anti-Intellectualism in American Life.*[8] While it's hard to discount a certain dumbing down of culture (manifest these days in political anti-rationalism, the denial of science, etc.), these authors also evoke an all-too-familiar yearning for a "literate" and "polite" American society of a lost era. Such is the function of nostalgia as it shapes the past from a current context. The U.S. never has been an especially well-read society, owing as much to educational and economic inequity as any lack of motivation. And as many of the aforementioned authors point out, American society retains a certain ambivalence toward "high culture" and other trappings of aristocracy, even as many envy the cleverness and ingenuity of "self-made" wealth.

On many scores, the rush to judgment about paint-by-number kits, coloring books, and related items reveals its own form of intellectual laziness. These products have weathered relentless critical assaults since their introduction in the 1950s, when the term "by-the-numbers" first became a synonym for a mechanical absence of creativity. Ironically, studies of consumers have shown that even early purchasers of the kits *knew perfectly well* what they were doing and harbored no illusions about channeling Leonardo da Vinci or Rembrandt in the hours they spent on number paintings. If anything, consumers were affirming the value of art in a way that made sense to them. As one writer observed, "paint-by-number with the apparatus of fine art made light of the authority of experts, who in this discourse enjoyed every advantage except being in on the joke."[9]

Leisure pursuits acquired heightened significance in the post-war decades, as a rapidly growing U.S. middle class suddenly found itself with extra time on its hands. How one spent spare moments became a marker of social standing, and before long cultural critics would assert that working people were wasting their idle hours on mindless activities like watching TV or painting by numbers. At a time when work itself was becoming increasingly routinized, the proper use of re-*creation* (recreation) could serve as an antidote, it was argued. While high culture might be one option, the prevailing prescription for most of America was to expend time off in a "useful" or economical way—as in household repair and do-it-yourself projects (for men) or home decoration (for women). Gendered aspects of hobbies also became manifest in such male-oriented consumer products as model airplanes and ships, with parts numbered much in the same way as painting kits geared toward women. As home economist Janet K. Smith would write, "An artistic home means more enjoyable living."[10] With brand titles "Craft Master" and "Masterpiece," paint-by-number kits seemed to bridge the hobby/art divide. And they were insanely popular. It is reported that, by 1954, more "number paintings" could be found in U.S. homes than original works of art.[11]

In his 1899 *The Theory of the Leisure Class*, Thorstein Veblen explored the cultural aspects of modern social stratification. Obviously, money and jobs played a role. But how people displayed their status became increasing important, especially in capitalistic societies. Veblen is best known for naming the "conspicuous consumption" and leisure practices of the privileged. More than merely showing the "visible success" and implicit dominance of the upper class, conspicuous consuming had to be non-utilitarian to distance the manager from the common worker, as Veblen put it, manifesting "a conspicuous exemption from productive labor and in the evidence which this exemption affords of their masters' wealth and power."[12] Paint-by-number violated these boundaries in all sorts of ways: first, by collapsing distinctions between art and non-art; next, by elevating inauthentic number paintings over genuine displays of wealth; and, finally, by turning audiences into producers of culture.

While paint-by-number may have subverted conventional notions of art and art-making, it ironically conformed to certain traditions its critics may not have recognized. In his landmark 2001 "Paint-by-Number" exhibition at the Smithsonian Museum, curator William L. Bird would note that canonical figures such as Leonardo often assigned numbered sections of paintings for completion by helpers.[13] Frequently, the training of assistants also involved copying the master's work. This same pedagogy of practiced emulation would prevail in nineteenth-century art education in the U.S., particularly the "learning by doing" movement to democratize drawing skills among mechanics and tradespeople. In the 1980s, the revered J. Paul Getty Trust would launch its "Discipline-Based Art Education" (DBAE) initiative using similar methods. In an effort to legitimize art education in schools, DBAE offered a curriculum that measured students' progress based on their abilities to replicate masterpiece examples. Even today, this same step-by-step teaching method is found in the thousands of "learn-to-draw" (or paint) books available in bookstores.

Given this history and context, Bird found much more in paint-by-number than critics did. Of course, the kits simplified and mechanized the process of artistry, making it "easier" for people to make things. But degrees of aesthetic class consciousness also percolated among the millions who spent hours with the kits, as mentioned above. Besides this subversive element, many critics failed to acknowledge the personal experience of creativity, which people discovered in the paint-by-number process, not to mention a newfound respect and admiration for art itself. If anything, a paint-by-number Leonardo on a living room wall represented an unprecedented "conversation" between audience and artist—"performed" in the paint-by-number process. And believe it or not, the effect often was transformative. As one customer put it, "A tree used to be just a tree to me. Now I see as many as 10 colors in a single tree."[14] Besides this, entire subgenres of paint-by-number would emerge as people took the kits to new levels, crafting elaborate picture frames, altering prescribed color schemes, or otherwise painting "outside the lines."

Artists themselves would periodically engage the paint-by-number phenomenon, either in the form of satirical critique or in gestures of dialogue with popular culture. Andy Warhol's 1963 "Do it Yourself" paintings were based on an actual Venus Paradise brand drawing kit, and were made in the same period as his famous Campbell Soups can images. In the 1980s, artist Paul Bridgewater created his "Abstract Paint-by-Number Kit" series, as a statement against public misperceptions of modern art's apparent simplicity. The next decade saw the 1997 release of the book *Painting by Numbers: Komar and Melamid's Scientific Guide to Art*, documenting an international poll of art and home décor preferences in which the artists purportedly surveyed two billion people.[15]

Of course, the story of paint-by-number continues—and not only in today's booming market for grown-up coloring books. Cultural forms do not exist in a vacuum, and those that become widespread phenomena tend to serve multiple purposes. Paint-by-number was (and remains) more a medium than a specific idea, which partly explains why it has resisted easy analysis. With roots in American ideals of selfhood and homemade goods, this mechanistic medium allowed its users to find personal expression in previously unknown ways. In this, painting by number shares much with popular forms of digital image making now ubiquitous in a mobile phone age.

Revaluing the Popular

Cultural boundaries may be troublesome in their exclusions, but also reassuring to those who benefit from them. This rarely goes well in the long run. The infamous high/low divide may have sprung from the presumptions of a social elite, but it also created plenty of antipathy among "ordinary" people. It wasn't the typical consumer who saw much of a problem with number painting or other forms of everyday creativity. The squabbles came from distanced observers, anxious that the blurring of cultural boundaries would create confusion, envy, exploitation, or, in some cases, a fraying of traditional values. From conservatives came worries that activities like paint-by-number would dilute the significance of art while diminishing its authenticity. Left-leaning cultural critics argued that commercially produced hobbies and crafts distracted people while taking their money. In nearly all instances, such generalized (and often negative) assessments came from those observing everyday creativity from the outside.

Concerns over popular culture began to grow in the early decades of the twentieth century, when the confluence of industrial capitalism and new media technology began to affect a rapidly growing U.S. Prior to this period, intellectuals and political leaders had seen the everyday preferences of ordinary people as unworthy of serious discussion, and of little consequence to the nation's well-being. In an age of minimal cultural philanthropy, both wealthy and poor people followed their respective diversions. While each group certainly knew what it *didn't like*, neither side paid much attention to the other. For its part, government tended mostly to care about the official culture of certified masterworks, majestic architecture, and other artifacts of national heritage and pride. If folk culture did attract any interest, it was seen as an eccentric expression of local or ethnic otherness. In overall terms, this disregard of what most Americans liked and consumed in everyday life eventually would be seen as the elitist construction it was. But this would take a concerted effort from activists and renegade scholars to push back against widespread beliefs that most people's tastes lacked agency or value.

The lack of attention to popular culture also resulted from a generalized disregard of domestic life as an area of study. Gender inequities cast a huge variety of creative activities by women as irrelevant or out-of-bounds until the contemporary women's movement would revisit history. Even as female artists and writers were excluded from galleries and publishing houses, other artistic genres of a practical kind similarly were undervalued. Sewing, needlepoint, embroidery, quilting, weaving, rug making, basketry, porcelain painting, ceramics, beading, cooking, gardening, floral arranging, collaging, scrapbook making, and home decorating constitute a partial list of what are now considered creative activities. In many ways, the story of photography has a similar history—in that the medium generally was considered "non-artistic" in the nineteenth century. When George Eastman introduced roll-film Kodak cameras in 1888 with the slogan "You Push the Button and We Do the Rest," millions of Americans suddenly found a novel means of preserving memorable moments. But the camera was seen more as a mechanical device than a means of creative expression. Commercial imaging and photo-journalism dominated the medium, even as figures such as Alfred Stieglitz would struggle to legitimize photography as art.

In the background, a generalized anxiety began to creep into Western societies, with worries emerging in the late nineteenth century about the fitness of bourgeois populations seen as the centerpiece of national economies and strength. Nervous colonial powers in Europe needed capable workers and leaders to sustain their empires. And the U.S. was developing its own global ambitions. Amid a host of strategies for population management, many Western countries began to think seriously about public standards of physical health, mental ability, and morality. By 1900, the new sciences of evolution and statistics gave rise to eugenics movements promising to improve the human species, as secular concepts of moral hygiene would take their place alongside latent religiosity. And perhaps not surprisingly, concern would rise over what some saw as the negative potential of popular culture.

New forms of media often got a lot of the blame, with printing itself causing a certain uproar in nineteenth-century America, when cheaply produced pamphlets, broadsides, and magazines began appearing in cities. High-minded moralists would express concerns over working-class youth and new immigrant populations who read, for example, the extremely popular *National Police Gazette*, first published in 1833, which often covered prize fights and reported on crime. The New England Watch and Ward Society launched a campaign against such "degrading" magazines for "manifestly tending to corrupt the morals of youth."[16] When movies came along in the early 1900s, new panics arose over such "coarse" and "vulgar" fare as Sigmund Lubin's *Chinese Massacring Christians* (1900) and George Méliès' *The Last Days of Anne Boleyn* (1905). It bears mentioning that early movies were especially popular among new non-

English-speaking populations in growing urban centers, for whom the visual language of film functioned as an alternate form of literacy.

Presumptions of cultural standards do not square well with American values of free expression, open markets, and the fabled "democracy" of the country, especially when it comes to how people think. Here again, U.S. ideas of liberty and equality come into conflict with singular definitions of value and purpose. History has shown that working out such matters of public priorities and policies is the purpose of politics. And while most people don't think of "culture" as a "political" matter, often it became a topic of heated partisan debate. Sometimes people forget that the direct regulation of speech (whether in words, texts, or images) is forbidden by the First Amendment of the U.S. Constitution, except when such speech violates copyright, libel laws, or national security. Hateful opinion, pornography, and unpopular political views remain present in the mediascape as one aspect of America's open society, their presence also held in place by the fundamental supply-and-demand premise of democratic capitalism. It might even be said that the circulation of bad taste proves that democratic culture is working.

Not everyone agrees with this, of course. The expansiveness of commercial free speech in American society has prompted condemnations of popular culture from some quarters, especially in contrast to the purportedly non-commercial enterprise of the public museum and symphony. By the middle of the twentieth century, academics and public intellectuals would unify in their condemnation of mass-produced culture as the purveyor of many of society's ills: from selfishness and greed to superficiality and prejudice. Campaigns arose against sex and violence in movies, which would fail when subjected to legal scrutiny, but nevertheless result in periodic self-regulation moves by the entertainment community itself. From 1930 to 1968, the industry's Motion Picture Production Code enforced moral guidelines on movie content until it was replaced with similarly self-imposed film, television, and video-game rating systems still in use today.[17] Ultimately, the massive proliferation of viewing choices enabled by cable TV and the later effect of internet media would weaken further any efforts to contain media contents.

This raises the uncomfortable question of consumer desire—a matter that any discussion of cultural meaning must ultimately confront. Over and over, campaigns against media violence, erotic content, and artistic expression have overlooked the demand side of the equation, much as prohibitions against alcohol and drugs have done. In this respect, matters of public taste and preference become even more a matter of "culture" (in the sense of how a society thinks) than of any specific works or genres. This ongoing dialogue between the senders and receivers of cultural messages has been the concern of communications studies for several decades now, giving rise to a more nuanced understanding of how audiences accept, reject, or question messages, often attaching ideas of their own or exploring the meanings in their minds.

Efforts to analyze popular culture began in earnest shortly after World War II, initially in the form of literary studies and the emerging discipline of sociology. Theories of language and signification looked more closely at why people liked particular media and recreations, as social scientists explored the ways traditions and institutions shaped cultural habits. As this was taking place, a new mood of interdisciplinarity also began to infuse academic pursuits of many kinds, hence bringing new attention to topics preciously overlooked by traditional areas of research. As new programs and departments devoted to gender and ethnicity began to emerge in the 1960s and 1970s, centers looking at media and culture also appeared. Among these was the Centre of Contemporary Cultural Studies in Great Britain (CCCS), a nation long concerned with England's legacy of class division and colonialism. The CCCS would become famous for launching the field of "cultural studies," notably in the early scholarship of Raymond Williams and Stuart Hall.

"Culture is ordinary, in every society and in every mind," Williams wrote in a gesture of defiance against the so-called "cultured" manners, tastes, and superior sensibilities of the British upper class. Contrasting tradition with change in the development of social culture, William stated, "A culture has two aspects: the known meanings and directions, which its members are trained to; the new observations and meanings, which are offered and tested." In this spirit, the CCCS reclaimed the very idea of culture and recast it with a radical breadth as "the slow learning of shapes, purposes, and meanings, so that work, observation and communication are possible," but also, in Williams' words, "The testing of these in experience, the making of new observations, comparisons, and meanings: to mean a whole way of life—the common meanings; to mean the arts and learning—the special processes of discovery and creative effort."[18]

Synthesis was the foundation of cultural studies, as much as anything else—a principle very much in keeping with theories of relationality and intersectionality that were sweeping across academic thought at the time. In this new context, the unexplored territory of "everyday culture" engaged scholars interested in such matters as race, gender, sexual orientation, national identity, immigration, consumerism, and domestic life, among others. Initially at least, one strand of cultural studies sought to broaden notions of creativity beyond the parameters of artist practice, as exemplified in scholarship exploring the musical tastes, fashion habits, and group behaviors of specific "subcultures," with a frequent emphasis on the changing preferences of young people. Sometimes this work romanticized such cultural expression, attributing a certain autonomy to the sense of identity accruing from consumer purchasing decisions. Partly in reaction to a generation of criticism condemning the marketplace, cultural studies pointed to consumer choice as a skill or an application of knowledge, as well as an antidote to feelings of powerlessness and alienation. For example, Paul Willis attributed a certain empowerment in the presentation of the self through an artful display of clothing, personal grooming, home décor, and shared music.[19]

To Willis, the "symbolic creativity" expressed in consumerism played a central role in the way people engaged the world, and staked out a territory to call their own. Willis wrote that

> symbolic work and creativity mediate, and are simultaneously expanded and developed by, the uses, meanings, and effects of cultural commodities. Cultural commodities are catalyst, not product; a stage in, not the destination of, cultural affairs. Consumerism now has to be understood as an active, not passive, process.[20]

Willis saw symbolic creativity in clear juxtaposition to high culture's "categories of exclusion" that "have no real connection to most people or their lives." He wrote that art institutions like museums "exhaust everything else of its artistic content. If some things count as 'art,' the, rest must be 'non-art.' Because 'art' is in the 'art gallery,' it can't therefore be anywhere else."[21]

Despite its obvious populist appeal, this formulation had a number of shortcomings—notably its overgeneralization of the workings of both popular culture and fine art. Shoppers may get a self-affirming boost when picking out a new pair of jeans, but just as often they are responding to a promotional pitch or unknowingly choosing from a merchandizer's preselected inventory. In his book *The Practice of Everyday Life*, Michel de Certeau took a more balanced view in discussing the give and take of consumerism and spectatorship.[22] Individuals indeed may exert a certain control over what they buy and choose to see, de Certeau wrote. But this autonomy remains partial and not always helpful. As de Certeau observed, it's important to dig a little deeper into the ways audiences derive meaning from images or commercial products, taking into account "what the consumer 'makes' or 'does' during this time and with these images. The same goes for the use of urban space, the products purchased in the supermarket, the stories and legends distributed by the newspapers, and so on."[23]

Other contradictions between generalities and specifics would haunt cultural studies as it traveled from Britain and took hold within American academia. As late as the 1990s, overarching premises of class-based social critique continued to put pressure on the particularized politics of identity—often creating frictions within cultural studies as it emerged alongside (or sometimes overlapping with) more focused programs in African-American Studies, Asian/Asian-American Studies, Chicanx/Latinx Studies, Gender Studies, and LGBTQ Studies, among others—not to mention multidisciplinary programs addressing such topics as media, migration, translation, and transnationalism. For this reason, it sometimes proved difficult for cultural studies to escape the gravitational pull of intellectual habits and biases. As Paul Gilroy commented:

> Cultural studies may be a more or less attractive candidate for institutionalization according to the ethnic garb in which it appears—the question of whose culture is being studied is a pressing one, as is the

issue of where instruments which will make that study possible are going to come from.[24]

Finally, the radical interdisciplinarity of cultural studies soon raised suspicions that it simply stretched itself too thin, lacking the in-depth analyses of conventional scholarship. Over time, this pushed cultural studies to more closely examine what counted as "academic research," the epistemology of disciplinarity, and questions over the functions of hybridity, coalition, analogy, affect, or other markers of "affinity" among knowledge forms.

Are We All Artists Now?

It's often said that the internet makes everyone an artist. Just pick your platform and Facebook, Instagram, Pinterest, Tumblr, or YouTube will provide you a creative "voice," along with a billion other people. Technology remains a magic world in today's America—more than ever tied to U.S. ideals of progress, innovation, and endless growth. No longer merely a "computer" phenomenon, it's hardly an exaggeration to say that the mobile phone has become today's most ubiquitous form of aesthetic production (user texts, photos, and videos) and consumption (music, games, and apps). Of course, phones are but the latest iteration of culturally transformative media technologies, manifest in earlier introductions of cable networks, television, radio, and movies. These consumer novelties fall within the broader patterns of change brought about by other "technologies," broadly writ, which transformed the industrial era into the information age. Yet amid this the ongoing evolution, it's also important to consider the potential downsides of innovation and to consider whose interests really get served by America's uncritical belief that technology always makes life better.

As with popular culture, "technology" was ignored through much of Western history, and not taken seriously until after World War I. Beginning in ancient Greece, topics in philosophy, politics, and aesthetics took precedence over what were considered the everyday banalities of craft and technology. This set in place a long association of technology with crude utilitarianism, rather than the high-minded ponderings of poets and intellectuals. As the Western Enlightenment unfolded in the eighteenth century, technical matters became part of the "mechanical arts" (practical, materialist) rather than "fine arts" (creative, idealist). Romantic Era poets and painters particularly disdained technology's growing influence in the 1800s—as the machines came to embody progress and what later would be termed "modern" impulses. Isolated movements would protest the mechanization of certain jobs, such as the famous Luddite protests of British textile workers. But for the most part, industrial technology was seen as a healthy and positive "extension of man," in a way of thinking later termed "technological instrumentalism." Reflecting on such marvels as the cotton gin, locomotive, telegraph, and water mill, Thomas Carlisle would effuse in 1829 about the "Age of Machinery" making possible "what no individual hopes to accomplish single-handed and without mechanical aids."[25]

Throughout the nineteenth century, technological instrumentalism's neutral and positive image went largely unchallenged, even as industry's wondrous new machines were transforming ever more jobs into factory work. With the mechanization of labor, goods once made by hand became assembly-line products, as workers' lives conformed to the timeclock. Around 1900, the gleam of the machine began to tarnish as the effects of "technological systems" became increasingly apparent. In this atmosphere, a new brand of scholarship began applying scientific concepts to analyze how societies worked. The "sociology" practiced by Émile Durkheim, Karl Marx, and Max Weber found institutions and class formations where previous generations had seen only nature and kinship. While Durkheim would dissect the ways society itself operated like a machine, Marx would focus on ways technology affected production and exchange. Of the three, Weber most directly would describe technology as a force of its own with the power to change societies, especially as it becomes an institutionalized value. Weber saw the rationalist impulses of Western capitalism creating what he termed an "iron cage" of technological infatuation and bureaucratic artifice.

Historian Andrew Feenberg later coined the expression "technological substantivism" to describe the rising tide of critique that would emerge, beginning in the early twentieth century.[26] As the world was careering between wars, the excesses of fascism and capitalism would stimulate further debates over technology's benefits (efficiency, productivity, life enhancement) and drawbacks (job losses, mass consumerism, lethal weaponry). Technological substantivism became more urgent after World War II, with the nuclear arms race and the recognition of industrial pollution. Martin Heidegger argued that technology itself wasn't the problem, but rather what people chose to do with it. In his essay "On the Question of Technology," Heidegger famously stated that "the essence of technology is by no means anything technological."[27] Society's obsession with mechanics and gadgetry resulted from a dangerous deception, Heidegger wrote, inasmuch as "technology itself is a contrivance."[28]

The 1980s brought a fresh round of technological critique, triggered in part by environmental disasters in Chernobyl and Bhopal, but also the undeniable links of technology to nationalism, capitalism, militarism, and other forms of power inequity. Feminist scholars such as Sandra Harding and Donna Haraway would question the androcentric orientation of science and technology, as well as their tendencies to support Western epistemological conventions. From other quarters, the expression "digital divide" would emerge with recognition that not all groups had equal access to computers. By the end of the decade, it became apparent that developing nations often lacked the telephone infrastructure to support the emerging internet phenomenon—and that entire countries had been left out of the World Wide Web (on which 80 percent of traffic was in English).[29] The problem of so-called "e-waste" rose with the addition of outdated computers and used toner cartridges to landfill piled with discarded televisions. Much of this often-toxic material ended up in scrapyards near poor neighborhoods, with large amounts (up to 40 percent of U.S. output) being shipped to developing nations.[30]

Weighing in philosophically, Michel Foucault spoke of inequities resulting from "technologies of the self"—through which people internalized rules of expertise, responsibility, and normality.[31] In some ways, Foucault's premise conjoined the psychic realm of subjectivity itself to the intersection of technological systems at micro, meso, and macro societal levels. This analysis was taken even further by thinkers such as Gilles Deleuze and Bruno Latour, who developed models of viewing technology as a matrix of overlapping groupings and interconnected attitudes—infinitely complex, with the model of the network a guiding metaphor.[32] Ultimately, much of this thinking returned the conversation over technology to the dialectic of self and society, with aesthetics and creativity playing central roles.

More recent history has seen two major economic "tech booms" in the 1990s and 2000s, corresponding to respective expansions in internet and mobile phone diffusion. Despite popular beliefs that such frenzied growth has put more capability into people's hands, not everyone agrees that this necessarily translates into expanded artistry. These mixed feelings about digital creativity parallel previous concerns over the artistic value of media that "anyone" can use. Longstanding beliefs that art is a special substance made only by those with certain talents has meant that Instagram images, for example, get little credit as legitimate creative expression. Ironically perhaps, educators are among those arguing that if a process is too "easy," it simply can't be worthwhile. UCLA Children's Media Center Director Patricia Greenfield has written that the visuality of online experience has diminished the sort of imagination people use when reading, and that a wired world has weakened capacities for critical reflection.[33] Writing in *Educational Leadership*, public school technology administrator Doug Johnson similarly argued that teachers are being "lulled into the false impression that they have been developing creativity in students when using technologies (Flipbook, iMovie, Toontastic, etc.) that produce brilliant-looking results."[34]

Underlying such worries over artistic craft and authenticity, one sees an important shift taking place in American culture—a displacement of the "author-function" as typically defined. In the 1960s, scholars of language and media began looking at the role of audiences in the reading and understanding of texts. Theorists began to assert that while authors get the ball rolling in shaping the meaning of a text, readers play an equally important role.[35] People have different ways of interpreting what they experience in words, signs, or images. And because of this a given reader may believe (or not believe) what is being said, misconstrue the intended meaning, or even invent a new scenario or outcome from a given text. In this sense, authors and readers create meaning together in a back-and-forth dialogue, rather than a one-directional process. This premise also applies to viewers of media and purchasers of products, who similarly have their own ideas about what a commodity means or does.[36] Consumer capitalism always knew this, as it appealed to purchasers who "vote" with their money, even while being influenced by externalities such as advertising and social pressure.

What happens when the reader/consumer's ability to supplant the author/producer function is taken to the next step—and the reader actualizes its imagination by "talking back" in self-made new works? Combine audience authorship and capitalistic free choice with ubiquitous two-way internet communication, and you arrive at the interactive media-sharing, "Do-It-Yourself" (DIY) movements, and the second-hand marketplaces of the 2000s. Obviously, new tools such as Flipbook and iMovie are making homemade "amateur" media easier to produce, as open-distribution sites like YouTube are now known to everyone. It's no wonder this is triggering cultural anxiety on many levels, not only among pious educators. Peer-to-peer (P2P) file sharing of the type made famous by Napster in the early 2000s went mainstream with the advent of Spotify and P2P protocols like Bit Torrent. This upended the entire music industry, not only by undercutting copyright and purchasing, but also in flooding the internet with music from everywhere. These days, software like Acid, GarageBand, Logic Pro, Reaper, and Studio One makes it possible for any teenager to make a professional-sounding song. Meanwhile, the twenty-first century has also seen a massive online increase in homemade or recycled merchandizing through Etsy, eBay, and such creative product sites as ArtFire and ArtYah. Certainly, the recent recession deserves some of the credit for motivating this blurring of the lines between consuming and producing. But clearly technology has enabled something bigger in terms of personal expression.

All of these recent developments bespeak both an expansion of everyday creativity and an inversion of the top-down model of media production/distribution that characterized much of the twentieth century. While the democratizing potentials of the internet remain counterbalanced by mega-producers, pay-walls, and such internet behemoths as Facebook and Google, it is fair to argue that the online media-sphere has extended a continuing expansion of viewer choice and voice. What remains an open question is whether enhanced entertainment options, sharing capabilities, or social networking resources have any effect on the ways people regard each other or thrive in other aspects of their lives. Put another way, are any of these technological enhancements having a lasting or beneficial effect on American society as a whole? After all, material inequities still afford some people more access than others to the benefits of new media, and, if anything, disparities between rich and poor seem to be growing in the U.S. Meanwhile, race relations and gender politics continue to worsen along with a rising tide of anti-immigrant sentiment, as groups across the country are turning against each other with a vehemence (and violence) rivaling the most destructive nations in the world. The internet may have allowed people to connect in unprecedented ways. But why is it that so many Americans feel no better about each other?

Part of the answer clearly lies in seeing technology as both an enabler of existing impulses and their extension in potentially novel forms, for better or worse. That said, it's often difficult for people to hold conflicting ideas in their minds, as debates over art and creativity so often have shown. A famously polarized American society doesn't help matters, especially when economic pressure and political contention enter the picture. I'm not saying that a "middle" position or an endless compromise is the answer, since there certainly are instances when technological instrumentalists or substantivists get things right in their own terms. But like the matter of creativity, technology demands contextual analysis—with an awareness that its usefulness and meaning can change over time. History has shown how one era's technology can prove a blessing or a curse (or both) for subsequent generations. In fact, some of the more sanguine thinkers about the internet describe it as just such a mixed bag. All of this begs the question of whether technology is a manifestation of deeper values, as Heidegger observed—a tool that serves whatever master commands it. Whatever the case, it is clear that singular views simply don't apply to such a dynamic and multifaceted topic as technology, especially in a democratic society in which a plurality of perspectives purportedly reigns.

Creative Democracy

In the notorious 1972 "Documenta" international exhibition in Kassel, Germany, artist Joseph Beuys launched his well-known "Direct Democracy through Referendum" project—in which he portrayed society as a work of art in which every person can contribute creatively. Calling his premise "social sculpture," Beuys joined a rising movement for participatory art work in his often-quoted statement, "Every human being is an artist."[37] While the premise of community cultural work arose from many quarters during this period, Beuys is remembered as one of the most clear-minded early advocates for art as an egalitarian force. He stated:

> Only on condition of a radically widening of definitions will it be possible for art and activities related to art to provide evidence that art is now the only evolutionary power. Only art is capable of dismantling the repressive effects of a senile social system.[38]

In questioning the purpose and context of art, Beuys joined a growing critique of aesthetic formalism and commodification. This strand of "conceptual art" often is dated to Marcel Duchamp's interventions with "Readymades" (common objects like stools and bicycle parts displayed in museums) first seen in the early decades of the twentieth century. Later projects of this kind included the free-form participatory "happenings" staged by artist Allan Kaprow, the street performances of Laurie Anderson, and the ambient noise compositions of musician John Cage—all of which challenged conventional distinctions between artist and audience.

Such assaults on aesthetic convention presaged postmodern movements in the arts, which would emerge in the 1980s to question the premises of aesthetic value, while looking even more closely at the way art institutions conveyed ideology. Underlying much of this was an assault on the separation of art from everyday life, which many believed obscured the museum's role in politics, economics, and the shaping of ideology. Postmodern artists frequently staged their interventions inside the institution itself, appropriating the very spaces and motifs they found objectionable. One example of this was the 1989 *Museum Highlights* project by Andrea Fraser, in which the artist parodied a docent at the Philadelphia Museum of Art.[39] Unsuspecting visitors would hear what initially seemed a typical talk—into which Frazer would insert comments about cultural elitism, prejudice, and capitalism. In large part, the impetus behind such critique projects was to turn the "public" function of art back onto itself, using audience encounters to highlight the museum as one component of a broader apparatus of educational, legal, religious, political, and corporate institutions.

Many of these arts activists drew inspiration from Antonio Gramsci's notion of "cultural hegemony," which theorized ways that ruling classes normalized oppression by manipulating public beliefs, perceptions, and expectations. To combat this hegemony, artists needed to move "out of the studio" and directly engage audiences on their own terms—employing public forums, media projects, and other non-traditional means. In keeping with principles of socialization, Gramsci would write, "Every relationship of hegemony is necessarily an educational relationship."[40] Obviously, Gramsci was not speaking simply of the kind of learning one associates with the classroom. He was describing a more general process through which people come to recognize and validate power. For artists doing community cultural work, this message had great appeal, especially for the many starting grassroots nonprofits or finding part-time work in schools. For decades, fractional employment in K–12 settings had become widespread in school districts wanting to avoid hiring full-time art teachers. Hence, the situation provided (and still offers) a perfect opportunity for civic-minded creative types. In the 1980s, I myself wrote dozens of articles about such artist/educators. These led to my book *Cultural Pedagogy*, which made the simple point that nearly every artist is a "teacher" in the broader sense of the term, just as many dedicated teachers also employ creativity in engaging students.[41]

Over time, this movement would evolve into what became known as the "social turn" in the arts. While initially driven by egalitarian impulses on the part of artists, the interest in community engagement later took a decidedly practical direction. As museum and symphony attendance plummeted in the 1990s, institutions began hiring staff and expanding education departments to build audiences, offering free-admission days or special programs to bring in groups of school children, the elderly, or the economically disadvantaged. University art schools with dwindling student populations similarly started outreach programs and K–12 partnerships, partially as recruitment tools, but also in many cases to prove their relevance to local financial donors. All of this put activist artists in a bit of a quandary, as their initial impulses to abandon traditional institutions and aesthetics seemed to be co-opted. Others contended that artists had won the day. This raised uncomfortable questions about selling out and the haunting effects of ideological absolutism in neoliberal times. Was it impossible to simultaneously escape a money-driven society and exist within it? Many would conclude that the best one can do is to remain vigilant, pick one's battles, and realize that cultural struggle is an ongoing process.

Not helping matters in the 1990s was a heightened politicization of art by headline-hungry members of Congress. Accusations were leveled against the National Endowment of the Arts (NEA), among other agencies, for sponsoring what some labeled "obscene," "blasphemous," or "anti-American" projects.[42] This led to the famous cancellation of an exhibition by the late photographer Robert Mapplethorpe scheduled to be shown at the Corcoran Gallery of Art in Washington, D.C. Christian fundamentalist groups such as the American Family Association joined the fray in condemning the show, largely on the basis of a few homoerotic images. While the legitimacy of the exhibition later would be validated on First Amendment grounds, the Mapplethorpe incident was followed by similar controversies. Artists fought back nationwide with "anti-censorship" and "artistic freedom of expression" protests and marches. But the damage had been done. What later would be termed "the culture wars" produced a chilling effect, as both public and private arts funders began worrying about the negative publicity and the threat of legal action. While financially harmful to the arts in the long run, these events gave further impetus to discussions of the "public" aspects of creative expression—and who can claim authority to speak for the American people on such matters.

Democracy soon became a central theme for many activist artists—as a way of unifying impulses of institutional critique, freedom of expression, and identity politics. In one influential text, theorists Ernesto Laclau and Chantal Mouffe coined the term "radical democracy" as a more inclusive *strategy* for achieving egalitarian and participatory ends. Laclau and Mouffe underscored the limits of procedural democracy, especially as it played out in liberal societies, representative governments, capitalistic economies, and the judicial infrastructures that held them in place. They saw the seemingly reassuring discourses of "free elections" and "civil rights" as providing but a partial address of what democracy means and whom it affects—typically leaving many citizens failing to receive its advertised rewards. Ineffective laws governing civil rights, equal pay, and anti-discrimination were some of the most obvious examples of a system in which citizens unwittingly voted against their own interests, often supporting the very powers that held them back. The argument put forth by

Laclau and Mouffe suggested democratic approaches to all aspects of life—inasmuch as dialogue and fairness were as applicable at the breakfast table as they were at the voting booth. This meant internalizing democracy as an element of connection between self and other. As Mouffe wrote:

> In this case citizenship is not just one identity among others—as in liberalism—or the dominant identity that overrides all others—as in civic republicanism. It is an articulating principle that affects the different subject positions of the social agent.[43]

The twenty-first century has seen a renewed interest in cultural democracy, as the concept has reached the level of national policy. While the U.S. continues to polarize over old and new forms of inequity, incremental progress has been made in establishing diversity as a normative value. Culture is seen as both a source of discord and a possible way to bring people together. Two recent documents from the American Academy of Arts and Sciences (AAAS) and Harvard College strongly supported the concept of common civic values as conveyed through the arts and humanities.[44] The AAAS's congressionally commissioned *A Crucible Moment: College Learning and Democracy's Future* directly linked civic education with national productivity, arguing for "full literacy" to make the nation "innovative, competitive, and strong," knowledgeable in "international affairs and transnational studies," and hence equipped for "leadership in an interconnected world."[45] In more political terms, "The Teaching of the Arts and Humanities at Harvard College: Mapping the Future" directly called for the "experience of liberty through the experience of art itself," while emphasizing the importance of "enthusiastic identification" with "transformative social movements" and abilities to critique "the mesmerizing, often dehumanizing force of powerful institutions."[46] While significant in their own right, these two documents reflected a growing recognition across academic disciplines that the expansive neoliberal agenda advanced in the past decade in culture, education, technology, and the arts needs to be met with a continuing effort to maintain the plasticity of a democratic society.

This chapter began by examining the democratic implications of paint-by-number as a creative medium at the intersection of popular craft and fine art—highlighting the cultural anxiety and consternation that such a boundary practice caused. Most criticism of paint-by-number came from people who never tried the hobby, but who nevertheless held with singular and exclusionary opinions about art, as well as stereotypical notions of American consumers. Lost in the discussion was not only a recognition of the expressive potential of the medium, but also the different ways people saw number painting—and still do with grown-up coloring books. Such failures to recognize the variability of American culture—and to value these differences—also lay behind conflicts over popular culture, art in schools, and technology.

This same dynamic has replayed itself and over in American culture: the difficulty of reconciling a set of overarching national ideals with the vast diversity and dynamism of the U.S. Achieving what one might term democratic creativity clearly calls for a kind of public pedagogy in the cultural arena and beyond. Social-turn activists have long asserted that artworks and cultural institutions are excellent candidates as sites for such activity, especially at a time when so many are sponsoring outreach efforts. Performances and exhibitions provide a unique form of public space where citizens can come together, often in face-to-face encounters, in a manner frequently lost in an era of social isolation and fragmentation. Moreover, such new public forums can be augmented by online virtual communities as well. The challenge ahead lies in finding ways to reinvigorate democracy within popular culture. It would seem that the space of everyday creativity is where such emancipatory sentiments reside, waiting to be activated and celebrated if an atmosphere of openness can be maintained.

Notes

1 "Study Reveals Global Creativity Gap," Adobe *State of Create* Global Benchmark Study (Apr. 23, 2012) www.adobe.com/aboutadobe/pressroom/pressreleases/201204/042312AdobeGlobalCreativityStudy.html (accessed Aug. 23, 2018).
2 "The Psychology Behind People Believing They Aren't Creative," Creative-Something.com (Sep. 10. 2012) http://creativesomething.net/post/31270547473/the-psychology-behind-people-believing-they-arent (accessed Aug. 23, 2018).
3 Kate Harrison, "The Adult Coloring Book Craze Continues and There Is No End in Sight," *Forbes Magazine* (Feb. 2, 2016) www.forbes.com/sites/kateharrison/2016/02/02/the-adult-coloring-craze-continues-and-there-is-no-end-in-sight/#560adc27643b (accessed Mar. 24, 2019).
4 Heather Tooley, "Adult Coloring Books Make $100 Million in Just 5 Months for Walmart and Apps Are Cashing in, Too," Inquisitr.com (Jun. 2, 2016) www.inquisitr.com/3017955/adult-coloring-books-make-100-million-in-just-5-months-for-walmart-and-apps-are-cashing-in-too (accessed Mar. 26, 2019).
5 Ibid.
6 Kelly Fitzpatrick, "Why Adult Coloring Books Are Good for You," CNN.com (Jan. 6, 2016) www.cnn.com/2016/01/06/health/adult-coloring-books-popularity-mental-health (accessed Mar. 23, 2019).
7 Adrienne Raphel, "Why Adults Are Buying Coloring Books (for Themselves)," *The New Yorker* (Jul. 12, 2015) www.newyorker.com/business/currency/why-adults-are-buying-coloring-books-for-themselves (accessed Mar. 23, 2019).
8 Charles Pierce, *Idiot America: How Stupidity Became a Virtue in the Land of the Free* (New York: Anchor, 2010); Mark Bauerlein, *The Dumbest Generation: How the Digital Age Stupefies Young Americans and Jeopardizes Our Future (Or, Don't Trust Anyone Under 30)* (New York: Tarcher/Pedigree, 2009); Al Gore, *The Assault on Reason* (New York: Penguin, 2008); Richard Hofstadter, *Anti-Intellectualism in American Life* (New York: Vintage, 1966).
9 William L. Bird, Jr., *Paint-by-Number* (Princeton: Princeton Architectural Press, 2001) p. 10.
10 Janet K. Smith, "Art and the Homemaker," *Journal of Home Economics* 41 (Nov. 1949): 498.

11 *Paint-by-number*, p. 3.
12 Thorstein Veblen, *The Theory of the Leisure Class: An Economic Study in the Evolution of Institutions* (New York: Macmillan, 1899) p. 27.
13 *Paint-by-Number*, p. 33.
14 *Paint-by-Number*, p. 97.
15 JoAnn Wypijewski, *Painting by Numbers: Komar and Melamid's Scientific Guide to Art* (New York: Farrar Strauss Giroux, 1997).
16 "The Police Gazette (1870s)," Dangerous Ideas: Controversial Works from The William L Clements Library (2009) http://clements.umich.edu/exhibits/online/bannedbooks/entry19.html (accessed May 31, 2018).
17 Kevin S. Sandler, *The Naked Truth: Why Hollywood Doesn't Make X-rated Movies* (New Brunswick: Rutgers University Press, 2007).
18 Raymond Williams, "Culture is Ordinary," *Resources of Hope: Culture, Democracy, Socialism* (London and New York: Verso, 1958) www.hu.mtu.edu/~jdslack/readings/CSReadings/Williams_Raymond_Culture_is_Ordinary.pdf (accessed Sep. 2, 2018).
19 Paul Willis, *Common Culture: Symbolic Work at Play in the Everyday Cultures of the Young* (Boulder and San Francisco: Westview Press, 1992) p. 1.
20 *Common Culture*, p. 18.
21 *Common Culture*, p. 1.
22 Michel de Certeau, *The Practice of Everyday Life*, trans. Steven Randall (Berkeley: University of California, 1984).
23 *The Practice of Everyday Life*, p. xii.
24 Paul Gilroy, "Cultural Studies and Ethnic Absolutism," in Lawrence Grossberg, Cary Nelson, and Paula A. Treichler. eds., *Cultural Studies* (London and New York: Routledge, 1992). p. 187.
25 Thomas Carlisle, "Signs of the Times," (1829) VictorianWeb (n.d.) www.victorianweb.org/authors/carlyle/signs1.html (accessed Sep. 7, 2018).
26 Andrew Feenberg, *Critical Theory of Technology* (Oxford: Oxford University Press, 1991).
27 Martin Heidegger, *The Question Concerning Technology and Other Essays*, trans. William Lovitt (New York and London: Garland, 1977) p. 4.
28 Max Weber, *The Protestant Ethic and the Spirit of Capitalism* (1905), trans. Talcott Parsons (London and New York: Allen and Unwin, 1930).
29 Holly Young, "The Digital Language Divide," *The Guardian* (n.d.) http://labs.theguardian.com/digital-language-divide (accessed Sep. 9, 2018).
30 Kate Campbell, "Where Does America's E-waste End Up" PBSnewshour.com (May 10, 2016) www.pbs.org/newshour/updates/america-e-waste-gps-tracker-tells-all-earthfix (accessed Sep. 9, 2018)
31 Michel Foucault, *Technologies of the Self*, ed. Luther H. Martin and Patrick H. Hutton (University of Massachusetts Press, 1988).
32 Gilles Deleuze and Félix Guattari, *A Thousand Plateaus: Capitalism and Schizophrenia* (Minneapolis: University of Minnesota, 1978); Bruno Latour, *Reassembling the Social: An Introduction to Actor-Network Theory* (Oxford: Oxford University Press, 2007).
33 Stuart Wolpert, "Is Technology Producing a Decline in Critical Thinking and Analysis?" UCLA Newsroom (Jan. 27, 2009) http://newsroom.ucla.edu/releases/is-technology-producing-a-decline-79127 (accessed Sep. 9, 2018).
34 Doug Johnson, "Power Up! Technology and the Illusion of Creativity," *Educational Leadership* (Apr. 2014) www.ascd.org/publications/educational-leadership/apr14/vol71/num07/Technology-and-the-Illusion-of-Creativity.aspx (accessed Sep. 9, 2018).

35 Stanley Fish, *Surprised by Sin: The Reader in Paradise Lost* (Cambridge: Harvard University Press, 1968); Roland Barthes, *Writing Degree Zero*, trans. Annette Lavers and Colin Smith (New York: Hill & Wang, 1968).
36 Hans Robert Jauss, *Towards an Aesthetic of Literary Reception* (Minneapolis: University of Minnesota Press, 1982).
37 Alan Antliff, *Joseph Beuys* (London and New York: Phaidon, 2014) p. 5.
38 Joseph Beuys, as quoted in Caroline Tisdall, *Art into Society, Society into Art* (London: ICA, 1974), p. 48.
39 Andrea Fraser, *Museum Highlights: A Gallery Talk,* Museum of Modern Art (1989) www.moma.org/collection/works/128768?locale=en (accessed Sep. 19, 2018).
40 Antonia Gramsci, *Selections from the Prison Notebooks*, ed. Quinten Hoare (New York: International, 1972) p. 33.
41 David Trend, *Cultural Pedagogy: Art/Education/Politics* (South Hadley, MA: Bergin & Garvey, 1992).
42 Bill Kenworthy, "Public Funding of Controversial Art," First Amendment Center (Sep. 13, 2002) www.firstamendmentcenter.org/public-funding-of-controversial-art (accessed Sep. 20, 2018).
43 Chantal Mouffe, "Citizenship and Political Community," in Miami Theory Collective, ed., *Community at Loose Ends* (Minneapolis: University of Minnesota Press, 1991) p. 79.
44 Donna Heiland, "The Role of the Arts and Humanities in Civic Learning and Engagement: The U.S. Debate," *Arts and Humanities in Higher Education* (Jul. 2015) www.researchgate.net/publication/281573815_The_role_of_the_arts_and_humanities_in_civic_learning_and_engagement_The_US_debate (accessed Sep. 13, 2018).
45 National Taskforce on Civic Learning and Democratic Engagement, *A Crucible Moment: College Learning and Democracy's Future* (2016) www.aacu.org/crucible (accessed Sep. 13, 2018).
46 "The Teaching of the Arts and Humanities at Harvard College: Mapping the Future," Harvard College (2013) http://artsandhumanities.fas.harvard.edu/files/humanities/files/mapping_the_future_31_may_2013.pdf (accessed Sep. 13, 2018).

Chapter 5

Creative Differences

How to Raise a Genius

Everyone knows the cliché "My kid could do that," which is still heard sometimes among contemporary museums-goers. Usually dismissed by insiders as hapless boorishness, the expression also conveys an unmistakable disrespect for the institutional authority of art. At a time when Pollocks and Rothkos and are fetching millions at auction, who can begrudge the non-art public a little moral outrage? Ironically, this irreverence also validates one of modern art's foundational messages (that viewers play an enormous role in the meaning of works) even as it challenges the legitimacy of the entire enterprise. On another level, the contentious assertion of any "kid's" ability speaks to the tension between the ubiquitous creativity of the human species and its differential recognition within various societies. Given this apparent contradiction, accounts of art and innovation must be seen as reflections of other values in both a positive and negative sense. Hence, every story of artistic genius and fame projects an opposite narrative of ordinariness and obscurity, although these latter tales are rarely told.

Artworks and other creative products always are made by someone, and often crafted with a particular purpose or audience in mind. In a perfectly egalitarian society, anyone could pursue an artistic livelihood—much as proponents of the new creative economy now suggest. But as sociologists have long pointed out, no social order anywhere has ever been immune to differentiation and the elevation of some groups over others. Even in primitive societies, functionalist hierarchies develop as tasks and leadership roles are assigned. In industrial times, conflict theories of stratification became apparent in the ways powerful groups acquire and maintain advantage. Culture plays a role by shaping the ways people interact around symbolic meanings, while also advancing normative attitudes toward age, education, ethnicity, expertise, gender, nationality, religion, and wealth.[1]

As a reflection of human culture, art has served historically as an index of social echelons—both in what works have depicted and the internal structure of the art world. In elemental terms, visual representation conveys differentials in the abilities of some to "gaze" upon and depict others. Within the Western canon, this gaze has taken many forms: the adulation of gods or heroes, the appreciation of the human body (typically female), the pleasure of pastoral landscapes, and the exoticism of unfamiliar people and places. Seen in this context, the history of Western aesthetics coveys a search for philosophical truth, but also a veritable catalogue of privilege and bias. Much of this inequity went unacknowledged until post-war revisionism began to excavate matters of identity and difference.

Complicating the rise of America's new creativity are the nation's mixed feelings about the concept of art itself. From the earliest days of the republic, the idea of high culture has been seen as anathema to the egalitarian goals of a democratic society. But then as now, the U.S. also aspired to greatness in all things, hence setting in place yet another set of contradictions that the nation has struggled to reconcile throughout its history. The philosophical underpinnings of aesthetic value precede the founding of the United States, of course. And the infamous high/low cultural divide is hardly as neatly drawn as it was once thought to be. As the preceding chapter has shown, such seemingly familiar and innocent terms as "creativity" and "art" turn out to be seen in wildly varying ways, complicated further by America's diverse and changing demographics, economic stratification, political polarization, and growing worries that "unity" may be impossible.

This chapter begins by taking "My kid could do that" in a literal sense, examining the phenomenon of art prodigies as a case study in America's peculiar tendencies of cultural ranking. Turning next to a review of Western aesthetics and taste, I then will chart creativity's journey into democratic capitalism and its more recent encounter with neoliberalism. Especially in the Western world, societies long have held conflicting views about whether creativity is a practical or a metaphysical activity, although the idea of art as its own separate sphere is just a little more than two centuries old. In this light, the recent shift of the new creativity along more instrumental lines is but the most recent "turn" in art's long and twisted journey. Keep in mind that aesthetic debates do not exist in a vacuum, but are always elements of a broader social context. In the U.S., the institutionalization of cultural philanthropy is a particular case in point in both its democratizing and stratifying aspects, especially as museums would enter the picture. This history will be juxtaposed to the rise of mass culture and its critiques (from intellectuals and artists alike). Finally, the art market itself will be examined, as well as attempts by government to mediate its influence.

The Prodigy Effect

Artist prodigy Arushi Bhatnagar produced her first work at the age of four months, and generated 52 paintings before most children are able to stand.[2] Receiving her inaugural solo exhibition in 2003 at the age of 11 months, Arushi's pictures immediately began selling, winning her the *Guinness World Record* title of "Youngest Professional Artist."[3] At one year of age, the "infant prodigy" Bhatnagar then won a state-level competition in India for her semi-abstract renderings of flowers and trees, reminiscent of the post-impressionism of Vincent van Gogh. By age six, Arushi produced an additional 3,000 works and was entertained by Indian President Pratibha Patil, who also purchased one of the paintings. Her parents insist that Arushi received no early training. While the Bhatnagar saga may stretch the limits of credibility, the story fits well within mythologies of child prodigies such as Mozart. Separating fact from fiction in these accounts always is a tricky matter, especially when age and ability are conflated with fame and money in the murky terrain of creative expression.

Even though every youngster may be creative, only a small number get tagged as gifted or brilliant. The child prodigy phenomenon is an extreme example of exceptionalism among creative types, and illustrative of how this intersects with broader patterns of social differentiation and inequity. There's a rather robust literature on how and why artistic genius occurs, but much less writing about where it goes once discovered or how prodigies fare later in life. Also under-discussed are matters of external influence, "stage parenting," and the public appetite for children with near-supernatural abilities. By definition, a prodigy is a youngster who demonstrates superior adult-level skills before the age of ten.[4] While popular culture often refers to such children simply as "gifted," most studies show that inherent talents rarely flourish without significant support. Mozart's father was renowned composer and musician Leopold Mozart, who also wrote the leading book of his day on playing the violin. As a baby, Mozart was surrounded by musicians, as his father continually rehearsed, wrote music, and spoke about the craft. The young Mozart's family compelled him to practice three hours each day. And while Mozart is said to have begun "composing" at the age of three, it was Leopold who transcribed his son's melodies and turned them into finished pieces.

This raises the biggest question in the ultra-young prodigy phenomenon: the role of parents, caregivers, and, in some cases, outside opportunists, in the construction of "genius." Even this is dependent on the social milieu in which the prodigy is channeled into recognized norms of aesthetic worth. And, of course, money and institutions often play a big role. Putting the matter of giftedness bluntly in *Scientific American*, Scott Barry Kaufman recently wrote, "There is *no such thing* as 'innate' talent. No one is born with fully developed traits."[5] Individuals may differ in their genetic endowments, but grooming a superstar takes a lot of work. "Michael Jordan didn't pop out dunking a basketball from the free throw line," Kaufman explained. In actuality, what really makes the prodigy phenomenon remarkable is the conjoining of so many factors that make it possible. For one thing, latent aptitude must be matched with the prodigy's interest and drive. Then talent must be recognized and nurtured by parents or caregivers, often with the expenditure of enormous time and effort. Instruction and education usually enter the picture next, hopefully followed by receptivity of the external world. All of this is contingent on geography, historical forces, timing and trends, socio-economics and cultural attitudes.

Consider the case of Aelita Andre. Born in Australia in 2007, the young Andre began painting at the age of nine months.[6] As the story goes, the child crawled onto one of the wet canvases of her artist father and began "smearing paints around" in a remarkably prodigious way. Within a year, Aelita had "completed" 60 of her own works, and her mother—also an exhibiting artist—soon engaged a gallery and dealer to display the child's abstractions in an exhibit entitled "Prodigy of Color." Before the girl's second birthday, her paintings began selling for as much as $25,000, as the family began marketing them on international trips. By age four she landed a show at a blue-chip Chelsea gallery in New York City, as the "Pee-Wee Picasso" received coverage by the BBC, *Time Magazine, The New York Post,* and *The Washington Post,* among others. Eventually, this began to ruffle a few feathers, with one critic asking, "Is this a joke at the expense of the artworld?" No less a figure than Germane Greer weighed in with an essay entitled, "Would You Pay $3,000 for a Painting by a Toddler?" accusing Andre's parents of manufacturing the entire affair.[7] Among other things, Greer noted that the child's works had been done on canvases with pre-painted flat backgrounds and seemed suspiciously neatly done—as though removed from reach at a strategic moment. Greer wrote that, in effect, Aelita's artist parents were "using her as a randomly programmed automatic paintbrush." Historian and critic Robert Nelson seemed to agree, writing, "If it is a child's work, it's not a child alone. We're happy to credit the child, but it begins with a parental concept."[8]

The prodigy phenomenon raises ethical quandaries, as well as child-rearing concerns. Not the least of these is whether the young "artist" can reasonably comprehend or otherwise choose the course that recognition might bring. Psychologists who study this say that, unless prodded to do so, most young children simply lack the focus to paint or play piano day after day. Brooklyn college psychologist Jennifer E. Drake, author of "How to Spot Artistic Brilliance," wrote that while "prodigies have an intense desire to draw," the compulsion is an end in itself, with the process itself providing satisfaction. Prodigies typically "don't have any interest in sharing their work," Drake added.[9] Exhibits, sales, and publicity are adult concerns—as are the motivations of those who purchase the work. In many cases, parental pressures to professionalize a child take a toll, or at least diminish the amount of time available for play, friendships, or other types of learning. Some have even argued that the narrowing of a child's experience to a single form of artistry shuts off the possibility of exploratory creativity of other kinds.

Taking a more reflective view, Cornell University psychiatry professor Andrew Solomon has placed prodigies within the broader category of "abnormal" children, around whom families must make accommodations. The word "prodigy" derives from the Latin *prodigium,* referring to a monster that defies the natural order, Solomon noted in this well-known book *Far from the*

Tree: Parents, Children, and the Search for Identity. [10] Throughout Western history, prodigies were thought to be possessed—with genius equated to madness since the time of Aristotle. "Like parents of children who are severely challenged, parents of exceptionally talented children are custodians of children beyond their comprehension," Solomon wrote.[11] Putting prodigies at one end of a spectrum of "difference," Solomon detailed the often-perplexing dilemmas facing the adults around them: seeking expert consultation, finding a new community of those with similar experience, addressing the quandary of mainstreaming, and protecting the child from exploitation. "Prodigiousness compels parents to redesign their lives around the special needs of their child," Solomon noted, adding that "brilliance can be as much of an impediment" as any developmental anomaly.[12]

Traditional education can be a problem as well, given the careerist pressures often felt by prodigy families. Public schools in the U.S. provide little help for creative virtuosos, devoting the lion's share of gifted education money to high-testing academic achievers. Some prodigy parents take the route of private tutors or homeschooling, although these options cost money or someone's time. And as elsewhere in U.S. culture, the prodigy phenomenon tends to reflect broader social inequities. The most recent report from the National Association for Gifted Children notes that federal law sets no specific funding levels for talented youth, with state spending ranging from $150,000 (Idaho) to $157.2 million (Texas).[13] In Los Angeles—considered by many a national epicenter of creative activity—gifted student funding does little for the creatively talented. Only 2 percent of the district's 1,087 schools offer programs for gifted artists, and, according to the *Los Angeles Times*, most that do are charter schools supported by private money in wealthier parts of town.[14] "A key factor contributing to the disparities is the ability of schools in more affluent areas to tap foundations and community members for help as district funds have dwindled," the paper explained.

What is the family of a young arts prodigy to do? Autumn de Forest seemed an average Las Vegas kindergartener until the age of five, when she picked up a brush as her dad was staining wooden furniture. By the time he turned around, Autumn had rendered something "like a Rothko."[15] Before long, the youngster was creating dozens of paintings in the styles of other well-known artists: Henri Matisse, Georgia O'Keeffe, Picasso, and Roy Lichtenstein—even creating what she termed a "Barbie-Warhol" like the famous "Marilyn" lithographs. In this respect, the child shared an ability of the kind often associated with musical prodigies: an uncanny knack for mimicry but with little interpretive depth. Papa de Forest built the child a studio and the rest is history, as they say. With over $7 million in sales, Autumn became one of the top child art prodigies worldwide, invited to the Obama White House, and even meeting Pope Francis. Helping matters was a bit of name recognition, as her family lineage included the well-known painters Lockwood de Forest and George de Forest. But this legacy didn't stop the child's parents from falling into the hands of one of the art world's most unscrupulous hucksters—Ben Valenty.

As much as anyone, Valenty deserves credit for today's bull market for child prodigy art. Valenty also is known for the string of lawsuits against him from clients claiming he stole fortunes from them. Before he discovered children's art, Valenty was hawking rare coins and precious metals in four failed businesses sued for over $11 million.[16] Later, Valenty's telemarketing scheme selling Disney memorabilia was shut down by the FCC for fraud. As the art market boomed in the first decade of the twenty-first century, Valenty began calling prodigies, initially making a name for himself with the work of leukemia survivor Olivia Bennett. Valenty then discovered Romanian immigrant Beso Kazaishvilli (the "young Dalí"). Most notably, the promoter used his child celebrities to launch extravagant charity events, such as his $40-million Pyramid of Hope in Northern California. Even though Bennett and Kazaishvilli both have charged Valenty with withholding their earnings, Autumn's family has stuck with him.

Clearly, the pejorative "my kid could do that" has assumed new meaning in today's art world, as growing numbers of promoters and investors seemingly are taking the joke seriously. Despite a market dip in 2007, overall art sales have doubled in the past decade to over $66 billion worldwide—with heightened demand putting pressure on the supply chain.[17] Art prodigies fit easily into the broader search for new inventory. While not all investors can afford to spend millions on a Picasso, growing numbers are willing to gamble on a newly discovered talent in the hope of a big payoff. Magazine editor Henri Neuendorf explained, "Artists whose work comes up at auction are increasingly younger, as investors seek to capitalize on the next star artist by purchasing their work early." And make no mistake about it, in many cases the quality of the work itself may be irrelevant. According to Los Angeles dealer Stefan Simchowitz, "99 percent of people who buy a piece of art do so because they believe it will have more value in the future."[18] Apparently aware of this, Valenty would tell investors that works like Autumn's were about to quintuple in price, but by acting quickly they could own them virtually for free.

Of course, behind much of this is the fascination with prodigies themselves. Most experts agree that the "reality" of childhood genius may be less an issue than a public hunger for magical thinking. This is evident in popular culture's adulation of young performers and athletes, as well as the scores of coming-of-age books and movies featuring adolescents with superhuman abilities. In a world of strife and economic hardship, child prodigies feed dreams of boundless possibility, aside from any merit or beauty in their accomplishments. Tufts University's David Henry Feldman, who is one of the leading experts on child genius research, asserts that prodigies must be seen as products of the societies that create them—functioning variously as aspirational role models, youthful reminders, ideological embodiments, or money makers. They certainly don't emerge from a vacuum, observed Feldman, adding, "You find prodigies where cultures care to look for them."[19] Unfortunately, the downside of a society that makes prodigies into superstars is that so much talent is never discovered at all.

Judgments of Taste

If creativity is something all people have to some degree, what distinguishes great art from everything else? It's a simple question, but one that most people can't answer—especially in an age when almost everyone sees beauty as relative, styles as changing, and expert opinion not always in line with popular views. It's hard to imagine anything like a universal standard of artistic quality, although plenty of voices will use terms like "vision," "uniqueness," or even "taste" to name that ineffable something, even as cynics counter that it's all marketing and hype. One early approach to this dilemma was spelled out by the ancient Greeks, as they struggled to understand how the "self" relates to the world around it. What would later be termed "aesthetics" began in part in debates over "idealist" and "realist" schools of thought. Put in basic terms, idealists think of the world as they'd like it to be and speak in terms of timeless universal truths. Realists look at things as they actually are and describe a world of changing meaning and circumstance. This Enlightenment Era distinction was dramatized in a famous 1511 fresco by the artist Raphael entitled *The School of Athens*, which depicted a gray-bearded Plato as an idealist (his hand pointing upward to heaven) and youthful Aristotle as a realist (gesturing to the ground).

Perception always was the devil of the idealist/realist divide—the junction point connecting mind and body, joining the self to everything else. A person predisposed to certain ways of thinking will apprehend "reality" in a particular way—often without being aware of the process. Genetics and life experience instill ways of making sense of what is seen, heard, and felt, often through organizational templates that psychologists call "schema."[20] Culture plays a big role is shaping schema according to what one has known or gathered from others. But most importantly, internal mental processes and external stimuli are always in a state of interaction rather than isolation. This means that seemingly solid perceptions of reality are always subject to influence. Things get complicated further when perceptions are rendered in symbolic form through language, images, sounds, or actions. This is one reason why discussions of art and other forms of representation remain so fraught with disagreement.

Plato and Aristotle both pondered the dilemmas of perception and representation at length, although they reached different conclusions. Keep in mind that, in ancient Greece, artists held the status of workers rather than eccentrics or geniuses, their "craft" abilities evaluated according to skill (*technē*) with which they fashioned works. In his famous "theory of forms," Plato asserted that artworks and writings offered approximations or copies (*mimesis*) of the "ideal" essences of a world that exists only in the mind. Because artists functioned as manipulators of these approximations and presented them to others, their sculptures and poems had the ability of persuasion—for better or worse. In this way, Plato asserted that the "purpose" of art, if there could be one, lay in encouraging people to idealistically seek transcendence from their earthly lives by contemplating the eternal cosmos. In contrast, Aristotle contended that "real" things in the world gave people the ideas in their heads, not the other way around. Truth came from what could be observed and its virtues could be propagated in artworks. In striving to represent the natural world, artists did much more than copying a virtual impression, Aristotle argued, since artworks invariably altered or improved on what was used as source material. Hence, the artist always took something very specific and real from the world and created a fresh version of it.

Despite their disagreements, both Plato and Aristotle agreed on the role of intention in the transformation of perception into visual or written form. Artists could steer viewers in one direction or another—and they should be cognizant of this in making their works. This even-handed philosophy came unraveled as time unfolded, with art and other forms of rendering falling prey to ideology, influence, and power. Little consensus between idealism and realism would emerge in subsequent conceptions of art, especially as theologies complicated matters by defining the natural world and the human body as "creations" of God. From the Grecian era through the Renaissance and beyond, "nature" was celebrated and copied in aesthetic enterprise. All manner of technique and apparatus were developed to these ends—mathematical formulas, special tools and brushes, mirrors, perspective devices—usually without any thought to the constructed character of the representational schema themselves. This is to say that the degree to which one method of painting was more "true to nature" or otherwise "real" remained a matter of interpretation, rather than an objective fact.

During the eighteenth century, the secularizing mood of the late Enlightenment allowed the work of art to assume a new idealist identity for which it is still widely known—although not without its discontents. An initial concept of aesthetics came from German philosopher Alexander Baumgarten, who described an immaterial "sensibility" of perception and feeling distinct from the material "reason" of empiricism and factuality.[21] Writing in his 1750 work *Aesthetica*, Baumgarten stated, "Science is not to be dragged down to the region of sensibility, but the sensible is to be lifted to the dignity of knowledge."[22] Others would refine and clarify this impulse to specify the workings of art, music, and poetry. Immanuel Kant gets most of the lasting credit for this—first from his assertion of an independent (non-theological) account of human motivation, and later in Kant's attribution of an "autonomous character" to the creative realm. In carving out a separate faculty for art appreciation, Kant introduced a fresh vocabulary for aesthetics as "disinterested contemplation" and "art-for-art's sake."[23] Key in Kant's model of disinterested contemplation was a distinction between the mental appreciation of beauty and bodily appetites for things like food or drink.

There is no underestimating the implications in all of this for the idea of "taste" and its related value system, especially as these would inflect upon social class. While Kant saw aesthetic judgment as a universally held aspect of mind ("valid for all men," as he put it), traditions of artistic patronage by the wealthy and powerful already had linked art appreciation to status, heritage, ethnicity, and schooling.[24] This didn't so much result in a tectonic shift in aesthetics, per se, as it did a branching off of some of creativity's more rarefied strains into what became known as the "fine arts." Essential to the Kantian definition of fine art was its utterly non-utilitarian character—a form of art wanting nothing but to be itself. From this would emerge a long lineage of artistic practices based on improvements of technique or formal qualities. Consonant with this aesthetic philosophy was a rigid scorn among artists, critics, and their academies toward functional aesthetic forms. The so-called "applied arts" of design, graphics, fashion, architecture, decoration, and industry later would evolve in separate trajectories, develop their own schools, and establish professions specific to each. As Leo Marx has written:

> The habit of separating the practical and the fine arts served to ratify a set of overlapping invidious distinctions between things and ideas, the physical and the mental, the mundane and the ideal, female and male, making and thinking, the work of enslaved and free men.[25]

Trying to make historical sense of the form/function divide, sociologist Pierre Bourdieu applied a structuralist lens in his landmark work, *Distinction: A Social Critique of the Judgement of Taste*. Bourdieu saw the idea of aesthetic value as a kind of meta-currency, in describing what he termed "cultural capital." Due to such differentiating factors as national heritage, levels of education, and socio-economic standing, powerful groups could access cultural capital more easily and were able to define it according to non-utilitarian norms. "Ordinary people expected objects to fulfill a function," Bourdieu wrote, "based on the continuity of art and life, which subordinates form to function."[26] He explained that, like the term "culture" itself, taste was appropriated by powerful groups and defamiliarized in an assertion "of the absolute primacy of form over function, of the mode of representation over the object represented."[27]

Such worries over fine art and aesthetic idealism have a history, dating to anti-authoritarian sentiments rising in the nineteenth century. After all, royals and pontiffs had commissioned art works for centuries to promote correct beliefs or instill citizen loyalty. German philosopher Friedrich Schiller took issue with Kant's rigid view of art as an independent domain apart from bodily impulse and desire. In his *Letters on Aesthetic Education*, Schiller saw art as a bridge between natural wildness and civilizing control, capable of instilling a positive social sensibility through what he termed an "aesthetic state."[28] In other words, Schiller located in aesthetic education what is now termed "socialization." Disagreements over aesthetics would continue and definitions of art would multiply. In the mid-1800s, Ernst Bloch wrote in *The Principle of Hope* of what he saw as the backward-looking tendencies of art appreciation.[29] Instead, Bloch described aesthetics as a way of imagining a utopian future through what he called "a not-yet conscious." Friedrich Nietzsche would take this idea further in his 1871 *The Birth of Tragedy from the Spirit of Music* in which he proposed a dialectic between a moralistic (Apollonian) "healthy-mindedness" of high art while defending the hedonistic (Dionysian) "folk-diseases" of everyday culture.[30] More to the point, Nietzsche saw in aesthetics a revolutionary power, capable of overturning regressive orthodoxies and reversing "the denial of life."[31]

Contested aesthetic views also emerged with the rise of "modernity" as both a metaphysical premise and a marker of temporal change. The philosophical stage had been set by artist Charles Baudelaire, who in 1863 wrote in *The Painter of Modern Life* that commonly held standards of "timeless" beauty had clear expiration dates.[32] Echoing Nietzsche's critique of traditional canons, Baudelaire spoke of the modern as "our time" and the artist as the *flâneur* ("passionate spectator") of current moment. This notion had obvious appeal in an age of rapid technological advancement and new forms of communications such as radio and movies. But not everyone was thrilled. Aside from those worried about a decline in traditional values were others questioning the wisdom of focusing exclusively on the "now" (or the future). A few decades into the twentieth century, critics also pointed to a chilling rationalism accompanying the modern impulse. Among others, Max Weber would note the decline of spirituality in the embrace of novelty for its own sake, particularly when modernity was linked to consumer goods and profit-making.[33]

Efforts to reconcile such divides date to America's early days, as the new nation struggled to free itself from old ways of thinking. While many repudiated the oligarchical ways of Europe, others in America worried about being seen as crude-minded reactionaries, who had abandoned a civilized continent. Ralph Waldo Emerson's 1837 address to Harvard College summed up the new nation's conflicted philosophical views. Emerson proposed a melding of both worlds in the figure of "The American Scholar"—an individual both forward-looking and historically wizened, independent yet socially engaged, informed both by "books" and "nature" in equal measure. "We have listened too long to the courtly muses of Europe," Emerson said, while cautioning, "The mind of this country, taught to aim at low objects, eats upon itself."[34] Unfortunately, Emerson's even-handedness could not long forestall the cultural stratification that would overtake the U.S. But at least initially, no clear boundaries between "high" and "low" culture existed as they are known today. Few distinctions were made between the creativity of a portraitist and a shoemaker. There was little differentiation within the "public" for creative works, or the manner in which they were made available. Audiences of all kinds bought tickets for prize fights and performances of Shakespeare; paintings were sold at public auctions and P.T. Barnum's collections of oddities were shown in what were called at the time "museums."

Simply put, in the America of the 1800s all public culture was commercial. But by the end of the century, a new series of polarizing shifts began occurring. With rising industry, the growth of cities, and changing philosophical outlooks, concerns over public health and morality grew in many Western nations. All sorts of methods for population management would come into use, justified in part by newly emerging patterns of class formation. While status and privilege no longer derived from birthright alone, they nevertheless could accrue in democratic societies from superior drive or talent. Even in an America of presumed equal opportunity, some ideas were simply beyond the grasp of ordinary people. Governance and leadership were best left to intelligent visionaries. And great art similarly became attributed to the newly recognized category of the "genius" mind. Before long, a resurgent mythology would begin to circulate about "masterpiece" creativity, as art assumed a magical "aura" of specialness. In this new atmosphere, a reverie for canonical literature and creative tour-de-force was imagined to inspire the general public with "the best that has been said and thought in the world."[35] Hence, despite earlier efforts toward amelioration, lines of demarcation again would be drawn between the easily purchased amusements of working people and the esoteric collectables of the rich.

The Museum of Money

"Every time we speak of the 'institution' as other than 'us' we disavow our role in the creation and perpetuation of its condition," the artist Andrea Fraser recently said. Known for her museum critiques, Fraser added, "It's not a question of being against the institution: We *are* the institution. It's a question of what kind of institution we are."[36] In these comments, Fraser was putting a principle of classical sociology in artworld terms. Social hierarchies don't merely arise from privileged attitudes, or even from the classifications they imply. It takes the collective entity of the institution to put such ideas into action and give them material form, whether one is talking about a gallery, guild, or a government. In capitalistic societies, moneyed interests tend to hold the most sway, overtaking the compensatory role of the state in obvious and not-so-obvious ways.

Beginning in the 1800s with what looked like a little well-meaning philanthropy, the interplay of big money and creative expression evolved into an unholy alliance, largely due to the air of ethical superiority that arts giving assumed. The origin of the term "high culture" is attributed to English poet Matthew Arnold, who spoke in his 1869 book *Culture and Anarchy* of "the disinterested endeavor after man's perfection" (much as Kant did) as a trait that could lift the masses from the banal pursuits of folk culture.[37] Arnold also was among the first to critique modernity with this line of thinking, urging listeners at his lecture "The Modern Element in Literature" to look beyond the mere contemporary to a "comprehensive, commensurate, and adequate literature" that encompassed the "vast spectacle of life, while craving for moral and intellectual deliverance."[38] This updated view of aesthetic idealism sat well with America's cultural elites, who, like Arnold, saw art as a force of social betterment enabled by spiritual discipline.

Not-so-coincidentally, this moral/aesthetic program emerged with the simultaneous rise of big business in Western nations. In many ways,

Arnold's thinking provided a blueprint for philanthropists who would fund museums and schools to bring art to ordinary people in the interest of building a new bourgeois society. Not to be forgotten is that all of these men saw their "cultural" efforts explicitly as means of taming the otherwise "anarchistic" tendencies of the unruly masses beginning to aggregate around the very factories upon which their fortunes often depended. Many U.S. philanthropists had battled labor disputes and strikes, often using violent tactics, at exactly the time they were burnishing their public personae with the patina of art. As Terry Eagleton explained in his book *The Ideology of the Aesthetic*, this conversion of "morality into style" gave material form to an overtly political program. In effect, this program generated "a precious form of intersubjectivity, establishing a community of feeling subjects, linked by a sense of shared capacities" while also unifying national populations "with all the authority of a law."[39] As with consumer culture, the appreciation of fine art provided the illusion of agency, but not the reality of actual freedom. More worrisome still, such an "aestheticization of politics" could be dangerous in its appeal to emotions over rational sensibilities.[40]

Put another way, the confluence of motivated philanthropy and idealist aesthetics simultaneously advanced faith in the "public good" of art while also neutralizing any ideological implications of the enterprise. Indeed, amid all of the posturing about the "disinterested" purpose of art museums, there was little awareness of whose *interests* such views might be promoting. Patronage doesn't come cheap, after all. And the moneyed elites behind sponsored creativity—whether in cathedral naves or aristocratic parlors—consciously or unconsciously demonstrated power with their largesse. Of course, art always had mirrored the societies from which it sprang, including the groups and institutions in charge. But the operations of aesthetics ran quite a bit deeper than this, as they inhered in the very methods of depiction employed and even the contexts in which works were conceived and subsequently viewed.

It's important to note that the concept of the *public* museum is not as old as many might assume. Before the 1800s, but a few public galleries and sculpture gardens existed in Europe, such as those of the Vatican, British Museum, and Louvre. Owing to concerns that working-class crowds might damage artifacts, some of these institutions required written applications for admission. In the mid-nineteenth century, America began to compete with other nations' history, science, and art museums—with U.S. openings of the Smithsonian (1858), the American Museum of Natural History (1869), and the Metropolitan Museum of Art (1870). While fueled in part by nationalist sentiments, the growth in American museums also embodied the democratic ambitions of citizen education more generally. In keeping with public policy views of the time, art museums were seen as providing both a source of important knowledge and an incentive for upward mobility. The enormous scale of the buildings themselves, the classical architecture employed, and the exquisitely detailed framing of works all gave patrons the impression of entering something akin to a palace or temple. Is it any wonder that so many people then (as now) would find the encounter with fine art to be an off-putting experience, alien to their lived experience?

By the beginning of the twentieth century, the American museum movement was in full swing—propelled by a combination of new wealth and rising anxieties over popular culture. Many in America's ascendant millionaire class already saw themselves as both natural stewards of public thought and role models for capitalistic achievement. Hence, the Golden Age of American philanthropy was born, generating foundations and non-profit cultural institutions by the likes of Andrew Carnegie, Andrew Frick Clay, J.P. Morgan, and John D. Rockefeller. Writing his widely circulated "The Gospel of Wealth," Carnegie would attribute his moralistic giving to the will of God, while excoriating social welfare programs as "indiscriminate charity so spent as to encourage the slothful, the drunken, the unworthy."[41] Divine intervention notwithstanding, a cooperative government also helped men like Carnegie by implementing changes in income tax codes to encourage philanthropy and the creation of private charities.

Most striking in America's early days of "museumifcation" was the amount of creative endeavor it excluded, not only to counterbalance an ascending commercial culture, but also in an expansive move to mark its own territory. Soon the designation of "non-artistic" would apply to the commercial, scientific, industrial, decorative, and otherwise "practical" applications of human expression, while also discounting the many ways artistry had functioned around the globe in custom, ritual, and the details of everyday life. But, of course, that had been part of the point of Western aesthetics for a very long time: a means of differentiating itself from the rest of the world even as it was asserting "universal" principles and beliefs. To their credit, European and American artists themselves soon saw this problem. Consequently, much of the subsequent history of aesthetic modernism addressed epistemological questions about the definition and function of art, even as much of that self-same art operated within the confines of "art" institutions and schools. Such is the dilemma of solipsism.

In many instances, this self-critique only widened the gap between fine art and popular culture. Avant-garde movements in Europe and United States quite literally said that forward-thinking intellectuals and artists were ahead of the general public. While sometimes the avant-garde was critiquing the takeover of aesthetics by the wealthy, its radicalism often gave much of the public an impression that artists were oddballs, extremists, or simply crazy people. Technology played its own alienating role. As new publishing and photographic methods enabled anyone to acquire cheap copies of artworks, the scarcity of "original" paintings and sculptures made them ever more rarefied and inaccessible to the average person. And as the sales prices for masterpieces went through the roof, long lines would be seen outside museums where uniformed guards protected the sanctified objects inside. Before long, entire governments—especially European countries such as Britain, France, and Spain—would compete in claims of greatness based on their artistic troves. Following World War II, the U.S. would claim its own place at this table in boasting the inherent "freedom" conveyed by its abstract expressionist painters.

This nationalistic fervor drove audiences to museums as repositories of artistic value—places where already-recognized masterpieces were collected, preserved, and displayed. This viewpoint didn't take into account the function of museums as certifying institutions, adding value to artworks by virtue of their acquisition by experts with superior knowledge. And while the general public occasionally grumbled over the difficulty of some museum exhibitions, there was a general consensus that a painting or sculpture *must* be important if it appeared in the Metropolitan or Smithsonian. In this way, museums began to operate in a dialogue with the market for artworks—simultaneously responding to value while also creating it. And, of course, museums depended on contributions from well-off supporters, making their connection to privileged interests even more explicit.

The close connection between museums and money became a blessing and a curse. Obviously, museums benefited from patron generosity, while in exchange often reciprocating with special privileges, board-of-director seats, public recognition, or even programs catering to sponsors' interests. In some instances, donor priorities conflicted with the museums' intellectual integrity or civic mission. But there was a far worse problem. Just over a century ago, art auction prices were shattered by the 1913 sale in Paris of Rembrandt's *Bathsheba* for approximately $50,000. Compare this to the recent sale in New York of Picasso's *Les femmes d'Alger* for $179 million.[42] Like real estate and gold, art proved a reliable asset for the prudent investor, with the Stanford Graduate School of Business reporting fine art investments showing an annual return of 10 percent for 40 years.[43] While this was great news for speculators, the hyperbolic art market proved a nightmare for museums—heightening their need for even wealthier contributors and driving intense competition for artworks that could pull in audiences. All of this changed the way museums thought about their priorities and purposes. In a story entitled "Why the Booming Contemporary Art Market Is Bad for Art Museums," the *Washington Post* pointed to staff lay-offs and other belt-tightening combined with an increasing push for publicity-getting blockbuster shows.[44] "Soaring prices mean museums simply can't keep up and must depend on donations to assemble portfolios of the best work, or they're priced out," the *Post* reported.

This drive to acquire ever larger storehouses of masterpieces was hardly innocent, for it even further narrowed the scope of what museums exhibited. This was pointed out decades ago in Douglas Crimp's 1980 *October* essay "On the Museum's Ruins," in which the critic joined a rising tide of postmodern artists in challenging "the museum's claims to represent art coherently."[45] In response, some museums made efforts to become more inclusive in what they exhibited—showing works by younger artists, women, and people of color.

This provoked outrage among conservative reviewers, who labeled the insurgent movement an "anything goes" abandonment of "quality standards." Drawing on the writings of Foucault, Crimp compared the guarded preserve of the museum to such "institutions of confinement" as the asylum and prison. Rather than remaining shut off to fresh ideas, Crimp proposed an updating of such museum staples of art appreciation as "tradition, influence, development, evolution, sources and origin" with more open and inclusive themes of "discontinuity, rupture, threshold, limit, and transformation."[46]

Since those days of postmodern controversy, many museums have diversified their offerings—somewhat in response to dwindling audiences, but also due to an increasingly globalized art world. Nevertheless, the entrenched canon of European and U.S. masterworks still reigns supreme in most museums, and continues to drive the art market as well. Here again, contemporary artists themselves have led efforts to push back—for example, Renée Green, Louise Lawler, Martha Rosler, and Fred Wilson (to name just a few). Such resistance also encouraged a growing literature about the implicit authoritarianism of museums and other mainstream arts institutions. Notable critics in this area include Benjamin Buchloh, Isabelle Graw, James Meyer, Brian Wallis, and John Welchman. Leading publishers have issued anthologies on this topic, including Routledge's *From Museum Critique to the Critical Museum* and *How Institutions Think: Between Contemporary Art and Critical Discourse* from MIT Press.[47] As Steven Dubin wrote in one such book:

> Museums have always featured displays of power: great men, great wealth, or great deeds. The emphasis could be on the spoils of war, victors in the marketplace, or man as the crown of creation. In all of these instances, museums have ratified claims of superiority.[48]

Perhaps the biggest driver of the institutional critique movement was the glaring fact that most artists never got shown in museums. This exclusivity has been part of what has kept so many artists poor, as detailed in Chapter 1. It's also what makes the promise of the new "creative economy" so alluring to some. Let's review the numbers. Like the U.S. economy in general, income gaps among artists widened dramatically since the beginning of the twenty-first century, with the vast majority turning to teaching or other "second jobs" to make ends meet. One recent study revealed that for every creative producer shown in a museum today there are over 650 who are not, with fewer than 300 of the nation's 200,000 artists receiving such exposure.[49] Statistics are even worse for actors, dancers, and musicians. Illustrating the influence of money in the visual arts, one-third of major museum solo shows in the U.S. featured artists represented by five blue-chip galleries: Gagosian Gallery, Pace, Marian Goodman Gallery, David Zwirner, and Hauser & Wirth. At New York's Guggenheim, the number was 90 percent. These numbers also are skewed in other ways. In its most recent index of big-ticket artists, the gallery world's Artprice listed but three women in its "Top 100 Living Artists."[50] The news on race and ethnicity is equally dismal. Aside from tokenistic nods to diversity, the visibility of people of color on museum walls and in leadership positions continues to be miniscule.

Mass Anxieties

Western culture likes to think in terms of opposites—of a world conveniently divided into mind and body, good and evil, self and other. The problem is that two-sided generalities usually break down when examined in detail. Such is the case with the tenacious high/low aesthetic dichotomy, which is better seen as an indication of cultural difference than any means of determining cultural worth or legitimacy. "To reduce all cultural criticism to the problem of quality is a symptom of the anxiety of contamination," wrote Andreas Huyssen in *After the Great Divide*. "The boundaries between high art and mass culture have become increasingly blurred, and we should begin to see that process as an opportunity rather than lamenting loss of quality or failure of nerve."[51] Let's pick up this anxious story in the decades just prior to World War II, when mechanically reproduced culture began to affect American society in unprecedented ways. In large measure, movies became the predominant form of entertainment during the Golden Age of Hollywood from the 1920s to the 1940s. At their height in 1939, U.S. movie studios (MGM, Paramount Pictures, Twentieth Century Fox, Warner Brothers) cranked out 400 films per year viewed by 90 million viewers (70 percent of the U.S. population).[52] By 1940, twice as many homes had radios (73 percent) as had telephone service, while sales of recorded music rose to 50 million units.[53] And with 42 million people reading daily newspapers, the overarching influence of mass culture was unmistakable, as more and more Americans were consuming the same products and messages.

Assaults on popular culture came from both conservative and liberals. Russell Kirk's *The Conservative Mind* was published just after World War II by no less a figure than T.S. Eliot, then an editor at Faber & Faber. Kirk's book was notable for its reactionary repudiation of both modernism ("change may not be salutatory reform") and egalitarianism ("civilized society requires orders and classes"). Drawing heavily from the writings of Edmund Burke, the book's "canons of conservativism" denounced populist aesthetics as an affront to "a transcendent order, or body of natural law, which rules society as well as conscience."[54] Conservatives found unlikely agreement from European leftist critics, initially worried about the use of media by fascist regimes. The manipulation of entire national populations could have dangerous consequences, they observed, with photography, radio, and film playing a big role. Market economies handed media to big business, which, in its one-directional messaging, unleashed powerful forms of persuasion, commodified popular aesthetics, and promoted negative consumer values. These arguments had their basis in

Karl Marx's theory of capitalism as a system of economic exploitation and mass confusion (creating "false consciousness").

Popular culture also had its supporters in certain quarters. In conservative circles, this stemmed from nostalgic views of European folk traditions or homespun Americana. Such thinking was very much in keeping with the Emersonian dualism discussed above, in which an independent-minded American sensibility struggled to retain contact with its origins. The other fondness for popular aesthetics was more socialist in character, emerging from the "people's culture" of worker movements and revolutionary projects. Certainly many filmmakers and artists of the early twentieth century were cognizant of the political possibilities of accessible media forms, among them Bertolt Brecht, René Clair, Salvador Dalí, Sergei Eisenstein, Vsevolod Pudovkin, and Man Ray—not to mention such intellectuals as Hans Richter and Siegfried Kracauer. Class and economic conflict often became central themes in movies because working people made up the bulk of the audience.[55] Despite later political censorship during World War I, workers and radicals continued to produce feature films and newsreels through the 1920s, even as the Hollywood film industry would blame industrial unrest on socialists or union organizers. Such radical sensibilities became more organized in the 1930s Film and Photo League, not to mention the activist groups that would emerge in the post–World War II period.

Taking a reflective view of these disputes, Henri Lefebvre observed in the 1940s in his *The Critique of Everyday Life* that modernization had centralized the organization of many societies, exacerbating people's feelings of existential alienation.[56] Lefebvre said that big business and government simply made individuals feel smaller. With the rise of mass communications, the emphasis of public policy and commerce largely shifted to a national level, as religion and high culture focused on a similarly abstract set of values. All of this put pressure on people to behave in prescribed ways, while ignoring the "ordinary" ways they invented themselves in daily life. This had negative consequences for democracy, of course, but also it eroded capacities for self-expression more generally. In this vein, Lefebvre wrote that "to reach reality we must indeed tear away the veil, that veil which is forever being born and reborn of everyday life, and which masks everyday life along with its deepest and loftiest ambitions."[57]

Nothing could have predicted the massive expansion of the cultural economy that new technology would soon enable—and along with it a novel form of national consciousness and a gigantic corporate infrastructure to reap its profits. Keep in mind that much of the above discussion occurred *before* the television age. And when TV finally *did* enter the picture, it played perfectly into the famous 1950s mood of social conformity and competitive consumerism. Less historically discussed has been the way that TV worsened individual feelings of insignificance and inadequacy, especially for those outside the norms of the era's fabled white middle class.

By the end of the 1960s, the domination of programming by a handful of large corporations (ABC, CBS, NBC) led Washington policymakers to step in—first, with the Prime Time Access Rule of the Federal Communications Commission (FCC), which sought to encourage local programming by limiting nationwide programming to four hours per night. Further efforts to democratize TV programming included the establishment of publicly funded programs to diversify and elevate what was available on TV: the Corporation for Public Broadcasting (CPB) and the Public Broadcasting Service (PBS). Further government support for culture came with the 1965 founding of the Institute for Museum Services (IMS) and the National Endowments for the Arts and Humanities (NEA and NEH). All of this had the paradoxical effect of leaving serious "high" culture in the hands of government and "low" culture to the marketplace. Throughout the history of American media, government largely has played this type of background role—deferring to industry on many matters of communications policy—chastened by the political heft of "free enterprise" and "free expression" evoked by corporations.

Creative Differences

"Do women have to be naked to get into the Met Museum?" asked a famous poster from the anonymous activist group the Guerrilla Girls, with the rejoinder, "Less than 4% of the artists in the Modern Art sections are women, but 76 % of the nudes are female."[58] Giving voice to a generation of complaint about art-world exclusion and sexism, the Guerrilla Girls project echoed ongoing critiques also leveled at the entertainment industry, which many asserted similarly propagated an objectifying "male gaze" while shutting women out of most behind-the-camera jobs.[59] Since the 1970s, concurrent criticism about race and ethnicity similarly would describe dehumanizing "white" or "colonial" forms of representation, seen in depictions of people of color or those from "other" nations as inferior, irrational, criminal, or inherently threatening—with threads of ageism, homophobia, and class bias often present.[60] While recent gestures of inclusion have been much publicized, any careful look at the overall composition and output of American culture shows that little has actually changed.

Democracy is premised on the participation of citizens in culture and governance, presumably in an egalitarian fashion. Through free speech and elections, the often tumultuous process of collective decision-making unfolds through the "voice of the people." Or at least that is the theory. But as everyone in America is painfully aware, equal participation in U.S. society is more a myth than a reality. Even as diversity has become the "new normal" in much of public discourse, and despite federal laws banning overt discrimination, longstanding patterns of bias persist (race, sex, religion, and class still leading the list), even as new categories emerge (disability, immigration status, and gender identity, for example). Such inequities are held in place by attitudes that reinforce social behaviors. Put another way, they are matters of American culture.

Or, more pointedly, they are aspects of "the politics of culture" in U.S. life—a concept receiving extensive scholarly attention in recent decades. As explained by Glenn Jordan and Chris Weedon in their anthology *Cultural Politics*, for a long time the connection between cultural and political matters was overlooked or disregarded.[61] Without doubt, many Americans still don't connect material inequities with enabling belief systems like competition and privilege. But as Jordan and Weedon further explained, forms of "cultural power" conveyed through art and entertainment allow certain groups to push selected attitudes as the "common sense" for the larger population. These politics also lay behind notions of "official culture" and the worldviews behind it. Put another way, cultural politics is what allows people to advance themselves and their interests—for example, in getting published or exhibited in a museum. Cultural politics determines the array of stories the nation tells about itself.

One of the most important of these stories concerns continuing inequity in a country purporting that everyone is "born equal" rather than into preordained stations of life. Remember that the nineteenth-century functionalist story for social stratification held that people had different skills (men are hunters). This later was updated with a conflict theory of inherent competition among groups (bosses exploit workers). As sociological theories, neither of these principles are discussed very much in public. And they no longer explain the vast complexity of social stratification. But if you think about it, functionalist and conflict theories persist nevertheless beneath nearly every justification for inequities in the workplace, the market, and elsewhere, particularly when they link to culturally transmitted beliefs and attitudes about human variation. Put another way, such views established inequity as a "natural" or somehow "normal" phenomenon. These pernicious myths have allowed old forms of stratification to perpetuate themselves, as habits of thought are transmitted from one generation to the next, circulated within or among groups, magnified in narrative form via media or artworks, and given material form in institutional practices.

Let's be blunt about America's past. It's no secret that the U.S. was founded and run for generations by wealthy able-bodied European men, most of whom also were Protestant heterosexuals. Their interests and worldviews set the tone for American culture. Much of the nation's legislative history has been consumed with attempts to correct the resulting unfairness through laws abolishing slavery, extending voting rights, legitimizing worker's unions, prohibiting discrimination, desegregating education, affording reproductive freedom, disability access, and marriage equality. In each case, a disenfranchised group had to push against the status quo in a public way, and very often in the domain of public culture. America's grand traditions of free expression purportedly made this possible. After all, the open exchange of contesting viewpoints lies at the very heart of a democratic society. But what has happened when some groups have had more access than others to the means of public communication? Consider these facts: men make up 90 percent of top living artist sales and 98 percent of the directors of big-budget movies.[62] Also, 89.5 percent of leading movie actors/actresses are white, as are 80 percent of museum staff and visitors.[63]

Representational disparities in the arts and entertainment industry have been widely documented and discussed for decades. But history is hard to reverse and markets seem to have minds of their own. Diversity is sorely lacking in the major institutions of culture, just as more attention is due to the changing interests of consumers. In 2016, the Academy of Motion Pictures and Sciences—which makes the famous Oscar award—responded to revelations that not a single non-white performer had been nominated in the prior year's Best Actor or Actress categories, and that women were poorly represented in production roles. One explanation was that the 6,261 voting members of the Academy itself were, according to the *Los Angeles Times*, "overwhelmingly white (89%) and male (73%)."[64] At the same time, a study from UCLA also revealed that the top executives at major Hollywood films studios were 94 percent white and 100 percent male, with the American Association of University Women (AAUW) adding that "a whopping 85 percent of films had no female directors, 80 percent had no female writers, and one-third lacked female producers."[65] Bowing to pressure, the Academy issued over 700 new membership invitations, largely to women and people of color, with a promise to double its diversity numbers within five years.

Most people think of the art world as inherently progressive in terms of race, gender, and sexual orientation—given common impressions of artists as liberal thinkers. Unfortunately, this isn't always reflected in what museums exhibit or how art is valued. Art markets tends to be lagging indicators of social change, with recognized names commanding the highest prices, and dead artists outperforming living ones. These dynamics have been roiling the art world for decades, where activism has been more visible but the institutional power structure is less centralized. Government agencies and trade organizations like the College Art Association have tried to help. The National Endowment for the Arts (NEA) took an early lead in promoting diversity through aggressive funding for emerging artists, women, and people of color. But because it is a federal program reliant on congressional appropriations, the NEA has been buffeted over the years by political pressures to temper such efforts.

As a government agency, the NEA "funds, promotes, and strengthens the creative capacity of our communities by providing Americans with diverse opportunities for arts participation."[66] Underlying this official charge has been the NEA's function in obviating (or at least lessening) the market imperative in determining aesthetic worth, often using the academic model of "peer review" in which artists and performers evaluated each other's efforts. Especially during the decades just following the NEA's 1965 founding, the agency gave money directly to artists and to hundreds of non-profit galleries, event venues, and publishers enabled by federal funding—in some regions supplemented by moneys from state and local arts agencies. Down somewhat from its highest budget in 1992, the NEA today divides its $146 million appropriation into over

2,300 grants, supporting "more than 30,000 concerts, readings, and performances and more than 5,000 exhibitions of visual and media arts with annual, live attendance of 33 million." An additional 360 million people see NEA-sponsored projects through broadcast and cable media.[67] NEA grants often serve an honorific function for recipients, who typically match the relatively small awards with an additional 90 percent of funding from other sources.

In addition to direct support for underrepresented artists and performers, the NEA has taken strong advocacy positions, as seen in the agency's *NEA Arts* magazine. A recent issue entitled "Telling *All* Our Stories" opened with the following statement:

> We know that the arts can be a powerful tool to tell our stories. And yet, one needn't look further than the Oscars controversy to know that certain stories remain more frequently told and celebrated than others. For example, only four percent of classical symphony musicians are African American. Just five percent of professional staff in art museums are people of color, with even fewer among senior management. In film and television, there are nearly five men for every woman working as a writer, director, or producer.[68]

Similar diversity advocacy comes from the Corporation for Public Broadcasting (CPB), which is the primary federal sponsor of public broadcasting stations (1,400 nationwide). In recent years, the CPB has operated on an appropriation of $455 million divided among 580 grantees. By comparison, it cost $478 million to make *Avatar* and $444 million for *Pirates of the Caribbean: On Stranger Tides.* [69] As the agency states, "CPB strives to support diverse programs and services that inform, educate, enlighten and enrich the public. Through grants, CPB encourages the development of content that addresses the needs of underserved audiences, especially children and minorities."[70]

All of this begs the question of what audiences want, and to what extent they exercise discretion in art and media consumption. With the demographics (and contents) of cultural consumption often varying throughout the American public, how is the overall population responding? Taking a measure of the complex array of commercial and government-sponsored cultural offerings—as well as the rising volume of consumer-generated material—federal agencies teamed up with the U.S. Census Bureau in recently issuing the first comprehensive "Survey of Public Participation in the Arts, 2002–2017" (SPPA).[71] Somewhat like the "creative economy," the SPPA took a broad view of art in its analysis, including such popular forms as movies and online media as well as more traditionally defined forms such as opera and visual art exhibitions. Predictably, the report showed broad growth in "new media" activity (online music, photography, etc.) and declining interest in such cultural standards as museum shows and symphony performances. Also unsurprisingly, younger and less educated consumers preferred the former, with retirees and college-educated cohorts trending toward the latter.

Music was the most popular consumer category, with nearly 79 percent of adults listening via radio, TV, internet, or CDs. Attendance at live concerts was somewhat less impressive, with classical music falling from 9.3 to 8.8 percent.[72] Overall attendance also dropped for "visual and performing arts events" from 35 to 33 percent, with 9.1 percent going to an art gallery or museum (2 percent decline) and 8.3 percent attending a play (12 percent decline). Movie-going was reported by 59 percent of respondents, although this included people seeing but one film a year. Gender made little difference in movie attendance, although age seemed a factor. From a high of 76 percent among 18–24 year olds, film viewing was but 44 percent among those over 65. High school graduates went to movies at a rate of 36 percent, with those finishing college attending at 65 percent.[73]

"Non-white and Hispanic Americans saw no declines in their arts attendance," the report stated, adding that "on the contrary, they even showed increases in some categories." That said, African-American viewing of art exhibitions (12 percent) was half of that of European-Americans (24 percent).[74] Similar disparities were evident at classical music and ballet performances, although African-American preferences rose in the jazz category (from 8 to 11 percent), as did Hispanics listening to "Latin, Spanish, or Salsa Music" (from 17 to 18 percent). Also SPPA said that "young adults and Hispanic Americans are among the most active" in social dancing.[75] Hispanic Americans also were the only group in which reading books did not decline (31 percent). "Adults who read at least one work of literature" dropped among African-Americans (from 42 to 40 percent) and Whites (55 to 52 percent), with poetry showing the greatest decline across all groups (down 19 percent).[76] Turning to art education, the report said that "roughly half of all adults had experienced some arts learning at some point in their lives. But disparities persist by gender, race/ethnicity, and general level of educational attainment."[77]

So where does this leave things? As discussed, the output of American cultural institutions remains skewed in predictable ways, albeit with some recent diversity efforts. Meanwhile, audiences slowly are moving away from what was once called "high culture," while also reading less, attributable to such factors as electronic media and the changing make-up of the U.S. population. But ticket sales and attendance statistics are not the whole picture. It's widely known that the twenty-first century also has seen a great expansion in the making of culture by those traditionally seen as consumers. In many ways, this is a logical extension of mental dialogue that always take place in any cultural activity. Former beliefs in a propagandistic one-directional flow of messages from producer to receiver may still be valid, but contemporary audience research has shown that people also think for themselves. And viewers do more than merely making program choices.

As a generation of TV offerings like *The Daily Show* has made self-conscious criticality a given in millennial society, more and more people are taking the means of media production into their own hands by making, posting, and otherwise stating their views. Audiences quite literally are "coming to voice" and talking back to mainstream culture (and its authority) much as activist bell hooks famously advocated.[78] Documenting some of this activity, the SAAP collected information on "Making and Sharing Art." The e-mailing, posting, or sharing of photographic images accounted for 27 percent, with an additional 15 percent of people saying they made images themselves.[79] While technology accounted for a lot of this, it might surprise many to know that dancing (at clubs, weddings, and elsewhere) was the biggest category of "making," accounting for 32 percent of respondents. Fiber-related activity was reported by 13 percent in the form of knitting, weaving, crocheting, quilting, and sewing. Much smaller numbers were seen in creative writing (6 percent), visual art (6 percent), dance (5 percent), playing a musical instrument (5 percent), and acting (1 percent).[80]

While these patterns of self-expression and user-generated content may seem new, its more accurate to say they are newly recognized. People always have expressed themselves through such everyday activities as home decorating, cooking, sewing, and snapshot photography, not to mention crafts like "Paint-by-Number" kits as discussed in Chapter 4. But such activities rarely have been considered "creative", let alone artistic endeavors. Much of the discussion above has considered the role of social construction in such designations, detailing hierarchical views of artistry (a reverence for prodigious genius, idealized beliefs in "great" art, a fascination with celebrity culture) held in place by what feels to many like an unassailable institutional structure. And, of course, the role of money, markets, and social differentiation cannot be understated.

In this chapter, cultural stratification in the U.S. has been traced to longstanding conflicts in Western epistemology and metaphysics, which translated into dominant aesthetic views predating the nation's founding. In turn, these philosophical traditions played themselves out in contradictions in American politics and economics, notably the difficulty of reconciling democratic egalitarianism and free-market competition. From all of this emerged a set of classifications assigning the authorship and stewardship of artistic works to some groups over others. Over time, obvious and not-so-obvious patterns of cultural inequity have fed lingering feelings of alienation within a population that nevertheless demands creative experiences. Of course, great art and genius are more than fabrications—and exceed the mere conjoining of social construction and material interest. Despite the admitted hype surrounding child prodigies or other people with "gifts," most Americans accept as self-evident the premises of differential talent and skill in creative pursuits, much as they do in athleticism and intelligence. In what follows I will examine the literature on exceptional creativity, both in historical and anecdotal terms. The next chapter also will look at the extensive scientific study attached to extraordinary achievements, especially as interest has been propelled by the creative economy in recent years.

Notes

1 Anthony Giddens and Mitchell Duneier, *Introduction to Sociology* (10th ed.) (New York: W.W. Norton, 2016) pp. 17–20.
2 Shailendra Bhatnagar, "Arushi Bhatnagar: World's Youngest Professional Artist," arushiarts (Mar. 12, 2012) http://arushiarts.blogspot.com (accessed Oct. 2, 2018).
3 Ibid.
4 Andrew Solomon, "How Do You Raise a Prodigy?" andrewsolomon.com (2012) http://andrewsolomon.com/articles/how-do-you-raise-a-prodigy (accessed Oct. 3, 2018).
5 Scott Barry Kaufman, "The Mind of the Prodigy," *Scientific American* (Feb. 10, 2014) http://blogs.scientificamerican.com/beautiful-minds/the-mind-of-the-prodigy (accessed Oct. 3, 2018).
6 Jacoba Urist, "What Makes a Child an Art Prodigy?" *The Atlantic* (Nov. 7, 2014) www.theatlantic.com/entertainment/archive/2014/11/what-makes-a-child-an-art-prodigy/382389 (accessed Oct. 3, 2018).
7 Germane Greer, "Would You Pay $3,000 for a Painting by a Toddler?" *The Guardian* (Jan. 29, 2009) www.theguardian.com/artanddesign/2009/jan/26/art-australia (accessed Oct. 3, 2018).
8 Ibid.
9 "What Makes a Child an Art Prodigy?"
10 "How Do You Raise a Prodigy?"
11 Andrew Solomon, *Far from the Tree: Parents, Children, and the Search for Identity* (New York: Scribner, 2012) p. 405.
12 *Far from the Tree*, p. 406.
13 *2014–2015 State of the States in Gifted Education*, National Association for Gifted Children and the Council of State Directors for the Gifted (2016) www.nagc.org/resources-publications/gifted-state/2014-2015-state-states-gifted-education (accessed Oct. 4, 2018).
14 Zahira Torres and Ryan Menezes, "Only 35 L.A. Public Schools Get an A in Supporting the Arts," *Los Angeles Times* (Nov. 2, 2015) www.latimes.com/local/education/la-me-lausd-arts-20151102-story.html (accessed Oct. 4, 2018).
15 Gabby Laurent, "Meet Art Prodigy Autumn de Forest, 14, Who Has Sold $7m of Paintings," *Financial Times* (Oct. 1, 2016) www.ft.com/content/1598a426-3880-11e6-9a05-82a9b15a8ee7 (accessed Oct. 5, 2018).
16 Thomas Fields-Meyer, "Hype, Art, and Lawyers," *People Magazine* 61, no. 18 (May 10, 2004). http://people.com/archive/hype-art-and-lawyers-vol-61-no-18 (accessed Oct. 5, 2018).
17 Katya Kazakina, "Art Market Nears Record with $66 Billion in Global Sales," Bloomberg News (Mar. 12, 2014) www.bloomberg.com/news/articles/2014-03-12/global-art-market-surged-to-66-billion-in-2013-report (accessed Oct. 5, 2018).
18 Henri Neuendorf, "Art Demystified: Why is Buying Art as an Investment Frowned Upon?" Artnet News (Aug. 4, 2016) https://news.artnet.com/art-world/art-demystified-art-as-investment-581129 (accessed Oct. 5, 2018).
19 "The Mind of the Prodigy."
20 Gillian Fournier, "Schema," PsychCentral (Jul. 17, 2016) https://psychcentral.com/encyclopedia/schema (accessed Oct. 22, 2018).
21 Terry Eagleton, *The Ideology of the Aesthetic* (Cambridge and Oxford: Basil Blackwell, 1990) p. 17.
22 Alexander Baumgarten, as quoted in Ernst Cassirer, *The Philosophy of the Enlightenment* (Princeton: Princeton University Press, 1951) p. 340.
23 Immanuel Kant, *The Critique of Judgement (Oxford World Classics)* (London and New York: Oxford University Press, 2009).

24 *The Critique of Judgement*, p. 51.
25 Leo Marx, "The Idea of Technology and Postmodern Pessimism," in Yaron Ezrahi, Everett Mendelsohn and Howard P. Segal, eds., *Technology, Pessimism, and Postmodernism* (Amherst, MA: University of Massachusetts Press, 1994) p. 14.
26 Pierre Bourdieu: *Distinction: A Social Critique on the Judgement of Taste*, trans. Richard Nice (Cambridge, MA: Harvard University Press, 1984) p. 32.
27 *Distinction*, p. 30.
28 Johann Christoph Friedrich von Schiller, *On the Aesthetic Education of Man in a Series of Letters* (1789), trans. Reginald Snell (New York: Dover, 2004).
29 Ernst Bloch, *The Principle of Hope, Vol. 1–3* (1984–1918) (Cambridge, MA: MIT Press, 1995).
30 Friedrich Nietzsche, *The Birth of Tragedy out of the Spirit of Music* (1871) (New York: Cambridge University Press, 1999).
31 Friedrich Nietzsche, *The Will to Power*, ed. Walter Kaufman (1901) (New York: Vintage, 1968) p. 853.
32 Charles Baudelaire, *The Painter of Modern Life* (1863) (New York: Da Capo, 1964).
33 Max Weber, *The Protestant Ethic and the Spirit of Capitalism* (1905) (New York: Norton, 2009).
34 Ralph Waldo Emerson, "The American Scholar" (1837) www.emersoncentral.com/amscholar.htm (accessed Aug. 25, 2018).
35 Matthew Arnold, *Culture and Anarchy* (1869) (New York and Oxford: Oxford University Press, 2009).
36 Andrea Fraser, as quoted in James McAnally, "A Call for a Collective Reexamination of Our Art Institutions," Hyperallergic.com (Oct. 12, 2016) http://hyperallergic.com/329506/call-collective-reexamination-art-institutions (accessed Oct. 12, 2018).
37 Matthew Arnold, "Preface," *Culture and Anarchy: An Essay in Political and Social Criticism* (New York: Macmillan, 1875) p. xxxvii.
38 Adam Kirsch, "Introduction to *The Modern Element*" (2008) http://poems.com/special_features/prose/essay_kirsch.php (accessed Mar. 21, 2019).
39 Terry Eagleton, *The Ideology of the Aesthetic* (London: Blackwell, 1975) p. 75.
40 Walter Benjamin, "Theories of German Fascism: On the Collection of Essays *War and Warrior*, edited by Ernst Jünger," *New German Critique* 17 (Spring 1979): 120–8.
41 Andrew Carnegie, "The Gospel of Wealth" (1886) www.carnegie.org/publications/the-gospel-of-wealth (accessed Jun. 4, 2019).
42 "Most Expensive Painting," *Time Magazine* (Sep. 9, 2015) http://content.time.com/time/specials/packages/article/0,28804,1917097_1917096_1917100,00.html (accessed Oct. 11, 2018).
43 Arthur Korteweg and Roman Kräussl Patrick Verwijmeren, "Research: Is Art a Good Investment?" *Insights* by Stanford Business (Oct. 21, 2013) www.gsb.stanford.edu/insights/research-art-good-investment (accessed Oct. 11, 2018).
44 Robert Elkind, "Why the Booming Contemporary Art Market Is Bad for Museums," *Washington Post* (Aug. 4, 2016) www.washingtonpost.com/posteverything/wp/2016/08/04/why-the-booming-contemporary-art-market-is-bad-for-art-museums/?utm_term=.72a02986c70f (accessed Oct. 11, 2018).
45 Douglas Crimp, "On the Museum's Ruins," *October* 13 (Summer 1980): 51.
46 "On the Museum's Ruins," p. 44.
47 Katarzyna Murawska-Muthesius and Piotr Piotrowski, eds., *From Museum Critique to the Critical Museum* (London and New York: Routledge, 2018); Paul O'Neill, Lucy Steeds, and Mick Wilson, *How Institutions Think: Between Contemporary Art and Critical Discourse* (Cambridge, MA: MIT Press, 2017).

48 Steven C. Dubin, *Displays of Power: Controversy in the American Museum from the* Enola Gay *to* Sensation (New York: NYU Press, 1999). p. 3.
49 Julia Halperin, "Almost One Third of Solo Shows in U.S. Museums Go to Artists Represented by Just Five Galleries," *The Art Newspaper* (Apr. 2, 2015) http://old.theartnewspaper.com/articles/Almost-one-third-of-solo-shows-in-US-museums-go-to-artists-represented-by-just-five-galleries/37402 (accessed Oct. 13, 2018).
50 Maura Reilly, "Taking the Measure of Sexism: Facts, Figures, and Fixes," *ArtNews* (May 26, 2015) www.artnews.com/2015/05/26/taking-the-measure-of-sexism-facts-figures-and-fixes (accessed Jun. 14, 2018).
51 Andreas Huyssen, *After the Great Divide: Modernism, Mass Culture, Postmodernism* (Bloomington and Indianapolis: Indiana University Press, 1986) p. ix.
52 "No. 1440. Selected Communications Media: 1920 to 1998," *Twentieth Century Statistics*, U.S. Census Bureau (1999) www.census.gov/prod/99pubs/99statab/sec31.pdf (accessed Oct. 15, 2018).
53 Nick Hunter, *Popular Culture 1940–1959* (New York: Raintree, 1960) p. 29.
54 James M. Wilson, "The Drama of Cultural Conservativism," *The Point* 3 (2012) https://thepointmag.com/2010/politics/the-drama-of-cultural-conservatism (accessed Oct. 15, 2018).
55 Nathan Godfried, "Labor," in Peter Simonson et al., eds., *The Handbook of Communication History* (London and New York: Routledge, 2013) p. 320.
56 Henri Lefebvre, *The Critique of Everyday Life: Volume 1*, trans. John Moore (London and New York: Verso, 1947).
57 *Critique of Everyday Life*, p. 57.
58 Guerrilla Girls, "Do Women Have to Be Naked to Get into the Met. Museum?" (1989) www.guerrillagirls.com/projects (accessed Oct. 17, 2018).
59 Laura Mulvey, "Visual Pleasure and Narrative Pleasure," *Screen* 16, no. 3 (1975).
60 Edward Said, *Orientalism* (New York: Vintage Books, 1978).
61 Chris Jordan and Glenn Weedon, "Introduction," in Chris Jordan and Glenn Weedon, eds., *Cultural Politics: Class, Gender, Race and the Postmodern World* (London and New York: Blackwell, 1995) p. 11.
62 Deanne Gertner, "Gender in the Art World: A Look at the Numbers," NINE dot Arts (May 31, 2017) www.ninedotarts.com/gender-in-the-art-world-a-look-at-the-numbers (accessed Oct. 16, 2018); Brent Lang, "Women Comprise 7% of Directors on Top 250 Films (Study)," *Variety* (Oct. 27, 2015) http://variety.com/2015/film/news/women-hollywood-inequality-directors-behind-the-camera-1201626691 (accessed Oct. 16, 2018).
63 Roxanne Gay, "The Oscars and Hollywood's Race Problem, *New York Times* (Jan. 22, 2016) www.nytimes.com/2016/01/24/opinion/the-oscars-and-hollywoods-race-problem.html?_r=0 (accessed Oct. 16, 2018); Betty Farrell and Maria Medvedeva, *Demographic Transformation and the Future of Museums, American Association of Museums* (2010) www.aam-us.org/wp-content/uploads/2017/12/Demographic-Change-and-the-Future-of-Museums.pdf (accessed Jun. 4, 2019). As a point of reference, the most recent (2018) U.S. population estimates put the demographics of the nation's population (327 million) in the following terms: female 50.8 percent, male 49.2 percent, white 76.6 percent, African American 13.4 percent, Hispanic/Latino 18.1 percent, Asian 5.8 percent. See Quick Facts: United States, Population Estimates (Jul. 1, 2018) www.census.gov/quickfacts/fact/table/US/PST045218 (accessed Jun. 18, 2019).
64 "The Film Academy's 2016 Class: An Exclusive Club Gets Much Bigger After #OscarsSoWhite," *Los Angeles Times* (Jul. 4, 2016) www.latimes.com/entertainment/movies/moviesnow/la-et-mn-academy-new-members-live-updates-20160629-htmlstory.html (accessed Jun. 4, 2019).

65 "The High Cost of Hollywood's Gender Bias," AAUW (Feb. 4, 2016) www.aauw.org/article/hollywoods-gender-bias (accessed Oct. 17, 2018).
66 The National Endowment for the Arts, nea.gov (2016) www.arts.gov (accessed Oct. 27, 2018).
67 National Endowment for the Arts Appropriations History (2016) www.arts.gov/open-government/national-endowment-arts-appropriations-history (accessed Oct. 17, 2016); NEA Annual Report 2015 (2016) www.arts.gov/sites/default/files/2015%20Annual%20Report.pdf (accessed Oct. 17, 2018).
68 "This Issue," *NEA Arts* 1 (2016) www.arts.gov/sites/default/files/nea_arts/FINAL_NEA_Arts_1_2016.pdf (accessed Oct. 17, 2018).
69 Charles Stockdale and John Harrington, "From 'Transformers' to 'Avatar', These Are the 50 Most Expensive Movies Ever Made," *USAToday* (Jul. 6, 2018) www.usatoday.com/story/life/movies/2018/07/06/transformers-avatar-50-most-expensive-movies-ever-made/762931002 (accessed Nov. 10, 2018).
70 "About CPB," CPB.org (n.d.) www.cpb.org/aboutcpb (accessed Oct. 17, 2018).
71 *Survey of Public Participation in the Arts, 2002–2017*, National Endowment for the Arts (2018) www.arts.gov/artistic-fields/research-analysis/arts-data-profiles/arts-data-profile-18 (accessed Oct. 18, 2018)
72 *Survey of Public Participation*, p. 13.
73 *Survey of Public Participation*, p. 23.
74 *Survey of Public Participation*, p. 2.
75 *Survey of Public Participation*, p. 43.
76 *Survey of Public Participation*, p. 25.
77 *Survey of Public Participation*, p. 43.
78 bell hooks, *Talking Black: Thinking Feminist, Thinking Black* (New York: Between the Lines, 1989).
79 *Survey of Public Participation*, p. 34.
80 *Survey of Public Participation*, p. 43.

Chapter 6

Divine Madness

The Crazy-Artist Myth

I've known thousands of creative people in my four decades in the arts—as friends, colleagues, students, and employees. I gravitated to art as a kid, completed an MFA degree, worked in arts non-profits, taught art in public schools and universities, and served as an art school dean, while writing hundreds of exhibition reviews. Like anyone else in this profession, I've always been aware of the crazy-artist stereotype, even as personal experience has shown me that artists are as sane as people in other fields. And yes, I've seen a few psychiatric hospitalizations and a couple of suicides, as well as more than a little eccentric behavior. But more commonly, I've found that because instability routinely is expected in creative types, they often must demonstrate *extra* measures of rationality and responsibility as they manage galleries, organize performances, write grants for funding, administer their finances, or negotiate the workplace.

Nevertheless, it's hard to imagine anyone unfamiliar with the crazy-artist stereotype. Popular culture loves to depict artists as moody, fanciful, self-indulgent, unreliable, childlike, rebellious, obsessional, delusional, fanatic, psychotic, or otherwise unhinged. Such ideas are rooted deep in Western culture, with Plato and Socrates both speaking of creativity as "divine madness" and Lord Byron famously telling fellow poets, "We of the craft are all crazy."[1] Clearly, artists themselves have been part of the problem, often reveling in their status as visionaries or outsiders. And most people can call to mind at least one famous artist or writer who has done something shocking (Van Gogh cutting off his ear) or committed suicide (Sylvia Plath, Jackson Pollock). Media accounts haven't helped matters either, glorifying the bizarre or self-destructive habits of creative types in Hollywood movies like *Synecdoche, New York* (2008), *Crazy Art* (2009), *Black Swan* (2010), *Being Flynn* (2012), *For No Good Reason* (2012), *Loving Vincent* (2017), *Van Gogh: Of Wheat Fields and Clouded Skies* (2018), and *At Eternity's Gate* (2018). Arguably, this negative attention may be one reason so many people don't see themselves as "creative" types, even as the new cultural economy rebrands artistry as an aspirational ideal.

Notice that already I've evoked a handful of famous names. This speaks to popular culture's strong association of mental illness and prodigious achievement. Craziness helps to explain the vagaries of creative talent, as well as the mysterious power that artworks can have on audiences. Societies often have handled the unfamiliar this way, sorting certain people and behaviors into categories and then assigning them a value. Habits of mind can drive such assessments, aside from anything the person in question may or may not do. Neutrality often goes out the window, as projection and prejudice come into play. And perhaps obviously, the focus on famous artists leaves out everyone not marked by greatness or "genius," while also discounting the universe of day-to-day creativity discussed in Chapter 4.

"The truth of the matter is that the *vast* majority of creative people are not mentally ill and, more importantly, the *vast* majority of those suffering from psychopathology are not geniuses," wrote psychologist Arne Dietrich in "The Mythconception of the Mad Genius."[2] Dietrich argued that the "creativity and madness binary" is both a false dichotomy and a sweeping overgeneralization. How can one speak of the "creativity" of actors, choreographers, composers, designers, filmmakers, installation artists, musicians, novelists, painters, poets, and web producers in a single breadth? "Madness" is an even more ambiguous concept, with scores of diagnostic categories and professional definitions, foisting the crazy-artist stereotype on everyone from manic-depressives, schizophrenics, and substance abusers, to autistic children, "savants," and seniors with dementia. Given this confusion, nuanced thinking about artistry and insanity has been relatively rare until recent decades. Partly, this is because mental illness remains so underreported and misunderstood within the population at large. In this vacuum, retrospective views of famously crazy figures become default values to prove a "natural" connection between prodigious artistry and mental illness.

Taken as a whole, the literature on creative madness paints a mixed picture at best. Research in the nineteenth century drew on primitive understandings of the mind and often was tainted by romantic preconceptions about artists and writers. Later attempts to prove the crazy-artist thesis often were poorly designed and ended up with contradictory findings. But public fascination never abated. A quick look at Amazon.com finds 9,436 books currently listed about "mad artists" and another 5,120 on the topic of "art and mental illness."[3] While most of this literature conforms to expected stereotypes, a modest discourse has emerged in recent years to challenge (or at least question) the crazy-artist construction, including books like Judith Schlesinger's *The Insanity Hoax: Exposing the Myth of the Mad Genius*, Simon Kyaga's *Creativity and Mental Illness: The Mad Genius in Question*, and Pascal Geilen's *Creativity and Other Fundamentalisms.* [4] Scrutiny seems to be growing within academia as well, with new journals appearing in the past decades in fields of art, education, and mental health, including *Creativity Studies* and *Journal of Creativity in Mental Health*, among others.

This chapter opens by tracing the myth of the crazy artist to its origins in the nineteenth century, as positivistic studies of genius and deviance coincided with a bourgeois and philosophical distancing of aesthetics from "reason." This intersectional *othering* of artists—as abnormal or unreasoning—shaped later views of creative madness in the fields of psychiatry and psychoanalysis. Discussion then turns to recent scientific findings on creativity—as well as the widely reported use of alcohol and other drugs by artists, writers, and other creative types. Oftentimes it seems that researchers have made up their minds before their studies begin. While the voluminous evidence of artistic wackiness seems hard to deny, is it possible that eccentricity or passion can get mislabeled in this process? Helping to sort this out in recent years have been advances in neuroscience, genetic mapping, and population research. Much more is becoming known about the workings of the brain, the body, and culture in the murky madness/creativity intersection. At the same time, parts of the puzzle still remain elusive.

The Art of Unreason

Western culture's fascination with crazy artists derives in part from the inexplicability of creativity itself. Sloppy definitions of artistry give the concept a generative or reactive aspect—often bespeaking an uneasiness over things unknown. Combine this with the emotional appeal of an image or song, and the stage is set for magical thinking. Plato drew an early distinction between creative inspiration and other kinds of intellect, likening artistry to a "form of possession" enabling a supernatural access to truth. "Whoever knocks on the door of poetry without the madness of the muses does not reach his aim," Plato said, adding, "The poetry of reason disappears before the poetry of the madman."[5] This familiar view of creativity also relates to the Platonic concept of "doxa," a Greek word meaning popular opinion or common belief. Plato spoke of the way doxa can fall prey to irrationality, and he worried that poets and artists could lead populations astray.[6]

Suspicions about creative madness deepened over time. With the rise of the Roman Empire came concerns about what were termed "melancholic" states of mind, with poets and other "men of genius" particularly prone. Cicero associated melancholy with excessive guilt, fear, or rage (sometimes attributed to bile in the body) making someone lethargic, "unsound," or "suspicious toward everything, and hating the company of people."[7] During the Middle Ages, themes of magical enchantment pervaded perceptions of insanity and artistry. With the rise of Christianity, religiously inflected views saw madness as punishment from above, with confession or exorcism often prescribed. Beauty and artistic vision had similarly supernatural explanations, often attributed to the hands of Satan or God. While artists themselves were not always seen as crazy, writers and painters often depicted the tormented and became associated with such ideas—as in Geoffrey Chaucer's *Canterbury Tales* and such works as Hieronymus Bosch's 1494 *Extraction of the Stone of Madness*, and Vittore Carpaccio's 1496 *The Healing of Possessed Man at the Rialto*.

In the Enlightenment Era, madness and creativity became conjoined as common enemies of reason. Mental illness was seen as a loss of rational faculties and a threat to public well-being, as creativity was partitioned from reason as its own often "irrational" thought process. Also underlying these views of madness and creativity was a specific Western view of the self as separate from the world around it. Rather than an integrated relationship with humanity or the cosmos, as seen in Buddhist or Hindu faiths, European culture prized the individual almost from the beginning. Artists and the "mad" were islands unto themselves. In his epic *Madness and Civilization: A History of Insanity in the Age of Reason*, Michel Foucault wrote of the era's rising preoccupation with behavioral disorders. As leprosy abated across Europe, France would begin conversion of its 2,200 sanatoriums into asylums for the insane—often casting a broad net in its definitions. A madman could be just about anyone falling outside accepted norms of reason or moral propriety. Such prejudice became the rationale behind the ultimate segregation and confinement of the mentally ill. Along with psychotics and the criminally insane, artists were often swept into this net, along with other aberrant types like vagabonds, prostitutes, and the disabled. Key in Foucault's formulation was the transformation of insanity from *wisdom* (as expressed in the works of Shakespeare and Cervantes) into a *disease* of unreason by medicine and psychiatry in the early nineteenth century.[8]

Bringing mental illness and artistry together in a U.S. context was Declaration of Independence signatory Benjamin Rush, who many name as the first American psychiatrist. In his book *Medical Inquiries and Observations upon the Diseases of the Mind* (1812), Rush advocated bloodletting, spinning patients on wheels, and what was termed at the time "moral therapy." In keeping with public health concerns, Rush also drew the following conclusions about artists:

> From a part of the brain preternaturally elevated, the mind sometimes discovers not only unusual strength and acuteness, but certain senses it never exhibited before … Talents for eloquence, poetry, music and painting, and uncommon ingenuity and several of the mechanical arts, are often involved in the state of madness.[9]

Explaining that he commonly found delusional behavior in creative people due to "the vague and distracting exertions of genius," Rush saw similar proclivities in adolescents, unmarried adults, farmers, and "people with dark hair."[10]

Consonant with the era's rising interest in the mind, scientists drew up optimal standards of psychological fitness. This launched the search for the "normal" person, following newly emergent fields of genetics and statistical analysis. Charles Darwin's principles of evolution had popularized beliefs in the possibility of human improvement by natural selection. It didn't take long for many in the nineteenth century to speculate that social improvement could be achieved by encouraging greatness—and weeding out weakness. Soon came studies of "genius" as well as "defective" traits. Meanwhile, statisticians were working out the relationships of norms and deviations, first in the physical and biological sciences, but later applying them to public policy. By the mid-nineteenth century, Belgian astronomer Adolphe Quetelet arrived at his formulation of *l'homme moyen* (the "average man") as a model of balance between extremes, from which he extrapolated a "mathematics of society" charting probable heights, weights, and measures of health. Of controversy at the time, Quetelet began looking at behavioral data—specifically crime rates—in asserting that the "natural laws" of averages should inform the civil law.[11]

While the field of psychiatry did not yet exist as known today, Quetelet's work greatly influenced studies of intelligence, criminology, and their possible genetic origins. Quetelet's readers included Émile Durkheim, Karl Marx, Florence Nightingale, and Edward Burnett Tylor. Another notable follower was English sociologist Francis Galton, a half-cousin of Darwin, who pressed to extend the new statistical methods into the realm of public policy. Within a decade of Quetelet's writing on norms and deviations, Galton devised methods for the systematic elimination of defectives from England, a population that included the "feebleminded," the deaf, and the blind. Galton's work had profound implications in "scientifically" validating latent prejudices against certain groups. It would take another century for the damaging effects of normativity to be systematically analyzed and understood, as it became clear that statistical mid-points, in fact, excluded the majority of values on most scales.

In his 1869 study *Hereditary Genius: An Inquiry into its Laws and Consequences*, Galton examined population data and the biographies of famous people. From this he devised a grading system ("A" to "G") for intelligence in animals and people, observing, "I presume the F class of dogs, and others of the more intelligent sort of animals is near commensurate with the F of the human race."[12] Galton distinguished "talent" (aptitude for specific tasks) from "genius" (capacity for originality and creativity). Like others in the period, Galton sought to prove that ability ran in families, finding this occurring for at least one generation, after which it diminished in what he termed "regression to the mean." This led Galton to advocate eugenic marriages among genius families, as well as practices of selective human breeding more generally. Notably, he put no women or non-whites in his "genius" category.

Focusing on mental illness, Italian physician Cesare Lombroso was a fervent believer in the inheritability of psychopathology, claiming that it appeared in physical features and defects (which he termed "stigmata"). Also known as the "father of positivist criminology," Lombroso wrote that "born criminals" could be identified on sight (through the emerging pseudo-science of phrenology), owing to "ape-like" features such as sloping foreheads or long arms. In researching his often-referenced 1891 *Man of Genius*, Lombroso pored over encyclopedias, biographies, letters, and writings by famous artists and authors in a search for repeated idiosyncrasy. And, of course, he found it, concluding that "insane painters and poets are so numerous" because "the imagination is most unrestrained when reason is least dominant. For the latter, by repressing hallucinations and illusions, deprives the average man of a true source of artistic and literary inspiration."[13] And after visiting museum portrait galleries and studying images of artists and writers, Lombroso concluded that they were typically small, pale, and emaciated—often with big ears and stammers. In a section entitled "Insane Genius in Literature," he wrote that "quickened by insanity … these men have less hatred of novelty, and more originality, than normal people."[14]

Others looking at genius biographies saw no insanity at all. In his 1904 *Study of British Genius,* Havelock Ellis scoured 66 volumes of that nation's *Dictionary of National Biography* to pick out entries he deemed as showing artistic or intellectual "ability of high order." Ellis eventually complied 1,030 files that he scrutinized according to heredity, health status, economic class and intellectual/creative distinction. The list was 95 percent male. Ellis stated that "we must put out of court any theory as to genius being a form of insanity," attributing the periodic appearance of mental instability to a range of environmental conditions, such as "impoverishment, rejection and persecution."[15] Ellis also took interest in other suspect groups of the era, and is perhaps best known for authoring the first objective study of homosexuality. In his 1897 book *Sexual Inversion,* Ellis detailed the practices of gay men and countered prevailing views that homosexuality was a disease, moral failing, or crime.[16] Notably, Ellis wrote, "There are certain avocations to which inverts seem especially prone. One of the chief among these is literature" and that such individuals "in the proportion of 56 percent possess artistic aptitudes."[17]

Romantic Intentions

All of this occurred as philosophy was separating creativity from other forms of thinking. In structural terms, the aesthetic principles of Immanuel Kant cast artistic pursuits as a particular form of *unreason*—sometimes good, sometimes bad, but always beyond most people's rational processes. Such thinking fed the romanticism of the nineteenth century, which further encouraged beliefs in creative madness. The writings of Samuel Taylor Coleridge, John Keats, and William Wordsworth, and others celebrated the quasi-magical powers of the imagination in generating originality and *ex nihilo* ("out of nothing") creativity. "Oh for a life of sensations rather than of thoughts," Keats would proclaim.[18] Gothic fantasy paintings by William Blake, Eugène Delacroix, and Francisco Goya further advanced emotionality as a valued form of experience. At issue was what many in the romantic era saw as a false dichotomy between philosophy and science. The title of Jane Austin's *Sense and Sensibility* summed up this opposition.[19] In the absence of a bridging terminology, Coleridge coined the term "psychosomatic" to speak of the intersection of body and soul, speaking of a space between uncontrolled insanity and a self-conscious exploration of unreason, writing, "My case is a species of madness, only that it is a derangement of the volition, and not the intellectual faculties."[20] Also motivating the romantics was a recognition that normative reason tended to pathologize dissent, alternative thinking, moral ambiguity, or any forms of non-standard behavior as "deviant" or mad. Already many contemporary views of artists had been set in place, laying the groundwork for beliefs about creative people as pathologically different from others. This added a mystical value to the "exceptional" character of artists' works while devaluing "normal" activities of most people.

With so much attention in the nineteenth century given to states of mind, it's hardly surprising that new theories of consciousness would emerge. As philosophers and artists pondered the contest of reason and unreason, the mentally ill received growing attention. Care for those considered mad largely was limited to the severely afflicted, who usually were confined in asylums or tended by family members at home. Aside from scattered beliefs that mental illness was a "nervous condition," medical knowledge of the brain was absent. Treatments were crude, often harsh, and generally ineffective. In this context, neurologist Sigmund Freud took up studies in 1885 with physicians using hypnosis to treat what was deemed the female condition of "hysteria." Freud's breakthrough came when he asked patient "Anna O." to speak while hypnotized, during which she made up stories and recounted forgotten "reminiscences." Anna O. subsequently reported feeling better from what she termed the "talking cure." This famous case led Freud to his theory of the *unconscious* and the importance of "sweeping the mind clean" in therapeutic dialogue.[21]

While Freud's discovery of the unconscious is regarded as his most significant contribution to modern thought, the idea had already been used as a literary device in the works of such writers as Shakespeare and Robert Louis Stevenson. Nevertheless, the unconscious proved an influential counterpoint to nineteenth-century rationalist beliefs that individuals could fully know themselves and their world. Freud later would formulate his concept of "neurosis" (nerve illness), which he attributed to anxieties over unresolved conflicts or traumatic memories. In *Civilization and its Discontents*, Freud famously generalized neurosis quite broadly, as a symptom of conflict between an individual's desire for freedom and society's demand for conformity. Freud said that customs and laws compelled people to repress primitive impulses, with feelings of discontent inevitably lingering in the mind.[22] As Western nations were careening between world wars, Freudian therapy provided both an explanation for individual anxiety and an avenue to self-improvement, especially for those who could afford to pay.

Psychoanalysis played an unintended role in the pathologization of artists during the early twentieth century. As Freudian analysis entered mainstream culture, its well-heeled clients also made up a large part of the museum-going public, within which art was recognized as a channel for otherwise socially unacceptable thoughts. Meanwhile, books such as Freud's 1899 *The Interpretation of Dreams* encouraged artists and writers to explore the unconscious, often with bizarre imagery.[23] To Dada and Surrealists such as André Breton and

Salvador Dalí, psychoanalytic theory was both a tool for aesthetic inquiry and a rationale for provoking a complacent bourgeois society. As Breton wrote in his 1924 *Manifesto of Surrealism*:

> It was, apparently, by pure chance that a part of our mental world which we pretended not to be concerned with any longer—and, in my opinion by far the most important part—has been brought back to light. For this we must give thanks to the discoveries of Sigmund Freud.[24]

Others found encouragement for fantastical imagery in the writings of Carl Jung, a former student of Freud, who later attracted his own following among artists using symbolism, archetypes, and mythological figures in their work. In the long run, Freudian analysis proved to be something of a fad, losing ground to the medicalized field of psychiatry from the mid-twentieth century onward, owing to advances in brain science and drug treatments for mental illness. The field of psychoanalysis may have gotten smaller, but some of its strains also grew more sophisticated.

In the late twentieth century, Jacques Lacan became a household name for artists and intellectuals. Beginning in the 1950s, Lacan advocated a "return to Freud" informed by then-emerging theories of structural linguistics. In proposing that "the unconscious is structured like a language," Lacan mapped (and diagrammed) the subterranean workings of the mind as a source of insight. Lacan also devised vivid metaphors that appealed to artists, such as his formulation of the "mirror stage" of child development.[25] In Lacan's mirror stage, the infant's realization of a "me" is cast in visual terms, when the child-subject first sees itself as an image-object. This "gaze" creates a template for subsequent self/other concepts, as the youngster realizes that it both sees others and is seen by them. While the gaze is inherently neutral, it is influenced by forces inside (genetics, psychology, and learning) and outside (history, culture, and power).

Gendered aspects of the gaze would prove especially influential among artists and intellectuals. Whereas Freud had sexualized gender around the idea of having or not-having a "phallus," Lacan framed the matter in terms of desire and lack, which are not always anatomically dependent.[26] Put this way, representations of the phallic gaze could be analyzed as expressions of power dynamics, as discussed in the scholarship of Juli Carson, Jane Gallop, and Avital Ronell, among others. Before long, Lacanian theory enabled culture and media to be seen in new ways, as artists examined who was looking (or making the representations), who was being depicted (and in what ways), through what kind of lens (due to privilege, bias, or exclusion), and enabled by what kind of systems (owing to institutions, social classes, or economies). Psychoanalysis would play a vital role in the discourses of identity and difference that would grow in the post-war decades.

Paralleling this activity around the psychological dimensions of creativity was rising fascination with artwork by psychiatric patients. Partly an artworld reification of the crazy-artist stereotype, the movement also reflected clinical speculation that artworks could serve as diagnostic tools. French poet and clinical analyst Marcel Réja is credited with the first book in this area. In his 1907 *Art by the Mad*, Réja matched drawings and paintings by patients with a series of symbols he attributed to mental illness.[27] At the same time, Réja attributed an exceptional "vision" to artists with psychiatric disorders, notably Edvard Munch, whom he described as evoking "universality by his unself-conscious non-allegorical allegories, his ability to pierce the exterior."[28] Countervailing views already were emerging in art criticism, however. The era's rising interest in psychological processes brought Alois Riegl of the Vienna School of Art History to write about the subjective role of the *audience* in the aesthetic experience. In formulating his famous principle of the "beholder's involvement" (later rephrased by Ernst Gombrich as the "beholder's share"), Riegl described the mental collaboration between viewer and artist, as well as the highly individualized character of meanings derived.[29] This raised the important question of whether "madness" resided in the work of art or in the projective imaginations of therapists and museum goers.

But there was no stopping the craze for what later would be called "outsider art." German psychiatrist and art historian Hans Prinzhorn took issue with madness-and-creativity interpretations as formulated by Réja and predecessors such as Lombroso. In his 1922 book *Artistry of the Mentally Ill*, Prinzhorn derided attempts to find diagnostic clues in artwork by psychiatric patients, and instead asserted their legitimate aesthetic value through what he termed a "disquieting feeling of strangeness."[30] Initially showcasing works by ten "schizophrenic masters," Prinzhorn later built a huge collection of works by artists he romanticized as naïve, unschooled, and otherwise uninfluenced by aesthetic conventions.[31] Comprising more than 5,000 images by 450 patients, the "Prinzhorn Collection" had an enormous impact on the art world of the post–World War I era. The influence of the Prinzhorn Collection helped set in motion a fascination with outsider art, later advanced as *art brut* (raw art) by Jean Dubuffet in the 1940s, as well as figures like Breton, Marcel Duchamp, Paul Klee, and Pablo Picasso. Before long, the scope of outsider art would expand to include work by children, folk artists, and so-called "primitives" from outside the Western world.

Psychiatric Autopsies

Crazy artists have fascinated mental health professionals for over a century. Emil Kraepelin is known in psychiatry circles for his pioneering work on manic depression, and its manifestations in creative personalities, notably observing that bipolar symptoms often had cyclical characteristics.[32] Writing in his classic 1921 book *Manic-Depressive Insanity and Paranoia*, Kraepelin said that creative people often exhibited a "heightened distractibility" and a "spinning out of the circle of ideas and a jumping off of others." He specified that:

> These may under certain circumstances set free powers which otherwise are constrained by all kinds inhibition. Artistic activity namely may, by the untroubled surrender to momentary fantasies or moods, and especially poetical activity by the facilitation of linguistic expression, experience a certain furtherance.[33]

Aside from artists and writers, Kraepelin said mental illness was common among socialists and Jews.[34] A strong proponent of eugenics and racial hygiene in pre–World War II Germany, Kraepelin also is remembered for his reconceptualization of Lombroso's notion of born criminality into a diagnosis of "moral insanity" manifest in such forms as alcoholism, delinquency, homosexuality, and thievery.

As seen in biographical genius studies, contradictory findings among psychiatrists appeared almost from the beginning. From 1928 to 1944, psychiatrist Adele Juda studied 19,000 German artists and scientists, noting that "eminent artistic talents were frequently found in combinations such as music and poetry, music and painting, and painting, sculpture, and architecture." Unlike Kraepelin, Juda concluded:

> There is no definitive relationship between the highest mental capacity and psychic health or illness, and no evidence to support the assumption that the genesis of high intellectual ability depends on psychic abnormalities. The high number of mentally healthy geniuses speaks against such a claim and repudiates the slogan "genius and insanity."[35]

Going further, Juda observed that serious mental illness rendered a talent person incapable of productive work. "Psychosis, especially schizophrenia, proved to be detrimental to creative ability," Juda wrote.[36]

Things got even messier in 1978, when American psychoanalysts Mildred and Victor Goertzel found a *lower* level of mental illness in creative people. Published in their *300 Eminent Personalities: A Psychosocial Analysis of the Famous*, Goertzel and Goertzel found less than a 2 percent occurrence of mental illness—a significantly lower rate than the general population.[37] Also looking at family history, the team reported that parents and siblings of accomplished intellectuals and creative people had similarly low rates of psychiatric problems. The Goertzels noted that many they studied had faced challenges as children, but had acquired a certain tenacity (now termed "grit") growing up in poverty, broken homes, or with physical illness. In an earlier work entitled *Cradles of Eminence* (1962), the Goertzels had observed that high achievers often had strongly disliked school (especially their teachers), had highly opinionated parents/caregivers, and grew up "feeling different" from others.[38] From this the

Goertzels concluded that creativity—as a mechanism of adaptability—was more help than hindrance to mental health.

But soon creativity would be re-pathologized by neuroscientist and psychiatrist Nancy Andreasen, a recipient of the National Medal of Science and a former President of the American Psychopathological Association. Beginning in the early 1980s, Andreasen studied creative writers first-hand at the respected Iowa Writers' Workshop—including such notables as John Cheever and Kurt Vonnegut. Andreasen did extensive interviews with her 29 subjects, examined their family histories, and wrote nuanced papers about her research. Among other things, Andreasen noted the slipperiness of creativity as a research category, distinguishing between what she termed "big C" (recognized artistic accomplishment) and "little c" (everyday creativity). "What does it mean to have 'created' something?" Andreasen asked, adding, "Can creativity in the arts be equated with creativity in science or business?"[39] Notwithstanding this complexity, Andreasen matched her creative cohort with controls—and her findings were dramatic. "A full 80 percent of the writers had had some kind of mood disturbance at some point in their lives, compared with just 30 percent of the control group—only slightly less than an age-matched group in the general population," she wrote.[40]

I first discovered this back-and-forth discourse in 1997, when I wandered into a talk entitled "Creativity and Madness" while a guest at a psychiatry conference. I had never heard of Kay Redfield Jamison, whose book *Touched with Fire: Manic-Depressive Illness and the Artistic Temperament* had already been feted within the psychiatric field. Jamison's paper seemed largely a compendium of historical work on creatives she had begun in the 1980s, and I was skeptical as she began to plod through the well-known eccentricities of Blake, Byron, Mozart, and Pollock, punctuated with quotes like Emily Dickinson's "Could it be madness—this?"[41] Later I would learn of the enormous regard Jamison had earned for her research (named a "Hero of Medicine" by *Time Magazine* and winning a MacArthur genius grant) and personal ethics (revealing her own mental illness at a time when such disclosure could end a physician's career).[42] As her "Creativity and Madness" talk would demonstrate, part of what gave Jamison's work such resonance was her experience as both doctor and patient.

Keep in mind that psychiatric diagnoses always have derived from both medical and behavioral evidence. According to the American Psychiatric Association, findings of illness generally hinge on both "*symptoms*, a person's subjective report of an abnormality, and *signs*, objective findings of abnormality."[43] Within this logic, Jamison's broader *Touched with Fire* project had drawn on the self-reporting of famous figures, their behaviors as historically documented, as well as her own interviews with 47 British artists and writers, in which she found a 38 percent rate of mood disorders. As customary in psychiatry, Jamison saw "findings" emerging when symptoms and signs seemed to reinforce each other repeatedly. At the same time, Jamison also noted some of the limitations of such an approach. Among other things, she observed that famous artists may not accurately reflect the entire creative population, that artworks may not in themselves reflect someone's ongoing state of mind, and that strange behaviors may be transient and do not always mean a person is crazy.

Perhaps most importantly, Jamison's *Touched with Fire* shifted the conversation from a generic assumption of "creative madness" to a more focused consideration of manic depression, while also serving to destigmatize the condition in its less severe (and more common) manifestations. As Jamison put it, "Many are unaware of the milder, temperamental expressions of the disease or do not know that most people who have manic-depressive illness are, in fact, without symptoms (that is, they're pathologically normal) most of the time."[44] Additionally setting her work apart, Jamison explored the prevalence of "co-morbidities" (the intersection of two or more conditions), such as frequent substance use/abuse among bipolar people.

Both beginning their studies of artists four decades ago, the credibility of Andreasen and Jamison remains more or less unquestioned in psychiatry—largely due to their broader research on other topics: memory, neuroimaging, and schizophrenia (Andreasen); addiction, bipolar illness, and psychopharmacology (Jamison). Both women often reference each other's work, and within the field their different methodologies are seen as mutually reinforcing. At the same time, Andreasen's and Jamison's studies were small and have been challenged from outside the mental health field for their unrepresentative sampling, questionable evidence gathering, vague categorization and poorly defined criteria. Among other things, both researchers' study cohorts comprised mostly well-recognized male, white, and middle-aged figures. Criticizing the two bodies of work in his book *Creativity and Madness: New Findings and Old Stereotypes*, Albert Rothenberg observed that "the need to believe in the connection between creativity and madness appears to be so strong that affirmations are welcomed and treated rather uncritically," calling Andreasen's treatment of creativity "inexplicit and misleading" and Jamison's sampling numbers "far too small to draw adequate conclusions."[45]

"Psychological autopsy" is the term used by psychotherapist Judith Schlesinger in describing historical research on already-proven cases. Schlesinger's book *The Insanity Hoax* took an interdisciplinary approach to the creativity and madness fracas, examining conscious or unconscious unfairness in designing studies, carrying them out, or even formulating the project in the first place.[46] Schlesinger noted the difficulty of accurately defining "artistry" or "creativity," as well as a tendency among some scientists to drift into speculation and guesswork when looking at the unfamiliar territory of aesthetics. This can enable what researchers term "confirmation bias," as investigators search for, interpret, or favor information that reinforces a hypothesis, while failing to seek or value contradictory evidence.[47] Also causing confusion has been what psychologists Amos Tversky and Daniel Kahneman have called the "availability heuristic" that happens when repeated references to something make it seem common. (If people or media keep saying artists are crazy, does that make it true?)[48]

Further disagreements would follow in madness-and-creativity discourse. In 1992, psychiatrist Arnold M. Ludwig published findings based on 1,000 reviews in the *New York Times Book Review*. This work would eventually be published in Ludwig's ambitiously titled, *The Price of Greatness: Resolving the Creativity and Madness Debate*. A cautious scholar, Ludwig noted that many creative people reported transient or ongoing "mental disturbances" *at some point* in the course of their lives, and that artists, composers, and writers were hospitalized in psychiatric units more often than non-artists.[49] But rather than drawing a cause-and-effect connection in these correlations, Ludwig would state that "the relation that exists is not between mental illness and creative expression, per se, but between the presence or absence of mental illness and particular forms of creative expression."[50] Hence, Ludwig concluded, "Although intriguing, speculations of this sort are justified only if it has been established that mental illness is common among the eminent. Today, this has yet to be established."[51]

Subsequent research has looked at the possible mental health *benefits* of creativity. "One of the best things a person can do to maintain health is to find opportunities for self-expression," wrote cognitive psychologist Mark A. Runco in his widely read textbook *Creativity: Theories and Themes: Research, Development, and Practice*.[52] Runco cited research showing that people's abilities to communicate emotions, for example, often help them in forming relationships, managing stress, and coping with depression. In turn, this has been shown to reduce the likelihood of illnesses ranging from substance abuse to heart disease. Runco added that creativity also functions as a "cognitive modulator" in helping people make sense of their experience. This follows decades of research on the psychological benefits of artistic "self-actualization," as set forth in the writings of psychologist Abraham Maslow. "The ability to express ideas and impulses without strangulation and without fear of ridicule turned out to be an essential aspect of self-actualized creativeness," Maslow wrote.[53] He added that creative people are more likely to exhibit a "lack of fear of their own insides, of their own impulses, emotions, thoughts. They were more accepting than the average."[54] Similar views had been advanced by psychotherapist Carl Rogers, who stated that "the concept of creativeness and the concept of the healthy, self-actualizing, fully human person seems to be coming closer and closer together, and may perhaps turn out to be the same thing."[55]

Further findings have supported the psychological benefits of artistry in the areas of art therapy and everyday creativity, as discussed in Chapter 4. In a recent meta-analysis entitled "Advancing the Clinical Science of Creativity," psychologists Marie Forgeard and Jeanette Elstein explored these therapeutic benefits, finding that "such interventions typically lead to small but statistically significant improvement on a range of psychological measures," and also reporting that young adults "were more likely to be engaged in creative activities than other activities when they reported feeling happy and active."[56] Forgeard and Elstein also highlighted the positive association of day-to-day creativity "in almost all areas of life—excelling at work, solving thorny interpersonal problems, managing painful emotions," adding that creative thinking takes place not only in "creative therapies" (e.g. art therapy), but to some degrees in all forms of psychotherapy.

Positive news also has come from the corporate sector. Harvard Business School professor Teresa Amabile strongly advocated what she termed "creativity in the wild" in her 1996 book *Creativity in Context*. Amabile and her team analyzed 12,000 journal entries by more than 200 people working on exploratory projects, finding that creative traits made people more open and passionate about their work, and less likely to become thwarted or discouraged. In Amabile's pragmatic view, creativity is a key intrinsic reward system in what she terms "task motivation," as opposed to extrinsic payoffs like money or promotions. Moreover, Amabile concluded that creativity afforded people an adaptability and resilience in the workplace—and was just as important as "talent" in achieving good outcomes.[57] Amabile's work fits within a burgeoning outpouring of books on "outside-the-box" thinking, "creative entrepreneurship," and the commodification/harnessing of such traits—to be explored fully in Chapter 9.

Neuroscience and the Demography

As the above discussion has illustrated, beliefs in creativity as madness have a long history in Western thought—often kept alive in the mental health field despite contradictory findings. To fully understand this, it's important to remember that such ideas never exist in a vacuum. More than matters of superstition or prejudice, popular beliefs in creative madness are held in place by a market system predicated on viewing artworks as rare, valuable, and beyond the capabilities of ordinary people. While some artists may revel (and even promote) themselves as bohemian outsiders, others do so because of the attendant professional attention and benefits. All of this contributes to the unfortunate rendering of most people's creativity as unimportant. Countering such views are recent scientific studies of the creative brain and data analysis of artistic people in national populations.

"Creativity and Madness: Yes, No, or Maybe?" was the title of a recent issue of *Frontiers in Psychology*, one of the largest and most cited journals in the field. Perhaps surprisingly, editor Anna Abraham introduced the collection by announcing that *none* of the 14 contributors advocated a resounding "Yes" verdict: "It is patently clear that the evidence to make a strong claim in the affirmative (all highly creative people have some form of mental illness; all people who have some form of mental illness are highly creative) simply does not exist," Abraham wrote.[58] It wasn't that all the researchers gave a resounding "No," as much as they found the question itself so empirically frustrating due to the enormous breadth of artistry and the numerous categories of psychiatric diagnosis. With most contributors citing the methodological sloppiness of past research (as discussed above), their more or less unequivocal consensus was to call for more precision going forward, with discrete areas of creativity studied for specific maladies, with properly identified control groups, checks for study validity, and mulitperspectival analysis. As neuroscientist Rex Jung summarized, "Many thinkers and researchers have found the creativity and madness seem somehow to be intertwined, but the signal is weak, the image blurry, and the propensity toward romantic stereotypes is high."[59]

Empirical efforts to focus the creative madness picture have gone in two directions during the past decade, as neuroscience has looked more closely at the brain and demographers have crunched data in countries that track mental health statistics. While issues of definition and social context sometimes have been lacking in such research, the findings nevertheless shed new light on what has proven to be a difficult and often contentious question. For example, neuroscientists have been able to differentiate among various forms of creativity, chart their operations in discrete areas of the brain, and note parallels or interactions with structures of psychopathology. Demographers have shown tendencies of people with certain forms of mental illness to gravitate to creative occupations, while also documenting the predilections of family members.

How have neuroscientists defined creativity? Most people know that the brain is akin to a computer central processing unit (CPU), interpreting sensory data from the central nervous system. The essential function of the brain is cell-to-cell communication (both within itself and throughout the body) through the action of neurons, which operate by electrochemical means. While much is made in popular culture of the different regions of the brain and their primary functions (left/right sides, cerebellum, cortex, etc.), none of the functions of the brain (intellect, senses, emotions) are completely independent of each other. This is where things can get tricky, especially in studying matters like creativity or mental illness, which often affect multiple regions to varying degrees. And since chemistry is always involved, variables like hormones, drugs, and underlying psychopathologies such as bipolar disorder or schizophrenia can complicate matters even further.

Cognitive scientist Margaret A. Boden looked at what she termed *combinational, exploratory*, and *transformational* patterns of thinking. Combinational creativity, as the name implies, is the process of putting together known ideas to arrive at something unfamiliar. It's the kind of creativity seen in visual collage (artworks, advertisements) or analogy-making (verbal, visual or musical). To Boden, combinational thinking tends to be grounded in a pre-existing cultural "conceptual space." Exploratory creativity uses existing stylistic rules or conventions "to generate novel structures (ideas or artifacts), whose possibility may or may not have been realized before the exploration took place."[60] Making a new painting or even speaking a sentence are examples of this, as the final product can't be completely anticipated. While combinational and exploratory creativity can be tracked by researchers, transformational creativity remains in the realm of theory. Transformational thinking can come by surprise, Boden wrote, "wherein the novel idea appears to be not merely new, not even strange, but *impossible*."[61] To achieve this, a pre-existing style or conceptual space must be altered or abandoned, as in playing chess by a made-up set of rules.

Creative thinking often differs from cognitive problem-solving in its non-linearity. Researchers Liane Gabora and Apara Ranjan wrote about this puzzle in a paper entitled "How Insight Emerges." They noted that people spend 15 to 50 percent of their time "mind wandering"—recalling and playing with existing ideas and ideas in an undirected manner—and that this is when "eureka" moments often occur. When consciously trying to solve a problem, the brain searches for relevant ideas through associative memory and often connects to nearby "spiky" representations. But when the process isn't successful and one stops concentrating, a background search often continues nevertheless across the "flat" representations of the brain. In a relaxed or inattentive state of mind, the search for big spikes diminishes and more subtle ones are likely to be found. As Gabora and Ranjan explain it, with flat processing "multiple items with overlapping distributions of micro features are evoked simultaneously. Thus, flat activation is conductive to forging remote associations among items not usually thought to be related, or detecting relationships of correlation."[62] Put another way, "In a state of defocused attention more aspects of a situation get processed; the set of activated micro features is larger, and thus the set of potential associations one could make is larger."[63]

Geneticists have continued to work on creativity, too—as they have since the days of Galton and Lombroso—and this is where correspondences to madness have been examined empirically. Contemporary scientists have become more prudent in defining "artistic" qualities than their predecessors, nowadays focusing on specific traits they can measure—for example, novelty-seeking and neural hyperconnectivity—which correspond to certain symptoms of mental illnesses. According to Harvard psychologist Shelley Carson, these traits are influenced by measurable levels of dopamine and serotonin in the prefrontal and subcortical brain. Many artists use "novelty-seeking" in their creative work, professional trajectories, or personal lives. As Carson wrote, "Internal rewards (via the dopamine system) for seeking novel aspects of the environment or novel stimuli may provide the creative individual with intrinsic motivation and intellectual curiosity."[64] This reward system also is associated with substance abuse, hypomania, and mania, as well as the reduced inhibition of the schizophrenic and psychosis prone. Hyperconnectivity (the overactive neural linking of brain areas) is a frequent component of schizophrenia, but also is reported by many artists as an enabler of the defocused flat-actualization process. Specific genes have been linked to both novelty-seeking and hyperconnectivity, giving them a degree of inheritability.

But genes don't guarantee anything in these matters, beyond the latent potential for a characteristic or talent. While asserting a "shared genetic vulnerability model of creativity and psychopathology," Carson also noted that compensatory "protective factors" often accompanied the same genes.[65] In this light, she cautioned against drawing quick conclusions, especially since many creative people value their traits and often benefit from them. In such cases, Carson stressed the importance of awareness of potential problems, much as is often suggested with genetic propensities for alcohol abuse. In more general terms, Carson also said that creativity might help non-artists with certain mental health problems, since "art, music, or writing therapies may improve symptoms of psychopathology by increasing protective factors associated with creativity."[66]

Is it possible to sort out these contradictory findings? Common sense would suggest that studies of artistic populations might provide some hard facts. Unfortunately, studies trying to do this have become bogged down in many of the same ways discussed above, with researchers looking for predefined answers or getting confused by murky definitions of "creativity" and "madness." Once again, the question emerges of how to distinguish the creativity of some people from that of everyone else, especially at a time when so few "artists" get paid for doing so. By looking at creative jobs? Counting gallery shows? Asking people what they do in their off hours? This gets even more complicated when one recognizes that huge numbers of people have mental health problems, in a wide variety of forms—and that many of these are unreported or untreated. Despite the presumed "proof" that recent demographic studies have offered, the reality is that such methods often reinforce existing stereotypes. The perpetuation of madness-and-creativity misconceptions is even more reason to question the broader social, economic, and political incentives behind them.

Population studies have looked at the numbers behind creativity and madness, most recently in Sweden and Iceland, where data on psychiatric hospitalizations and artistic professions is systematically collected. A widely publicized Swedish study drew on statistics from that country's national health system, using data collected from 1973 to 2013 on 1.2 million patients and their relatives. It compared evidence of mental illness with what people reported doing for a living, defining as "creative workers" authors, composers, designers, musicians, performing or visual artists, and university professors. Lead researcher Simon Kyaga found bipolar disorder 1.35 times more likely in creative fields, although there seemed no increased occurrences of anxiety disorders, substance abuse, or suicidality.[67] Authors were the exception, with schizophrenia and depression more pronounced, along with a 50 percent higher suicide rate. The study also found that creative work was more common among relatives of those it studied. Critics faulted the 40-year analysis for using inconsistent diagnostic criteria, such as the now-outdated European ICD-8 system, as well as overly broad definitions of illnesses. Also, the work categories seemed too vague, even to Kyaga himself, who admitted that mentally ill people sometimes have trouble in "normal" jobs and gravitate to creative fields so they can work alone. Besides this, according to creativity expert Keith Sawyer, "Occupational choice is the result of many things unrelated to creativity: Educational qualifications, opportunity, awareness, location of the job, salary, family values, and social commitments."[68]

Schizophrenia and bipolar illness were the big findings of an Icelandic study released in 2015, which, as in Sweden, looked at national mental health and job data. Funded primarily by the deCODE Genetics pharmaceutical company, the study of 86,292 people compared people with genetic indications of schizophrenia and bipolar disease with creative workers, cross-checking this with membership lists of trade organizations of actors, dancers, musicians, visual artists, and writers.[69] The study found that creatives were 25 percent more likely to carry genetic vulnerabilities for the illnesses. And within the organization memberships it found a 17 percent prevalence.[70] As with Sweden, skeptics disagreed with certain conclusions. Clinical psychiatrist Alan Manevitz wrote, "The authors don't necessarily define what kind of creativity they are talking about. There is a difference between people who may identify themselves as being creative, and people who actually work in creative professions."[71] Despite their shortcomings, the Swedish and Icelandic population studies point in a now-familiar direction: that *some* creatives seem prone to *some kinds* of mental illness, within the specific countries studied—which, empirically speaking, lack the demographic diversity of the U.S. No comprehensive mental health studies of artistic people have ever been done in America, perhaps because doing so in a reliable way remains such a daunting prospect. But there is plenty of data on psychiatric illness and substance abuse in the population at large. The numbers may surprise some readers.

According to the National Institutes for Mental Health (NIMH), 26 percent of Americans (78 million) suffer from a diagnosable psychiatric condition, with depression and anxiety disorders topping the list.[72] The crazy-artist stereotype plays into America's denial of its mental health problems by diverting attention toward an already-suspect group. Driving this diversion is the discomfort or outright stigma attached to behavioral disorders. In *Illness as Metaphor*, Susan Sontag wrote about the ways contemporary society reverses ancient views of sickness as a reflection of the inner self. These days, the inner self is seen as actively causing sickness—through smoking, overeating, addictive behavior, and bad habits—which people not-so-coincidentally often associated with creative or bohemian types. "The romantic idea that disease expresses the character is invariably extended to assert that the character causes the disease—because it is not expressed itself. Passion moves inward, striking within the deepest cellular recesses," Sontag wrote.[73] Put in these terms, the suffering person is blamed for the illness. Such conceptual slippage often occurs with behavioral disorders, which is one reason that stigma attaches so easily to mental illness.

For at least 50 years, psychiatric conditions have been recognized as biological illnesses by the fields of medicine and mental health, every bit as "real" and potentially life-threatening as cancer or cardiovascular disease. But because mental illness is not always visible, it remains misunderstood and often goes untreated. UCLA's Dr. Andrew Leuchter has stated:

> 15 million people in this country at any given time suffer from depression, but only about a quarter of these people seek and receive adequate diagnosis and treatment. One of the biggest reasons for this is that people do not want to be labeled as having a mental disorder.[74]

American Psychiatric Association President Dr. Paul Summergrad added, "It's very important that we stop seeing these illnesses as faults and blames, and see them as what they are, medical conditions, genetic conditions, brain disorders which require appropriate diagnosis, treatment, care and support."[75]

Matters get worse when a substance abuse problem accompanies a mental health condition, which it does an astonishing 43 percent of the time. The U. S. Department of Health and Human Services has reported that 9 million Americans have co-occurring mental health and substance abuse problems.[76] Obviously, none of these people want to talk about their conditions in any way. "Few academics have dared to engage this topic," writes Dennis Schepp, "to protect their reputations, if not their income."[77] Also, this silence doesn't begin to address the personal experience of the individual with the mental health condition. Lost in debates over privacy, non-disclosure, and professional standing is the emotional toll on the person with the disease. Being identified as "eccentric" may hold a certain caché in creative circles, but it doesn't change the reality that the label makes life harder for some. America's relentless emphasis on happiness and healthiness tells everyone that being anything less than "okay" is unacceptable and unmentionable.

It's clear to see how damaging the crazy-artist stereotype can be in all of this—lumping all creative people into a stigmatized category with the potential to do great harm. And if the population studies described above do apply to the U.S., they indeed suggest that artists and writers are at least as prone to psychiatric conditions as the general population, and slightly more so with certain illnesses. A handful of writers have spoken about these matters, most notably Jamison and Andrew Solomon. Jamison's *An Unquiet Mind* and Solomon's *Noonday Demon* systematically detail what life is like with bipolar disorder and chronic depression, respectively.[78] Jamison likened the struggle of hiding her condition to living in a perpetual "war against myself."[79] Of course, there has been no shortage of famous writers and artists with similar diagnosis, many of whom died from their illnesses: Chantal Akerman, Arshile Gorky, Ernest Hemingway, Georgia O'Keeffe, Yukio Mishima, Pablo Picasso, Sylvia Plath, Jackson Pollock, Anne Sexton, to name but a few.

Artists on Drugs

Drugs and alcohol add yet another dimension to the madness-and-creativity stereotype, especially when they result in fatalities. While such cases represent but a tiny fraction of the creative population, the attention they receive feeds public beliefs that artists are prone to excess and self-destruction owing to inherent psychopathology, character flaws, or having too much money. Some of this has been reinforced by the creative community in artwork and films celebrating drug experiences, and especially by writers talking about booze. Recent science seems to suggest that creativity and substance abuse sometimes link to the same genes. At the same time, popular mythologies about the excesses of artists draws attention from the everyday use of drugs and alcohol in the general population, especially if one looks at the burgeoning pharmaceutical industry.

Like creativity and madness, the concept of drugs resists easy categorization. The ancient Greeks may have expressed this best in the idea of the *pharmakon*, which refers to a substance being both a remedy and a poison. Almost any drug, food, or activity can function this way—helpful in certain amounts and circumstances, but also potentially dangerous or lethal. American society finds itself caught in confusion between the two extremes of the pharmakon, with a conflicting array of beliefs and ideologies making matters worse. Then there is the specter of habitual use—sometimes termed addiction—which also gets misunderstood. If you think about it, American culture glorifies its culture of workaholics, shopaholics, and exercise fanatics, "hooked" on its favorite foods, TV shows, or computer games. Like many "addicts," the U.S. lives in a certain degree of denial about its guilty pleasures—just as it also neglects its own health. Creative people get caught in the vortex of public confusion over all of this.

Creative people do *not* have a unique tendency for substance use, alcoholism, or addiction. But they sometimes share certain vulnerabilities found elsewhere in the population. As with conditions like bipolar disorder, artists and writers often have certain novelty- and sensation-seeking genes that many non-artistic people also have. One of these genes influences the "signaling of the neurotransmitter dopamine for pleasure and reward," according to Johns Hopkins physician David Linden, author of *The Compass of Pleasure: How Our Brains Make Fatty Foods, Orgasm, Exercise, Marijuana, Vodka, Learning, and Gambling Feel So Good*. Genetic variations suppress the dopamine system in some people, making them more prone to risk-taking, compulsivity, and overindulgence. Even within this group, low-functioning D2 receptors account for less than 40 percent of any substance use. Variables such as upbringing, social environment, and drug availability complete the picture. This explains why the intoxicant-infused culture of Hollywood's wealthy seems to have more people with substance abuse (even though no empirical studies have ever been done). As Linden concluded, "The link is not between creativity and addiction per se. There is a link between addiction and things which are a prerequisite of creativity."[80]

But let's not understate the relationship of drugs to creative states of mind. As discussed earlier, the brain's abilities to make new connections can be enhanced in unfocused or relaxed moments, as both neuroscience and anecdote have shown. Add to this the varied functions of alcohol, dissociative drugs, hallucinogens, marijuana, and stimulants on neurotransmitters and perception, and the attractiveness of drugs to artists and writers become clear. At the same time, it's important to realize that very few successful writers or artists report being intoxicated while creating or performing their work. Somewhat like mental illness, significant intoxication tends to reduce one's ability to be creative. A small number of actors and musicians and actors use alcohol or other drugs to help them perform, often to reduce the anxiety of appearing before audiences. And rock stars like Keith Richards and David Lee Roth were famous for stumbling around on stage. At the same time, few fans realize that well-known figures can experience stage fright like anyone else, with such mainstream classical music names as Pablo Casals, Arthur Rubenstein, Luciano Pavarotti admitting the problem.

There also is the chemistry of addiction itself to consider. On one hand, most people (over 90 percent) manage to drink, smoke marijuana, or take pain prescriptions without dependence or abuse—even though all people have some degree of potential for addiction. Everyone knows that drugs like opiates, nicotine, and alcohol are riskier than others. With such substances, the action of the drug raises the body's tolerance by changing the sensitivity of brain receptors, requiring more of the drug to achieve the same effect. Tolerance by itself is not addiction, but certain drugs more readily produce dependence as the absence of the drug causes withdrawal effects. People with low-functioning D2 receptors are more likely to experience this. Again, lots of people have this characteristic besides creatives. Non-artists use drugs and alcohol, moderately or to excess, in surprisingly high numbers. According to the National Institute on Drug Abuse, 52 percent of Americans drink, 20 percent smoke cigarettes, and 10 percent regularly smoke marijuana or use illegal drugs.[81] Of the U.S. population, 8.6 percent of people are classified as alcoholics or drug abusers, with 30 percent of men and 16 percent of women also reporting recent binge drinking.[82]

Public attitudes toward drug use are mercurial. Before the twentieth century, what we now call addiction was largely written off as a matter of personal idiosyncrasy or "habit." It wasn't until the early twentieth century that Western nations like Britain and the U.S. began to regulate products, mostly in the interest of maintaining quality standards in consumables.

Western literature is full of writing about drugs, praising or condemning drugs, books written by authors on drugs, or simply works referencing drugs as a path to something transcendent or unknowable. Many of the rationales and myths about artistic drug use are found in this canonical literature. Thomas De Quincey's *Confessions of an English Opium-Eater* often is credited as one of the earliest frankly pro-narcotics books, launching the tradition of addiction literature in Europe. De Quincey discussed both the positive and negative effects of his addiction to opium and to laudanum (an alcohol and opium mixture), as well as the widespread use of such substances in Victorian culture. *Confessions* is often cited for its poetic descriptions of intoxication: De Quincey's delight in wandering the streets of London while high on a drug that "stealest away the purposes of wrath" and "givest back the hopes of youth … Thou hast the key of paradise, O just, subtle, and mighty opium!"[83] De Quincey's account of opium's irresistible appeal typifies nineteenth-century views that substances possess their own agency.

Following such beliefs, negative depictions of alcohol and drug use became more prevalent—especially with the emergence of drug laws and temperance campaigns. Émile Zola's *L'Assommoir* (1877) recounted the story of a working-class couple in Paris whose lives were ruined by progressive alcoholism.[84] Translated into English as *The Drunkard, The Drinking Den*, and *The Gin Palace*, Zola's book directly equated alcoholism with the new "masses" of low-income urban dwellers. The nineteenth century's quasi-medicalization of addiction led to a theory known as "degeneration"—the view that biological and moral factors combined with toxic social influences could trigger a downward cascade within an individual—which would continue in subsequent generations. In its crude determinism, the concept portrayed addiction as an acquired characteristic of people and their families. Such Darwinist beliefs contributed significantly to early twentieth-century abstinence movements, with addiction seen as a contagion infecting certain populations. This led to a U.S. constitutional ban on alcohol beginning in 1920.

By the mid-1930s, America discovered that Prohibition couldn't stop a nation determined to keep drinking. Soon arose the figure of the great American writer-alcoholic: F. Scott Fitzgerald, Ernest Hemingway, Tennessee Williams—all the way through Charles Bukowski, John Cheever, and Raymond Carver, among others. These men celebrated their booze, blackouts, and bad behavior in a way that introduced an invigorated sense of purpose in addiction. Without doubt, America's embrace of alcohol received periodic license in times of war and economic stress as well. Heavy drinking always had played an important role in military culture as both a stress reliever and bonding ritual. All of this provided degrees of social sanction for habitual substance use, the normative aspects of which in 1950s America were depicted in the *Mad Men* (2010–2015) TV series. While substance/use abuse often became fused with competitive masculinity, the ethos of the hardworking alcoholic also complimented broader ideals of self-determination and creative individualism.

Junkie: Confessions of an Unredeemed Drug Addict by William Burroughs appeared in 1953 as a mass-market paperback, in obvious homage to De Quincey. Chronicling Beat-era use of heroin and other drugs, *Junky* was among the books condemned by Senator Joseph McCarthy for its "appeals to sensuality, immorality, filth, perversion, and degeneracy."[85] The work drew on Burroughs' own life on the street, reading like a travel guide to the American underworld. *Junkie* didn't so much valorize substance abuse as it detailed the stresses of supporting a drug habit, its daily discomforts, and the sickness of withdrawal. Owing to Burroughs' frankness as a writer, the book was peppered with philosophical reflections on drugs, the most eloquent of which spoke of his subjective experiences in various states of intoxication. "Junk is not a kick. It is a way of life," Burroughs wrote.[86] In this, Burroughs was not simply describing a "lifestyle." He also was describing the long-term internalization of addiction as a form of transformation and self-knowledge.

Extending the theme of insight through drugs, Aldous Huxley published his *The Doors of Perception* in 1954, following his mescaline use in the American Southwest. Huxley cautiously characterized his account as a limited and quasi-scientific experiment with the drug—which he felt could create "temporary bypasses" to alternate realities of the kind "certain people seem to be born with." Huxley wrote that through these gateways flowed "something more than, and above all something different from, the carefully selected utilitarian material from which our narrowed, individual minds regard as a complete, or at least sufficient, picture of reality."[87] In different ways and using different drugs, Burroughs and Huxley charted what would emerge as an important dichotomy of drug use, which further would elaborate itself in the following decades. As Burroughs drew a picture of substance use as a consuming identity and one-way street, Huxley introduced the idea of utopian "experimentation" with drugs and their non-addictive use.

Everyone knows how drug use would be romanticized in the decades of the 1960s and 1970s, notably in the psychedelic writings of Carlos Castaneda and Timothy Leary, and frequently expressed in music of the era. If this writing shared a common theme, it lay in extending Huxley's philosophical approach to intoxicants as a form of inquiry as well as antiauthoritarianism. The very language of psychedelic "tripping" or "experimentation" suggested a purposeful aura of drug use as a search for meaning. While Leary's famous mantra of "Turn on, tune in, drop out" would be critiqued as inherently solipsistic, it also spoke of the era's yearning for change—manifest in student activism, civil rights struggles, and in uprisings in cities around the world throughout the 1960s.[88] In "An Essay on Liberation," Herbert Marcuse would write, "Today's rebels want to see, hear, feel new things in a new way: they link liberation with the dissolution of ordinary and orderly perception." Casting drugs as both metaphor and actuality, he wrote that "the 'trip' involves the dissolution of the ego shaped by the established society—an artificial and short-lived dissolution."[89] Marcuse regarded drug use as a component of what he termed "the new sensibility" emerging to confront the "one-dimensional society" of consumer capitalism and media consolidation.

Until this point, substance abuse appeared in popular culture only in movie or TV crime dramas—whether this meant hard-drinking detectives, secret agents, or the drug-addled underworld. One early exception was Otto Preminger's *The Man with The Golden Arm* (1955). Starring Frank Sinatra as a heroin addict, the film often is cited as the first Hollywood feature to directly address illegal drug use. Within the film noir genre, a few movies about alcoholism also appeared in this period—such as *The Lost Weekend* (1945) and *The Days of Wine and Roses* (1962). But, in general, depictions of drinking served to reify alcohol use. Moderate consumption would be treated in media and the public mind as a pleasurable and disinhibiting social lubricant, sometimes cast as a quintessentially American (Joe Six-Pack) way of having a good time. Between 1965 and 1974, *The Dean Martin Show* aired 264 episodes featuring the visibly intoxicated Martin as the quintessential rascal drunk. Later series like *Cheers* (1982–1993), which was set in a tavern, would likewise minimize the downside of excessive boozing in the character of the always-loveable Norm. The six-decade James Bond movie franchise (1962–present) featured the frequently drinking, but never intoxicated, spy in a relentless pursuit of his signature martini ("shaken, not stirred") through 26 films. A key element in these depictions was the uniform presentation of alcohol as a mostly positive, harmless, and, above all, normal diversion. Then there was the Budweiser effect. By the end of the 1990s, American beer companies were spending $596 million annually on televised sports advertising.[90]

Eventually, images of intoxication had to change. Hollywood's own drug and alcohol problems partly account for this shift. The cocaine epidemic of the 1990s took a heavy toll on the entertainment industry, with many celebrities succumbing to overdose or other health problems. And, of course, many in the industry simply drank. The combined death toll included Chris Farley, Phillip Seymour Hoffman, Margaux Hemingway, Whitney Houston, Michael Jackson, Heath Ledger, Anna Nicole Smith, Ike Turner, and Amy Winehouse—although these are simply the more well-known figures. While news reporting has tended to treat these tragic deaths as isolated incidents, growing numbers of performers have come forward to disclose their addictions. The list includes Drew Barrymore, Ed Begley Jr., Judy Collins, Jamie Lee Curtis, Robert Downey Jr., Edie Falco, Colin Farrell, Lou Gosset Jr., Melanie Griffith, Ed Harris, Richard Lewis, Lindsay Lohan, Kelly McGillis, Nick Nolte, Tatum O'Neal, Katey Sagal, and Martin Sheen.

None of this public attention has helped the crazy-artist stereotype—if anything exacerbating public confusion about "drugs" and who consumes them. Given the prevalence of unwellness in the American population, almost everyone takes "drugs" or medications of some form or another, either on a short-term or long-term basis—and the line gets very messy with over-the-counter medications, natural remedies, and nutritional supplements. Seven in ten Americans currently take at least one prescription of some kind.[91] Twenty percent use five or more medications regularly. The bestselling drugs in the U. S. are Lipitor (cholesterol-lowering), Nexium (heartburn), Plavix (blood thinner), Advair (asthma), and Abilify (depression/anxiety).[92] But one-a-day vitamins also are a kind of "drug," and some argue an unhealthy one for many people. Without doubt, drugs constitute one of America's biggest industries. Limiting the discussion to the "pharmaceutical industry" itself, expenditures for medicines account for over $1.2 trillion of the world economy, one-third of which is spent in the U.S.[93] And business for drug makers has been huge, with the number of prescriptions written in America expanding by 61 percent in the past decade to 4 billion per year.[94] Money spent by consumers has risen by 250 percent from $72 billion to $250 billion. This isn't counting the nearly $500 billion spent each year.[95]

Everyone is implicated in America's "drug problem"—from everyday consumers of prescription medications, to the pharmaceutical industry promoting real and imagined ills, to the media system that promises quick fixes to problems, to an economic system that keeps everyone on edge. Throughout these pages, I've discussed the general sense of anxiety that continues to pervade U.S. culture, an absence often tinged with underlying "emptiness."[96] In her book, *Crack Wars: Literature, Addiction, Mania*, theorist Avital Ronell discussed how contemporary society deals with its worries, often seeking out objects on which to project them.[97] Drugs become easy fear-objects, as do so many popular figures of threat: the terrorist, the mental patient, the illegal immigrant, the sexual deviant, the homeless person, and so on. But because drugs are both ubiquitous and personal, Ronell said that they resonate in people's deepest thoughts about health and happiness, sickness and misery—and have done so throughout history. Ronell wrote, "Drugs resist conceptual arrest. No one has thought to define them in their essence, which is not to say 'they' do not exist," adding:

> Everywhere dispensed, in one form or another, their strength lies in their virtual and fugitive patterns. They do not close forces with an external enemy (the easy way out) but have a secret communications network with the internalized demon. Something is beaming out signals, calling drugs home.[98]

Like creativity and madness, drugs and addiction go right to the metaphysical heart of American culture—speaking of ineffable forces that provoke wonder and amazement, but also often come laden with mystery and fear.[99] Heidegger wrote about this phenomenon when he said that creative and addictive urges were rooted in the instability of human existence—what he termed the "thrownness" of being.[100] Heidegger asserted the inevitability of a certain degree of anxiety in life—especially in the face of the alienation human subjects experience in the face of the "world" of things different from itself. He wrote that this latent anxiety was a necessary driver of desire, motivation, and, ultimately, the source of freedom itself. One only ever perceives "freedom" in relation to an oppositional force, Heidegger pointed out—explaining that ideas like art, madness, and intoxication (and other technologies) upset the necessary tension either by obfuscation, revelation or both. This suggests something elemental about the human culture of creativity, madness, and drugs—so deeply engrained that it is almost unspeakable.

In this chapter, I have detailed the imperfect search to explain the complexities of art-making, its relationship to mental disorders, and the substances sometimes associated with both. The resulting confusions have perpetuated stereotypes of artists as quintessentially different others, with their creative abilities beyond the reach of "normal" people. What began centuries ago in superstitions about divine madness and ethereal muses evolved into fears that creativity could threaten reason—that how people arrived at "truth" itself was somehow at risk. In a way, the discourses of mental illness and narcotic have swarmed around these same anxieties, suggesting that science or philosophy might somehow find a way to explain the matter for good. As suggested in the figure of the pharmakon, the enigmatic confluence of remedy/poison in artistry, mental illness, and drugs can be traced to the very function of language in its awkward articulation of philosophy, especially as inherited from history's archive. Creativity and madness speak of absence and presence, life and death, past and future—and the elusive craving for something that seems always just beyond reach.

Notes

1 Ian Sample, "New Study Claims to Find Genetic Link Between Creativity and Mental Illness," *The Guardian* (Jun. 8, 2015) www.theguardian.com/science/2015/jun/08/new-study-claims-to-find-genetic-link-between-creativity-and-mental-illness (accessed Nov. 2, 2018).
2 Arne Dietrich, "The Mythconception of the Mad Genius," *Frontiers in Psychology* 5 (July 2014): 8.
3 Search results in "Books," Amazon.com (Nov. 12, 2017) www.amazon.com/books-used-books-textbooks/b/ref=topnav_storetab_b?ie=UTF8&node=283155 (accessed Nov. 13, 2018).
4 Judith Schlesinger, *The Insanity Hoax: Exposing the Myth of the Mad Genius* (Ardsley-on-Hudson, NY: Shrinktunes, 2014); Simon Kyaga, *Creativity and Mental Illness: The Mad Genius in Question* (London and New York: Palgrave Macmillan, 2014); Pascal Geilen, *Creativity and Other Fundamentalisms* (Amsterdam: Mondriaan Fund, 2013).
5 Paul Feyerabend, "Creativity—A Dangerous Myth," *Critical Inquiry* 13 (Summer 1987): 700.
6 Plato, *Gorgias*, trans. Donald J. Zeyl (Indianapolis: Hackett, 1987).
7 George Kazantzidis, "Cicero on Melancholy," in William Harris, ed., *Mental Disorders in the Classical World* (Leiden and Boston: Brill, 2013) p. 263.
8 Michel Foucault, *Madness and Civilization: A History of Insanity in the Age of Reason*, trans. Richard Howard (1961) (New York: Vintage, 1988, 1972) p. 16.
9 Benjamin Rush, as quoted in Kay Redfield Jamison, *Touched with Fire: Major Depressive Illness and the Artistic Temperament* (New York: Free Press, 1993) p. 53.
10 Benjamin Rush, quoted in Ronald T. Takaki, *Iron Cages: Race and Culture in 19th-Century America* (Oxford: Oxford University Press, 1990) p. 32.
11 Gustav Jahoda, "Quetelet and the Emergence of the Behavioral Sciences," National Center for Biotechnology Information (Sep. 4, 2015) www.ncbi.nlm.nih.gov/pmc/articles/PMC4559562 (accessed Nov. 7, 2018).
12 Francis Galton, *Hereditary Genius: An Inquiry into Its Laws and Consequences* (London: Macmillan, 1969) p. 36.

13 Cesare Lombroso, *Man of Genius*, ed. Havelock Ellis (Turin: Turin, 1891) p. 205.
14 *Man of Genius*, p. 161.
15 Havelock Ellis, *A Study of British Genius* (1904) (Ithaca: Cornell University Press, 2009) p. 191.
16 Havelock Ellis, *Sexual Inversion* (1897) (New York: Nabu Press, 2010).
17 *Sexual Inversion*, pp. 293–4.
18 John Keats, "Letter to Benjamin Bailey, November 22nd, 1817," john-keats.com (n.d.) www.john-keats.com/briefe/221117.htm (accessed Nov. 4, 2018).
19 Jane Austin, *Sense and Sensibility* (1811) (New York: Collins Classics, 2010).
20 Samuel Taylor Coleridge, "Letter to Joseph Cottle, 16 September, 1826," in Allan Ingram, ed., *Patterns of Madness in the Eighteenth Century: A Reader* (Liverpool: Liverpool University Press, 1998) p. 221.
21 "Sigmund Freud on Anna O. Case," freudfile.org (n.d.) www.freudfile.org/psychoanalysis/annao_case.html (accessed Nov. 4, 2018).
22 Sigmund Freud, *Civilization and Its Discontents* (1930) (New York: Norton, 2010).
23 Sigmund Freud, *The Interpretation of Dreams: The Complete and Definitive Text* (1899) (New York: Basic Books, 2010).
24 Athanansia Nikolakouli, "The Influence of Sigmund Freud's Theory of Dreams on the Movement of Surrealism," *Manoeuvresto* (2012) http://manoeuvresto.blogspot.com/2012/03/influence-of-sigmund-freuds-theory-of.html (accessed Nov. 5, 2018).
25 Jacques Lacan, "The Mirror Stage as Formative of the I Function as Revealed in Psychoanalytic Experience" (1936) in *Écrits: The First Complete Edition in English*, trans. Bruce Fink (New York: Norton, 2002) pp. 76–82.
26 Jacques Lacan, *The Four Fundamental Concepts of Psychoanalysis* (1973) ed. Jacques-Alain Miller, trans. Alan Sheridan (New York: Norton 1998) p. 203.
27 Marcel Réja, *L'Art Chez les Fous* (Paris: Société du Mercure de France, 1907).
28 Sue Prideaux, *Edvard Munch: Behind the Scream* (New Haven: Yale University Press, 2007) p. 169.
29 Margaret Olin, "Forms of Respect: Alois Riegl's Concept of Attentiveness," *Art Bulletin*, LXXI, no. 2 (June 1989) www.collegeart.org/pdf/artbulletin/Art%20Bulletin%20Vol%2071%20No%202%20Olin.pdf (accessed Nov. 15, 2018); E.H. (Ernst Hans) Gombrich, *Art and Illusion: A Study in the Psychology of Pictorial Representation* (New York: Bollingen Foundation, 1960).
30 Hans Prinzhorn, *Artistry of the Mentally Ill: A Contribution to the Psychology and Psychopathology of Configuration* (1922) trans. Eric von Brockdorff (New York: Springer-Verlag, 1995).
31 Alan Beveridge, "A Disquieting Feeling of Strangeness: The Art of the Mentally Ill," *Journal of the Royal Society of Medicine* 94, no. 11 (Nov. 2001): 595–9.
32 Emil Kraepelin, *Manic-Depressive Insanity and Paranoia*, trans. R. Mary Barclay (Edinburgh: Livingston, 1921).
33 *Manic-Depressive Insanity and Paranoia*, p. 17.
34 Andreas Ebert and Karl-Jürgen Bär, "Emil Kraepelin: A Pioneer of Scientific Understanding of Psychiatry and Psychopharmacology," *Indian Journal of Psychiatry* 52, no. 2 (Apr.–Jun. 2010): 191–2.
35 Adele Juda, "The Relationship between Highest Mental Capacities and Psychic Abnormalities," *American Journal of Psychiatry* 106 (Oct. 1949): 84.
36 Ibid.
37 Mildred Goertzel, Victor Goertzel, and Ted Goertzel, *300 Eminent Personalities: A Psychosocial Analysis of the Famous* (San Francisco: Jossey-Boss, 1978).

38 Mildred Goertzel and Victor Goertzel, *Cradles of Eminence: A Provocative Study of the Childhoods of over 400 Famous Twentieth-Century Men and Women* (Boston: Little, Brown, 1962).
39 Nancy Andreasen, "Secrets of the Creative Brain," *Atlantic* (Jul./Aug. 2014) www.theatlantic.com/magazine/archive/2014/07/secrets-of-the-creative-brain/372299 (accessed Nov. 9. 2018).
40 Ibid.
41 Kay Redfield Jamison, *Touched with Fire: Manic-Depressive Illness and the Artistic Temperament* (New York: Free Press, 1993) p. 49.
42 Kay Redfield Jamison, *An Unquiet Mind: A Memoir of Moods and Madness* (New York: Vintage, 1996).
43 Abraham Nussbaum, *A Pocket Guide to the DSM-5 Diagnostic Manual* (Arlington, VA: American Psychiatric Association, 2013) p. 3.
44 *Touched with Fire*, p. 6.
45 Albert Rothenberg, *Creativity and Madness: New Findings and Old Stereotypes* (Baltimore: Johns Hopkins University Press, 1990) p. 21.
46 *The Insanity Hoax*, p. 117.
47 "Confirmation Bias," ScienceDaily (n.d.) www.sciencedaily.com/terms/confirmation_bias.htm (accessed Nov. 11, 2018).
48 Amos Tversky and Daniel Kahneman, "Judgement under Uncertainty: Heuristics and Biases," *Science*, 185, no. 4157 (Sep. 27, 1974): 1124–31.
49 Arnold M. Ludwig, "Creative Achievement and Psychopathology: Comparisons Among Professions," *American Journal of Psychopathology* 26 (1992): 330–56.
50 Arnold M. Ludwig, *The Price of Greatness: Resolving the Creativity and Madness Debate* (New York: Guilford, 1995), p. 100.
51 *The Price of Greatness*, p. 128.
52 Mark A. Runco, *Creativity: Theories and Themes: Research, Development, and Practice* (New York: Elsevier, 2014) p. 121.
53 Abraham Maslow, "Creativity in Self-Actualizing People," in *Toward a Psychology of Being* (New York: Wiley, 1968) p. 24.
54 "Creativity in Self-Actualizing People," p. 26.
55 Carl Rogers, *On Becoming a Person: A Therapist's View of Psychotherapy* (Boston: Houghton-Mifflin, 1961) p. 57.
56 Marie J.C. Forgeard and Jeanette G. Elstein, "Advancing the Clinical Science of Creativity," *Frontiers in Psychology* 5 (June 2014): 30.
57 Teresa Amabile, *Creativity in Context* (Boulder: Westview, 1996) p. 34.
58 Anna Abraham, "Madness and Creativity: Yes, No, or Maybe?" *Frontiers in Psychology* 5 (June 2014): 5.
59 Rex Jung, "Evolution, Creativity, Intelligence, and Madness: 'Here Be Dragons,'" *Frontiers in Psychology* 5 (July 2014): 34.
60 Margaret A. Boden, "Creativity as Neuroscientific Mystery," in Oshin Vartanian, Adam Smith, and James C. Kaufman, eds., *Neuroscience of Creativity* (Cambridge, MA: MIT Press, 2013). p. 6.
61 Ibid.
62 Liane Gabora and Apara Ranjan, "How Insight Emerges in a Distributed, Content Addressable Memory," in *Neuroscience of Creativity*, p.27.
63 "How Insight Emerges," p. 29.
64 Shelley Carson, "Creativity and Psychopathology: A Shared Vulnerability Model," *La Revue canadienne de psychiatrie* 56, no. 3 (Mar. 2011): 144.
65 Ibid.
66 "Creativity and Psychopathology," p. 145.

67 Simon Kyaga et al., "Creativity and Mental Disorder: Family Study of 300,000 People with Severe Mental Disorder," *British Journal of Psychiatry* 199, no. 5 (Oct. 2011) www.nationalelfservice.net/cms/wp-content/uploads/2011/10/Nov-BJPsych-creativity-and-mental-illness.pdf (accessed Nov. 17, 2018).
68 Keith Sawyer, "Creativity and Mental Illness: Is There a Link?" *Huffington Post* (Nov. 2, 2012) www.huffingtonpost.com/dr-r-keith-sawyer/creativity-and-mental-ill_b_2059806.html (accessed Nov. 17, 2018).
69 "Does Creativity Share a Genetic Root with Schizophrenia and Bipolar Disorder?" *Neuroscience News* (Jun. 8, 2015) http://neurosciencenews.com/genetics-creativity-schizophrenia-2105 (accessed Nov. 17, 2018).
70 Laura C. Mallonee, "Icelandic Scientists Find Link Between Creativity and Mental Disorders," *Hyperallergic* (Jun. 11, 2015) http://hyperallergic.com/213620/icelandic-scientists-find-link-between-creativity-and-mental-disorders (accessed Nov. 17, 2018).
71 Alan Manevitz, "Study Shows Link Between Creativity and Psychiatric Disorders" (Jun. 12, 2015) www.amanevitzmd.com/blog/Study-Shows-Genetic-Link-Between-Creativity-and-Psychiatric-Disorders (accessed Nov. 17, 2018).
72 "Developmental Disabilities Increasing in the U.S.," Center for Disease Control www.cdc.gov/features/dsdev_disabilities (accessed Jul. 30, 2018); "Alzheimer's Facts and Figures," Alzheimer's Organization (2017) www.alz.org/alzheimers_disease_facts_and_figures.asp (accessed Jul. 29, 2018).
73 Susan Sontag, *Illness as Metaphor* (New York; Vintage, 1977) p. 45.
74 Dr. Andrew Leuchter, "The Stigma of Mental Illness," Intelecom (Feb. 12, 2018) www.youtube.com/watch?v=NSXIX77bD3U (accessed Mar. 25, 2019).
75 "APA Reaches Out in Wake of Robin Williams' Death," Psychiatric News (Aug. 14, 2014) http://alert.psychnews.org/2014/08/apa-reaches-out-in-wake-of-robin.html (accessed Oct. 15, 2018).
76 "Rates of Co-occurring Mental and Substance Use Disorders," Substance Abuse and Mental Health Services Administration (n.d.) www.samhsa.gov/co-occurring/topics/data/disorders.aspx (accessed Oct. 15, 2018).
77 Dennis Schepp, *Drugs: Rhetoric of Fantasy, Addiction to Truth* (Dresden and New York: Atropos, 2011) p.11.
78 See *An Unquiet Mind* and Andrew Solomon, *Noonday Demon: An Atlas of Depression* (New York: Touchstone, 2002).
79 *An Unquiet Mind*, p. 3.
80 David Linden as quoted in David Biello, "Is There a Link Between Creativity and Addiction?" *Scientific American* (Jul. 26, 2011) www.scientificamerican.com/article/is-there-a-link-between-creativity-and-addiction (accessed Jun. 15, 2019). See also David Linden, *The Compass of Pleasure: How Our Brains Make Fatty Foods, Orgasm, Exercise, Marijuana, Vodka, Learning, and Gambling Feel So Good* (New York: Penguin, 2012).
81 "National Survey of Drug Use and Health," National Institute on Drug Abuse (NIDA) (2016) www.drugabuse.gov/national-survey-drug-use-health (accessed Nov. 23, 2018).
82 "Alcohol Facts and Statistics," National Institute on Alcohol Abuse and Alcoholism (2016) www.niaaa.nih.gov/alcohol-health/overview-alcohol-consumption/alcohol-facts-and-statistics (accessed Nov. 23, 2018); "Drug Facts—Nationwide Trends," NIDA (2015) www.drugabuse.gov/publications/drugfacts/nationwide-trends (accessed Nov. 23, 2018).

83 Thomas De Quincey, *Confessions of an English Opium-Eater*, ed. Terry Milligan (New York: Penguin Classics, 2003) p. 82.
84 Émile Zola, *L'Assommoir*, trans. Margaret Mauldon (Oxford: Oxford University Press, 1995).
85 Nathaniel Rich, "American Dreams, 1953: *Junky* by William S. Burroughs," *The Daily Beast* (June. 23, 2013) www.thedailybeast.com/articles/2013/06/27/american-dreams-1953-junky-by-william-s-burroughs.html (accessed Sep. 8, 2018).
86 William S. Burroughs, *Junkie: Confessions of an Unredeemed Drug Addict* (New York: Ace, 1953) p. 43.
87 Aldous Huxley, *The Doors of Perception* (1954) www.erowid.org/psychoactives/writings/huxley_doors.shtml (accessed Sep. 10, 2018).
88 Timothy Leary, *Flashbacks: A Personal and Cultural History of an Era* (Los Angeles: Tarcher, 1983) p. 253.
89 Herbert Marcuse, "An Essay on Liberation" (1969) www.marxists.org/reference/archive/marcuse/works/1969/essay-liberation.htm#s2 (accessed Sep. 10, 2018) n.p.
90 "Alcohol Advertising in Sports on TV," Campaign for Alcohol-Free Sports TV (2003) www.cspinet.org/booze/CAFST/QuickFacts.pdf (accessed Sep. 9, 2018).
91 "Nearly 7 in 10 Americans Take Prescription Drugs," Mayo Clinic (Jun. 19, 2013) www.mayoclinic.org/news2013-rst/7543.html (accessed Jul. 29, 2018); "Alcohol Use," U.S. Centers for Disease Control and Prevention (2014) www.cdc.gov/nchs/fastats/alcohol.htm (accessed Aug. 31, 2018).
92 "The 10 Most Prescribed Drugs," WebMD (2011) www.webmd.com/news/20110420/the-10-most-prescribed-drugs (accessed Sep. 13, 2014); "U.S. Pharmaceutical Sales—2013" (Feb 2014) www.drugs.com/stats/top100/2013/sales (accessed Sep. 13, 2018).
93 "Global Pharmaceutical Industry—Statistics and Facts," statista.com (2018) www.statista.com/topics/1764/global-pharmaceutical-industry (accessed Jun. 15, 2019).
94 "Nearly 7 in 10 Americans Take Prescription Drugs"; "Alcohol Use."
95 "Fact Sheet: The Use of Over-the-Counter Medicines," www.bemedwise.org/press_room/sep_2003_fact_otc.pdf (accessed Jul. 30, 2018); "What America's Users Spend on Illegal Drugs", Rand Corporation (2012) www.rand.org/pubs/research_reports/RR534.html (accessed Sep. 1, 2018); "Americans Spending More on Alcohol, Drinking Less," Huffington Post (Dec. 31, 2012) www.huffingtonpost.com/2012/12/31/americans-alcohol_n_2389375.html (accessed Sep. 1, 2018).
96 Gabor Maté, *In the Realm of Hungry Ghosts: Close Encounters with Addiction* (New York: North Atlantic, 2010) p. 272.
97 Avital Ronell, *Crack Wars: Literature, Addiction, Mania* (Urbana and Chicago: University of Illinois Press, 2004).
98 *Crack Wars*, p. 51.
99 *Crack Wars*, p. 13.
100 Martin Heidegger, *Being and Time* (1953) trans. Joan Stambaugh (Albany: SUNY Press, 2010).

Part III

Creative Industries

Doodling on your iPad is one thing, but most people experience creativity through organized interfaces like museums, schools, or websites. Even the seemingly untrammeled activity of posting a selfie or sending a tweet is mediated by a corporate gatekeeper—often tracking what you do and selling your data by the gigabyte. A long literature has analyzed and often criticized the role of institutions in managing creativity: the way it's taught in schools, filtered by galleries, commodified in entertainment, and promoted through how-to books. Technology also has played a role in making works accessible, from the printing press to the cell phone. And, of course, there is money as both an enabler and incentive, whether this means philanthropic giving or individual buying. This section looks at organized creativity as a dynamic and dialogical phenomenon, rather than a simple transfer from "makers" to "consumers." As such, "industrial" culture and creativity are tempered by historical context, political ideology, aesthetic preferences, and the ever-present tensions of democratic capitalism—especially as neoliberalism reshuffles wealth from bottom to top.

Responding to the creativity crisis, economists have begun scrutinizing how markets affect innovation. And researchers have reached some startling conclusions. Most Americans believe that competition drives ingenuity, that better ideas are rewarded in the marketplace, and that everyone benefits from this process. Now these common-sense assumptions are being questioned. People buy on emotion and are creatures of habit, often favoring what they "like" or what feels familiar. Responding to this, companies find it less costly to repackage existing products while also focusing on what is known to sell. This is why the purportedly "free" market seems stuck on churning out variations on essentially the same toothpaste and similar commodities. It does this because neoliberal fiscal policies favor short-term profit-taking over long-term investment.[1] In an age of "financialized capital," shareholder dividends get prioritized over the development of new ideas or products. This slows innovation while diminishing the value consumers receive. The obesity-inducing junk-food industry illustrates this trend. Hence, a growing list of mainstream experts agree that it is time to "rethink capitalism."

American public education was born in the industrial age. And in seeking to serve factories, schools came to resemble them, as Paul Willis noted decades ago in *Learning to Labour.* [2] If you think about it, this makes sense in organizational terms. How else can you educate millions of youngsters without rules, standards, tests, and metrics? But as John Dewey first pointed out, this approach often runs counter to the very values of experiment and creativity families say they value and want. Or do they? Chapter 7, "Factories of Knowledge: Why Schools Kill Creativity," looks at this question. Anxious times makes parents less open to subjects like art and music as they worry about kids' future earnings or college prospects. It doesn't help that Americans always have been suspicious of intellectual eggheads and too much book learning, especially as seen in the reactionary populism of recent years. All of this puts schools in a double bind, with creativity offerings usually suffering as a consequence. This chapter combines classic thinking on creative play and learning by figures such as Gregory Bateson, Melanie Klein, Jean Piaget, and Donald Winnicott, with historical views of schooling by Alfie Kohn, Jonathan Kozol, and Thorstein Veblen.

Sequels are the movies Americans love to hate. Even as they gripe about yet another *Avengers* or *Star Wars* installment, audiences flock to such films in record-breaking numbers. This contradiction opens discussion in Chapter 8, "Industries of Culture: Masterpieces vs. Dream Machines." While it's easy to blame this on a profit-driven media industry, the demand side of sameness can't be ignored. Anxious societies always have found comfort in familiarity and predictability, secure knowing that the hero wins or love always prevails. Aristotle first wrote about this in explaining the human hunger for stories with clear beginnings and endings. The art world saw similar troubles brewing decades ago, with postmodernism zeroing in on the end of fresh ideas—or even originality for that matter—while also understanding that makers didn't fully control what audiences perceived. Reception theory would add that interpretations varied from viewer to viewer, with different meanings "floating" around them. This didn't stop critics from condemning mass culture for pushing bad values on consumers, hence setting off debates about the "effects" of creative works still continuing today. Beginning with early critiques of the culture industry by Theodor Adorno, Walter Benjamin, and Max Horkheimer, this chapter traces the issue through writings by Louis Althusser, Ben Bagdikian, Jacques Derrida, Hans Magnus Enzensberger, Laura Mulvey, Elaine Showalter, and Carole Vance.

It's been nearly two decades since Richard Florida's *The Rise of the Creative Class* became a runaway bestseller in announcing that one in three Americans worked in creative fields.[3] As noted in Chapter 9, "Creative Economies: 'Big Magic' or Empty Promises?" Florida managed this large number by including many quasi-artistic fields in his calculus. But he launched a revolution nevertheless. Online creativity quickly became part of the buzz, encompassing everything from blogging to online retail. City planners soon saw the gentrifying potential of the creative class, as factory neighborhoods become overtaken by bohemian cafés and internet start-ups. Countries around the world became newly invigorated to promote local culture and national heritage. Indeed, the most recent report on this from UNESCO estimates $2.25 trillion (yes, *trillion*) is generated annually worldwide through performing arts, radio, music, books, newspapers and magazines, film, television, architecture, gaming, and advertising.[4] Critics such as Luc Boltanski, Ève Chiapello, Pascal Gielen, Max Haiven, and Angela McRobbie see this creative explosion as an extension of neoliberal agendas, extending premises of bureaucratic rationalism first advanced by early sociologists like Karl Marx and Max Weber.

Notes

1 Michael Jacobs and Mariana Mazzucato, eds., *Rethinking Capitalism: Economics and Policy for Sustainable and Inclusive Growth* (Malden, MA: Wiley-Blackwell, 2016).
2 Paul Willis, *Learning to Labour: How Working Class Kids Get Working Class Jobs* (New York: Columbia University Press, 1986).
3 Richard Florida, *The Rise of the Creative Class* (New York: Basic Books, 2002).
4 International Confederation of Societies of Authors and Composers, *Cultural Times: The First Global Map of Cultural and Creative Industries* (New York: UNESCO, 2016).

Chapter 7

Factories of Knowledge

Why Schools Kill Creativity

Pablo Picasso once quipped, "Every child is an artist; the problem is how to remain an artist once they grow up."[1] In this often-quoted slogan, Picasso neatly summarized idealized views of the universally creative child and the uncreative adult. In a similar fashion, he would later write, "It takes a long time to become young." What is one to make of such laments? Nostalgia over a lost youth? A yearning to escape a pressurized grown-up life? Regardless of origins, it's impossible to deny America's ongoing infatuation with childhood creativity. In what follows I'll review some recent theory about this cultural obsession and its origins—contrasting the recent hype over the creative economy with the dismal state of creative education in schools and universities. At a time when so many voices are advocating "creative" approaches for everyone and in all things, a set of cultural contradictions seems to be pushing back.

This chapter begins with a look at America's embrace of the creative "inner child" as a symptom of the country's anxious condition. Looking backward to happy times tends to make people feel better, even if it distorts what actually happened. This fondness for childhood imagery and ideas is seen in nostalgia-driven advertising, retro product lines, and movie remakes, not to mention the burgeoning inner-child self-help industry. You might expect this emphasis on childhood would translate into better funding for K–12 education and maybe more support for the arts in the "creative economy." Unfortunately, things seem to be moving in the opposite direction, even as the U.S. falls behind other nations in innovation and creative output.

Things get even worse at the post-secondary level. In the past year, 69.1 percent of high school graduates pursued further learning at community colleges, technical schools, colleges, and universities—making such institutions central to both job preparation and further academic study.[2] Meanwhile, universities accounted for 91.5 percent of U.S. research and development projects, dramatically outpacing the corporate sector in medicine, science, technology, humanities, and the arts.[3] Regardless of one's view of the creative industries, there is little question that institutions of higher learning drive American innovation, while enlightening citizens on everything from management to musicianship. The question is whether the monetary imperatives of the knowledge economy are shifting educational priorities away from humanistic purposes—and at what cost.

Without a doubt, higher education has taken a more "practical" turn in the past decade—frequently at the expense of the very values that drive creativity and social responsibility. Students raised in recessionary times are more grade-conscious, competitive, and vocationally motivated. Administrators are likewise driven by cost-cutting, metrics, and performance outcomes. Rising tuitions have increased student borrowing, as institutions have shifted teaching to low-paid adjuncts or graduate students. Utilitarian degrees and courses in engineering, business, and science have grown as programs in the humanities, arts, and social sciences have dwindled. Increasingly, observers both inside and outside the academy worry that this pattern may have negative consequences. And humanistic values may not be the only casualties. Evidence is growing that the commodification of learning tends to undercut the spirit of experiment and risk-taking needed for innovation.

Clearly, the new knowledge economy is changing the way Americans think and create ideas—often shifting priorities in a more competitive, businesslike, and rationalized direction. Is there a way to embrace this shift without sacrificing values of equality, justice, and mutual concern? It turns out that the future of idea production is being rewritten right now—and in unexpected ways. Not only are researchers seeing declines in novel thinking and creativity; in structural terms, academics in nearly all fields are recognizing that the very way ideas are organized and developed may also be causing problems. Businesslike attitudes toward knowledge as "property" fragment data and studies into academic disciplines, proprietary domains, and other "silos" of knowledge. This emphasis on the competitive ownership of information is getting in the way of progress. In an age of digital information, new forms of distributed research (and distributed teaching) are both a more efficient and more democratic way of maximizing intellectual potential. This is already taking place outside the halls of academe, where open-source software and other forms of information dissemination are occurring at a grassroots level in what is termed the new "sharing economy." Artists and others in creative fields may have new opportunities here—as they are already well versed in collaboration and are often acutely aware of their relationship with audiences. Add to this the growing ubiquity of online media-sharing, and the promise of shared creativity becomes apparent.

The Creative Inner Child

Western culture's recognition of child art dates to the eighteenth century, corresponding to evolving views of children as "blank slates" (tabula rasa) better served by nurturance and education than by discipline alone. At the same time, Enlightenment debates over individualism and personal autonomy were bringing considerable anxiety to the era, evidenced in worries that self-interest would overwhelm moral sentiments. This set the stage for the naturalism espoused by Jean-Jacques Rousseau in his book *Emile: Or, On Education*, seeing an inherent "goodness" in children, which becomes corrupted by adult desire and material want.[4] With the nineteenth century, views of "human nature" gave ways to theories of evolution and behavioral adaptation—owing in large part to the influence of Charles Darwin and Herbert Spencer. While the resulting rationalism eventually would make education more formulaic, an artsy transcendentalism would counterbalance American culture with an advocacy for an "educated imagination."[5] The Romantic Era writings of Ralph Waldo Emerson, Margaret Fuller, Henry Wadsworth Longfellow, and Walt Whitman advanced themes of emotion over reason and imagination over reality—setting in place a progressive tradition of push-back against the instrumentalist ethos of science and industry.

In the 1920s, Swiss psychologist Jean Piaget began charting children's "stages" of maturity, hence launching the modern field of child development.[6] Piaget saw "realistic" rendering as a learned ability rather than a natural inclination. In one famous study, Piaget asked a group of four-year-olds to draw familiar people or objects. He found that the images invariably had the same characteristics: drawn from memory rather than observation, exaggeration of certain salient features (faces, for example), and a disregard of perspective or scale. In other words, the images derived more from mental symbolism than they did conventional schema of visual representation. Piaget would note that at later ages children acquire the ability to "correct" their images to conform to normative depictions of reality. Later observations of so-called "feral" children (raised in the wild without human contact) found that such children often didn't speak or make pictures of any kind, further reinforcing the premise that language and "artistic" rendering were largely determined by culture.[7]

While Piaget concentrated on cognition and knowledge acquisition, one of his contemporaries, John Dewey, looked at motivation and communication. Dewey conceived what is now known as a "child-centered" approach to education, which saw a youngster's psychic investment as a key factor in learning. This would inform Dewey's views on aesthetics, notably his book *Art and Experience*, which advocated a process-oriented view of art-making.[8] Creativity was central to both thinkers as an expression of symbolic thinking (Piaget) and transformative imagination (Dewey), setting in place naturalist and progressive traditions that idealized children's capacities for self-expression. Over time, advocates for early childhood art instruction expanded the list of its presumed benefits: motor skills, language development, decision-making, inventiveness, cultural awareness, among others. Later studies found correlations between arts exposure and various kinds of intelligence. While arts groups would make much of research showing, for example, that playing a musical instrument helps with language acquisition, such studies often have been small or their findings relatively weak, owing in part to the difficulty of defining variables (i.e., What constitutes music?).[9]

The fields of psychoanalysis and child development had a lot to say about kids' creativity. Beginning in the 1920s, Melanie Klein analyzed children through their imaginative activities. In her "object relations" theory, Klein speculated that as youngsters played with toys or art materials, they often mimicked important (typically parental) relationships. Watching children as they fantasized with dolls and animals, made pictures, or acted out roles, Kline saw youngsters expressing feelings of infantile dependence, love, and anxiety—in effect, giving external form to feelings toward deeply valued figures ("internal objects") in their lives. From this, Klein advanced the concept of "projective identification" (also termed "projection") through which internal objects are split off and attached to another person, place, or thing—commonly seen, for example, when an adult unconsciously sees a boss or teacher in a parental role. Klein also saw projective identification as a pivotal aspect of artistic sublimation, with creative work another means of externalizing the internal.[10]

Pediatrician Donald Winnicott took theories of child development and creativity still further in his concept of "transitional objects," which he introduced in 1953.[11] Winnicott believed that as young children grow from a state of dependence to one of relative independence, they often imagine a symbolic substitute for the idealized "object" of the caregiver. Aside from providing a familiar comfort, the child's teddy bear or blanket is the youngster's first act of creative meaning-making. But as other theorists have pointed out, what gives this comfort object a unique significance is its *transitional* function—as a bridge between child and external world, between what exists and what does not, between "reality" and "imagination."[12] Winnicott said that the transitional object helped to open a "potential space" for both mother and child, a space of simultaneous connection and separation, "me" and "not-me," knowing and unknowing—and hence an arena of inquiry, exploration, and possibility. Winnicott would later generalize the concept of "transitional phenomena" and their ongoing importance throughout people's lives. He said that the act of attaching deep significance to "objects" (whether symbolized in things, people, images, or ideas) was an underlying aspect of science, religion, and human culture itself.

Winnicott's transitional object has continued to influence thinking about art and creativity. Bernard Stiegler would frame his discussion of sublimation in terms of transitional connections between internal states of mind and the outside world.[13] To Stiegler, the infant gets an initial motivation for "living" through the early bond with its caregiver, usually the mother. Later, the transitional object becomes a powerful bridge between the two, belonging to both but also to neither. As Stiegler wrote, the transitional object is

> both an external object on which the mother and child are *dependent* (losing it is enough to make this clear) and in relation to which they are

> thus heteronomous; and an object that, not existing but consisting, provides (through this very consistence) sovereignty to both mother and child: their serenity, their trust in life, their feeling that life is worth living, their autonomy.[14]

Whether people realize it or not, their transitional phasing remains in memory and consciousness, its template reactivated later in life and recognized in the image of the child. Or in Stiegler's words, "The transitional object does not only concern the child and mother: it is also, as the first *pharmakon*, the origin of works of art and, more generally, of the life of the mind or spirit in all its forms, and thus of adult life as such."[15] Put in these terms, the transitional object is the origin of creativity.

It's worth noting that object relations and transitional phenomena theories gained most of their public caché during the post–World War II era, its well-known "baby boom," and a less well-known upsurge in the mental health field. Dr. Benjamin Spock's *Commonsense Book of Baby and Child Care* sold more than 50 million copies in its advocacy of engaged and nurturing parenting, which included a heavy dose of transitional object theory.[16] Along with other child development figures of the era, Spock would literalize the transitional object as a comforting stand-in for the mother (or, more specifically, the breast). Spock's developmental and "permissive" methodologies put him at odds with proponents of traditional discipline and adult authority. But Spock's message of loving parenting also proved a popular antidote in a culture awash with Cold War paranoia and nuclear fear.

The concept of "play" loomed large in the 1950s and 1960s as an ideal for both children and adults. Theoretical work on imaginative play would continue, although the transitional space/object remains a central premise today. Transitional phenomena fit within more generalized views of play as ways that children formulate or test ideas in the "safe" space of imitation, pretending, and so on. Following Klein and Winnicott, psychologists observed that this imitative behavior begins very early in life. An important stage in infant development is the reciprocal "mirroring" of the caregiver and child (exchanging smiles, noises, touches, etc.) as a pre-linguistic way of connecting. Recent imaging science has found so-called "mirror neurons" that fire in the brain when animals act and then observe the same behavior in another. Such neurons play an important role in emotional recognition and the development of empathy.[17] As children learn to interpret behavioral cues and feelings through play, their capacities grow for relationship formation, collaboration, and social life more generally.

But as Picasso aptly pointed out, the pathways of play get more complicated as one ages. Always active in the imagination, play is reduced to an aspect of mind for most adults, perhaps lingering in fantasy, but usually sublimated into occasional use. One "plays" a sport, in part, as a way of acting out primitive competitive urges, or performs in a "play" as a socially sanctioned form of pretending. Anthropologist Gregory Bateson saw play as both a form of nascent creativity and a way for adults to test out ideas, especially when they face grown-up dilemmas or what he called "double-bind" situations.[18] Basing his thinking on observations of children and primates, Bateson envisioned play as a meta-environment in which oppositions can be *played* out in hypothetical or alternative scenarios. Roles can be invented or reinvented. Or games can be devised with rational or even irrational rules. Bateson's work on play became fertile ground for educational theorists, especially those studying play in learning environments. Even now, cultural theorists find Bateson's work helpful in explaining why some adults cling to childish impulsivity and emotionalism.

The ideal of the eternal inner child can be seen in the figure of Peter Pan and his mischievous ethos represented for over a century in stories and films.[19] First appearing in days of Piaget and Dewey, J.M. Barrie's 1902 story about "the boy who wouldn't grow up" was inspired by Barrie's older brother David, who died at the age of 13, and was memorialized by the Barrie family as a "forever boy."[20] Stories of enduring childhood have long attracted grown-up readers, and it's worth remembering that children's literature usually is written by adults. Peter Pan lives on in social behaviors, too—as famously seen in the life of Michael Jackson. For at least three decades, "Peter Pan Syndrome" has been used as a catch-all phrase for "man boys" who can't give up their childish ways.[21] None of this has been lost on the business world, in which adolescent-themed products keep appearing for what is termed the "Peter Pan Market."[22] Seen this way, the creative output of writers like Suzanne Collins, Bella Forest, Neil Gaiman, Veronica Roth, and J.K. Rowling only further encourages childhood regression, as Marvel superhero movies keep breaking box office records. Increasingly, stressed-out adult audiences are gobbling up stories about persecuted or damaged youth, who *magically gain power* to get romance, riches, or revenge. Is it any wonder that populist impulses of childish fantasy and wish fulfillment abound in such a symbolic economy?

Not that the inner child needs metaphoric support. Today's self-help industry leans heavily on youthful nostalgia, lost innocence, and a yearning for simple pleasures—with the very phrase "inner child" becoming its own cliché. Of course, stereotypes sometimes faintly connect to reality, albeit in distorted or exaggerated ways. You certainly don't need to explain this to merchandizers now cashing in on public demand to accept, awaken, discover, embrace, reclaim, recover, or rescue that little inner person you've lost. As *Marketing Magazine* put it, "Recalling a time of fewer responsibilities and a more carefree mindset, today's adults are looking for escapism from the ever-connected working world."[23] *Marketing* then enumerated not only the exponential rise in adult-oriented products that directly evoke childhood (neon sports shoes, goofy action movies, grown-up theme parks, etc.) but also a rising tide of nostalgia-based "retro" items for baby boomers like Dannon "Natural is Back" Yogurt, "Vintage" Heinz Ketchup, "Original" Lucky Charms, and the McDonalds "Classic" menu.

It goes without saying that the return to childhood has paralleled a gigantic upsurge of marketing to children themselves. Early in the television age, advertisers recognized the potential of children in household purchasing and began to target them as consuming subjects. And they were right. Today, children under 14 spend $40 billion on toys, computer games, and clothing, while influencing a staggering $500 billion in family buying each year.[24] This has led educators and consumer advocates to decry what Henry A. Giroux termed the "Disneyfication" of childhood in direct marketing to kids, advertising in schools, entertainment product placement, and a general encouragement of materialism.[25] Titles say a lot in works like Benjamin Barber's *Consumed: How Markets Corrupt Children, Infantilize Adults, and Swallow Citizens Whole*, Daniel Thomas Cook's *The Commodification of Childhood*, Peggy Orenstein's *Cinderella Ate My Daughter*, Alvin Poussaint's *Captive Audience: Advertising Invades the Classroom*, and Shirley Steinberg's *Kinderculture: The Corporate Construction of Childhood.* [26] Common among all of these has been a view of childhood as a concept changing over time. Once seen as a separate and protected developmental stage in life, childhood now seems a rehearsal for adulthood—even as grown-ups increasingly travel backwards into youth.

Might there be a silver lining in all of this obsession with childhood? Maybe so. Authors of the recent *Lifelong Kindergarten* assert that so-called "grown-up" values aren't always so great—and that a bit of youthful spirit "is exactly what's needed to help people of all ages develop the creative capacities needed to thrive in today's rapidly changing society."[27] Scholars looking at these trends have begun to see some positive themes besides a mere regression to childishness. As a demographic, behavioral psychologists say that kids behave differently from adults—and not in ways you might expect. Obviously, children are accustomed to dependency and the feelings of security that accompany it, especially when very young. But because of this, young children in particular also are less aggressive than grown-ups, according to researcher Fletcher Kenway.[28] Compared with adults, youngsters favor cooperation over competition when left to their own devices. Anyone who has watched parents screaming at a T-Ball or pee-wee soccer game will tell you that it isn't the kids who so desperately need to "beat" the opposition. Looking at the inner-child craze from the outside, it's clearly doing *something* for people crushed by worries over finding and keeping a job, supporting a family, or otherwise surviving in today's "Anxious America." On balance, the inner-child craze carries with it a resounding value that's hard to deny: *kindness*. All of this complicates any analysis of child creativity enormously—putting it on a feedback loop that promotes children's imagination on its own terms while also using creative products to promote childlike fantasy to adults. It's little wonder that this causes mixed feelings.

Art in Schools?

America has contradictory attitudes toward creativity in schools, which have only become more complicated with the buzz over the "creative economy." Parents see art as a great idea for toddlers and generally beneficial for students of all ages. Yet art has become nearly non-existent in middle and high schools, where a lean economy favors only the most "practical" subjects. Before they can read and write, most youngsters are encouraged to develop their communication and motor skills by painting, singing, or performing skits. It's hard to imagine a daycare facility or preschool without arts activities of some kind. But childhood doesn't last forever and creative expression is soon sidelined. Hence, there is an inverse relationship in American education between age and arts education, especially in public schools. As children grow older, they received less and less creative instruction. This fundamental reality seems at odds with the widely acknowledged need for creativity in American society.

This really isn't a new problem. Since the 1980s, contemporary "education reform" movements have downplayed arts instruction in middle grades and high schools in a reactionary privileging of "basic skills"—triggered by worries about U.S. students falling behind foreign competitors in science and math.[29] Meanwhile, education at all levels continues to orient itself along readily measurable and quantifiable lines. The recession of the 2000s obliged schools to prioritize efficiency and "accountability" as money got tight, with a heightened emphasis on "outcomes" defined by rigid norms and standards. Budgets became linked to aggregate test scores as competition between schools intensified. Teacher performance evaluations got more stringent, as a generation of K–12 students also experienced education in more utilitarian terms. At colleges and universities, institutional scale typically has afforded more learning options of all kinds, including creative kinds. Unfortunately, university art programs have been shrinking for decades as the landscape of U. S. higher education—typically seen as the gateway to careers and adult life—has become increasing stratified, often distributing creative education along all-too-familiar divides between disadvantage and privilege.

Statistics say much about public views of creativity in education. Surveys show a large majority (88 percent) in the U.S. believing "the arts are part of a well-rounded education for K–12 students," according to Americans for the Arts.[30] But this support quickly vanishes when money gets tight and core subjects like reading, math, and science get discussed. Such has been the case in every era of austerity, with music, dance, and visual art joining physical education in being cast as educational "frills." Also keep in mind that K–12 schools receive 90 percent of their funding from local and state governments, often from property taxes.[31] This creates what educational scholar Jonathan Kozol famously termed "savage inequalities" in school funding. Illustrating his point, Kozol contrasted the $3,000 spent per child in Camden, NJ, with the $15,000 per child allocated nearby in Great Neck, NY.[32] Such inequities directly affect which kids will or will not receive arts education, with economically disadvantaged and racial/ ethnic minority students often getting less. Hence, according to the most recent data, African-American and Chicanx/Latinx students are now half as likely to have access to art as their European-American peers.[33]

Then there is the recession and its aftermath. Budget cuts in schools became one of many reminders of an uncertain future for the Millennial generation socialized in an age of financial worry. Overall arts education spending in American public schools has fallen by 20 percent in the last decade. According to the U.S. Department of Education, most schools put a positive face on this in reporting "some" arts instruction for all students. Elementary schools place a higher priority on the arts, with 94 percent offering music instruction and 84 percent teaching visual art.[34] Dance and drama stand at 3 and 4 percent respectively. Often these numbers are a bit murky, since over 90 percent of schools count creative activities *within* academic classes such as history and English as "arts" instruction. When standalone arts classes are offered, it generally boils down to one arts class per week of less than an hour, taught by a single "specialist" teaching an average of 24 separate courses per week. Speaking recently to one such teacher working in California's Orange County, I learned that she taught 30 such classes. "I wouldn't call it 'art' education, really," she said. "I give them a break from their other classes. It's more like art therapy than anything else."

There is even less art instruction in public secondary education, even though most high schools report such classes "available" in the visual arts (93 percent), music (90 percent), drama (48 percent), and dance (14 percent).[35] Due to student demand, almost all school districts require at least some arts education in their curricula. Yet at the same time, federal policies like the No Child Left Behind Act (NCLB) and the Common Core State Standards Initiative heavily emphasize reading and math over other subjects—in both cases excluding art from recommended requirements.[36] But the story doesn't end there. Part of the purpose of NCLB was to improve educational outcomes by defunding low-performing schools (as defined by standardized tests) and allowing students to leave them, creating pressures in struggling schools to reduce even further their offerings in subjects like social studies, foreign languages, and art. Owing to these combined factors, but 57 percent of schools make coursework in the arts a graduation requirement, and in 7 in 10 such schools this means only a single course credit. Across the creative disciplines, courses rarely are taught by instructors with degrees in the arts. In both elementary and secondary education, many schools also bolster their arts instruction statistics by including after-school offerings like band practice or programs co-sponsored by community groups.

In response to this dire situation, advocates for art education have strengthened their arguments. "When we fail to provide access to arts and culture for students, we put them at a severe disadvantage not just now, but also in a ripple effect that will continue the rest of their lives," wrote Donna Cooper,

Executive Director of Public Citizens for Children and Youth.[37] Studies over the last decade have shown that art in schools improves attendance and graduation rates while also contributing to academic success in reading, science, and math.[38] A Johns Hopkins report entitled *Neuroeducation: Learning, Arts, and the Brain* documented how music and the visual arts "rewire" students' minds in ways that increase their attention spans and even raise IQ scores.[39] All of this comes in addition to rising vocational arguments for K–12 arts education in the new creative economy.

Factories of Knowledge

For 20 years, I have been teaching large arts and humanities general education courses at the University of California, Irvine. These 400-student classes are part of the undergraduate "breadth requirements" common in most colleges and universities, and hence draw enrollments from across the academic disciplines. At UC Irvine, this means that most of the class comprises science, technology, engineering, and math (STEM) majors. Aside from an orientation to more practical fields, I've noticed a clear shift in student attitudes in recent years—a heightened preoccupation with grades and rankings, combined with growing anxieties about future earnings. Many of my colleagues see this as well, often disparaging students more concerned with GPA metrics than learning itself, while increasingly behaving more like consumers of educational commodities. I take a more sanguine view.

Bear in mind that many of today's college students grew up during the Great Recession, when families of all incomes had money worries. With scant knowledge of a world before 9/11, it's little wonder that polls show Millennials expecting lower earnings than their parents, seeing the United States on a downward spiral, and believing the two-party system as fatally flawed.[40] Rising income inequality doesn't help matters, especially at UC Irvine where 6 in 10 students get financial aid and half are the first in their families earning a college degree.[41] Because of this, Irvine has been cited by the *New York Times* as the country's leading "upward mobility engine"—making the campus a national model of what public higher education can do.[42] But it's still not a cakewalk for degree seekers. As at most public universities in America, the majority of Irvine's full-time students also work at jobs to make ends meet.[43]

Higher education translates into higher wages. According to the U.S. Department of Education, people with four-year degrees earn roughly twice that of high school graduates.[44] Given these financial pressures, it's no surprise that college education is seen as a commodity. Almost all of the students I encounter are serious, hardworking, and focused. They want rational outcomes, high grades, and clear metrics. Most of all, they are driven to succeed—in a nation where struggle is expected and competition has been called the "state religion." In the minds of many, there is a Darwinian inevitability in seeing life as a contest, so much so that it is often seen as a natural instinct. This is manifest in a culture valorizing personal achievement, aggression, and America First—values reinforced in the ideologies of business, entertainment, celebrity, sports, and militarism.

But educators have long observed that competition can be dangerous when pushed too far—for the simple reason that a system producing "winners" always yields a larger pool of "losers." Alfie Kohn wrote in his book *No Contest: The Case Against Competition (Why We Lose Our Race to Win)* that Americans are caught in a vicious circle in which individual anxieties and structural conditions reinforce each other. Children are conditioned for a world of presumed scarcity, based on the following contradictory ontology: "If I must defeat you in order to get what I want, then what I want must be scarce," Kohn stated, explaining that when "competition sets itself as the goal, which is to win, scarcity is therefore created out of nothing."[45]

Kohn argues that the real lesson instilled by competition is personal inadequacy. Wins tend to be short-lived moments of self-satisfaction derived from external evaluation, implying that one's character rises in proportion to the number of those beaten. The transitory character of such winning means that any gain is fragile and contingent on the outcome of the next contest, setting off a repeating cycle, until one ultimately fails. The external character of the evaluation also can make young people feel they are not in control of what happens to them, as researcher Carole Ames has noted. Ironically, the very sense of autonomy that competition purports to instill is diminished by the anxieties that go along with it. Feelings of agency can become weakened even among successful students, but it takes a greater toll on those who fail. This tends to produce lower achievement in both groups, along with a plethora of esteem-related problems.[46]

Internalized competition is but one side of the equation. For the better part of a decade, professors and students alike have bemoaned the growing "corporatization" of universities, as bottom-line administrative thinking has encroached on high-minded idealism. Complaints have come from across the U.S. about skyrocketing tuitions, huge lecture courses, and growing numbers of low-wage occasional lecturers. Exacerbated by recessionary belt-tightening, a new philosophy taken over higher education—with numbers and budgets increasingly driving curriculum and research priorities. Humanities departments shrink as business programs grow, partly in response to student career worries. All of this has paralleled a continuing movement toward "accountability" in public education—with K–12 teachers finding themselves obliged to "teach-to-the-test" or risk losing their jobs. Competition for grades in science and math has superseded such "frills" as art education for most of the nation's kids.

"When Universities Try to Behave Like Business, Education Suffers," read a recent headline in the *Los Angeles Times.*[47] "For most of U.S. history, it was understood that universities, whether public or private, operated under a model distinct from business," the paper reported. But a shift took place in the 1980s and 1990s, as American culture became enthralled with marketplace values. "Until then, the private sector wasn't the model for the public sector," the *Times* reported, adding that "the prestige of the private sector now requires imitation by the public sector." Students seem to be losing out in this new environment. "They're not only saddled with an increasing share of the direct costs of their education," the *Times* stated, "but are offered a narrower curriculum as universities cut back on supposedly unprofitable humanities and social science courses in favor of science, engineering and technology programs expected to attract profitable grants and the prospects of great riches from patentable inventions."

The effect of corporatization on academic labor has been devastating. In 1975, over 70 percent of instruction was done by full-time professors—experts in their fields and committed to careers as professors. Today that ratio has reversed—with 70 percent of teaching delivered by adjunct faculty (non-tenure track)—with minimal experience, no job security, and often less commitment to the institution itself. Most adjunct instructors work multiple jobs to subsist—with over 50 percent earning less than $35,000 per year and 80 percent getting no health insurance.[48] Universities increasingly see adjunct teaching as a less valued enterprise than the highly compensated "research" mission of full-time faculty. This isn't just bad news for job-hungry young PhDs and MFAs. A recent study from the University of Southern California has shown that "students who take more classes from contingent faculty have lower graduation rates and are less likely to transfer" from two-year to four-year institutions.[49] *Forbes Magazine* similarly reported that "such faculty are less student-centered in their teaching, have less contact with students outside of class, and spend less time preparing for classes."[50]

Instructional declines and labor abuses are but a few symptoms of university corporatization. And this problematic trend is hardly a secret. Recently, the American Association of University Professors (AAUP)—the nation's largest organization of college educators—published an analysis of the shift from higher education as a "public investment" to the rising "private enterprise" model. "These changes reflect the neoliberal faith that free markets would restore productivity," the document stated.[51] But the AAUP asserts that privatization has had the opposite effect. With rising costs and narrowing academic options, colleges and universities have seen a steady decline in student applications—even though the overall population of high school graduates has grown. Pressures to avoid debt and to begin earning are some of the reasons, with low-income students attending traditional colleges at 10 percent lower rates than a decade ago.[52]

Concerns about corporatized higher education go back a century, evidenced in Thorstein Veblen's 1917 *The Higher Learning in America: A Memorandum on the Conduct of Universities by Business Men*. Even then, professors across the country worried about eroding educational values and a tightening of bureaucratic management. Veblen saw universities losing their status as protected preserves for "the cultivation and care of the community's highest aspirations and ideals," operating in the "apprehension of what is right and good," and "controlled by no consideration of expediency beyond its own work." With an eerie prescience, Veblen warned of the incursion of a rising business rationality in which "the men of affairs have taken over the direction of the pursuit of knowledge," while effecting a "surveillance of the academic work exercised through control of the budget."[53]

Corporatization intensified with the creative destruction of the 1970s and 1980s. No less a publication than *Time Magazine* expressed concern over this in a feature entitled "How Universities Turned into Corporations." *Time* described this as a period when "policymakers began to view higher education more as a private good (benefiting individual students) than as a public good (helping the nation prosper by creating better educated citizens)."[54] Others would join in noting the social consequences of this shift. In his well-known 1977 *Learning to Labor: How Working Class Kids Get Working Class Jobs*, Paul Willis documented how education came to reproduce class stratification rather than equalization. Examining vocational "tracking" in Britain, *Learning to Labor* drew comparisons between school and the workplace, likening teachers to job supervisors who paid students with grades rather than money. "There is no obvious physical coercion" in such a disciplinary model, but rather what appears to be a "degree of self-direction," Willis wrote.[55] Not that the students were oblivious to any of this, with many opting to push back. One of *Learning to Labor*'s key insights was its documentation of what later would be termed student "resistance," often manifest in oppositional attitudes, disengagement, and even intentional failure at school.

For his book *School is a Factory*, Alan Sekula went into California community colleges, interviewing and photographing students in vocational training programs. Accompanying one set of images, Sekula wrote:

> Three welding students pose for a portrait. They hope to graduate into jobs with metal fabrication shops in the area. Their instructors act like bosses, supervising the action from a glassed-in office. This apprenticeship program, like public education generally, is supported by taxes that fall heavily on working people and only lightly on corporations. Spared the cost of on-the-job training, local industry profits from the arrangement. Social planners also like the idea that vocational courses keep unemployed young people off the streets and dampen discontent.[56]

As a former Creative Arts dean at one of California's leading community colleges, I don't mind admitting that what Sekula wrote certainly is partly true. But such public two-year schools also serve other purposes. For many low-income students, community colleges offer a viable entry point into higher education, especially if they intend later transfer to a four-year school—which four in ten of students indeed do. The problem is with the remaining 60 percent—often underprepared by prior schooling and not able to afford further study even at state-funded universities—who see vocational programs or immediate work as their only options.[57]

The educational stakes are only rising in today's "knowledge economy"—an expression referencing both the decline in traditional manufacturing jobs and the rising role of expertise in a competitive job market. As with any highly desired consumer product, laws of supply and demand are putting new pressures on knowledge and pushing prices up. A recent report from the Council of State Governments entitled "America's Knowledge Economy," urgently warned public officials that "short-term" tendencies to reduce education funding only cripple "long-term" economic growth and prosperity.[58] The business press is beginning to voice similar concerns. "Education costs have soared over the past few decades leaving many potential students out in the cold," stated *Forbes Magazine.* [59] Citing statics from the College Board, *Forbes* reported that costs of higher education have risen an average of 5.2 percent every year since 1994—or more than double the rate of inflation. Annual tuition and fees for in-state students at a public university now stands at $39,508, with out-of-state students paying $97,690. By comparison, private universities cost an average of $135,010. Add the costs of housing, books, and supplies, and the price tag is even higher.

Student debt has become the new normal—in keeping with changing attitudes toward credit itself. Generations ago, the idea of being in debt or "falling behind with bills" was seen as a moral or social ill. But things have changed with the rise of consumer credit and the aggressive marketing of companies such as Visa and MasterCard. Breaking national records every year for the past two decades, total indebtedness for higher education now stands at $1.31 trillion. Outstanding loans have more than doubled since 2009 according to *Bloomberg News*, which observed, "No form of household debt has increased by as much since then."[60] And the toll of the loans is terrifying—with one-quarter of those owing now in default or at least 90 days late on their required payments. Making matters worse, student loans have been excluded from bankruptcy protection since 1998—thus condemning the current generation to a lifelong obligation unknown to their parents. Obama administration financial experts worried about the long-term consequences of this, predicting that the loans could soon slow the U.S. economy. Even President Trump has likened the debt to an "anchor" holding down young Americans—although his administration continues cutting federal programs to help student borrowers. Low-income students suffer the most—as they enter the workforce with less freedom to choose employment and more pressure to look for the biggest paycheck.

As schools have become costlier, pressures have grown to get the best value. Last year, UCLA broke national records for undergraduate applications, with more than 124,000 students seeking admission for a freshman class of 9,200—translating to a 7.2 percent acceptance rate. Similar (but less extreme) patterns are occurring across the country, pushing selectivity at prestigious public universities closer to that seen at Ivy League schools such as Cornell (12.5% acceptance rate),

Dartmouth (10.4%), and Yale (6.9%).[61] Students begin shopping for colleges as early as junior high, while struggling to optimize their chances through advanced classes and extracurricular activities. Nearly 50 percent of students see a high school counselor due to stress over this, according the American Psychological Association.[62] "Burn-out before college" is a rising phenomenon.

Meanwhile, business has boomed for SAT and ACT prep courses—much to the consternation of testing services. College Board President David Coleman thinks companies that offer SAT prep services are "predators who prey on the anxieties of parents and children and provide no real educational benefit."[63] Education experts have long argued that test prep providers exist not only because such high-stakes testing has failed students and colleges; they say the SAT and ACT provide poor measures of real academic achievement—and actually indicate nothing more clearly than family income. The two largest prep course providers, Kaplan and Princeton Review, charge $699 for a basic course, although some families pay as much as $1,000 per hour for private tutors or freelance college admission consultants. All of this has further stratified the college admissions process, while piling on costs before students even leave home.

Put all of this together and it's easy to see why my UC Irvine students are a little on edge. College degrees are now more expensive, competitive, and keyed to earnings than at any point in American history—so much so that many young people are buckling under the pressure. Universities seem unable to do very much to help because they themselves are a big part of the problem. Hence, amid an ever-tightening web of "financialized subjectivity," the current generation finds itself bound by the logic of capital within the very institutions of higher education that might be instilling values of humanistic wisdom and unbounded inquiry. Neoliberal culture promises them freedom and upward mobility, while supplanting other ways of looking at knowledge, work, or life itself. None of this bodes well for the current generation of college-age young people, much less the climate of experiment, risk-taking, and "creativity" so vital to innovation and new ideas.

College Art in Crisis

It might surprise many to know that no systematic studies exist of college- and university-level arts programs. This is partly due to the way art in higher education fragments into academic disciplines and professional training programs, as well as the complex array of public and private schools, community colleges and research universities, and the ever-expanding variety of for-profit entities and online learn-at-home opportunities. The National Center for Education Statistics (NCES) provides rough disciplinary percentages of bachelor's degrees earned by America's estimated 18.7 million college students, however. Of these, 5.1 percent graduated in the "Visual and Performing Arts" category, and another 4.6 percent in "Communications and Journalism." Larger breakdowns included "Business" at

19.4 percent, "Health Sciences" at 10.7 percent, and "Social Science" at 9.2 percent.[64] Beyond this, anecdotal evidence abounds of a decade-long decline in arts and humanities programs, described by many as a continuing crisis. The recession is partly to blame, with many students and their families simply opting for more surefire career paths, especially as college tuitions have risen.

On the other hand, college art has found new friends among creative economy advocates, with educators jumping on claims from people such as Richard Florida that 30 percent of today's jobs require creative skills.[65] Making the most of this, the National Endowment for the Arts (NEA) recently released a report entitled "The Arts and Economic Growth," compiled in partnership with the U.S. Bureau of Economic Analysis.[66] The document claimed that "arts and culture" contributed $704 billion to the U.S. economy (4.2% of GDP) and a whopping 32.5 percent of GDP growth in the past 15 years. This is *more* than sectors like construction ($619 billion) and utilities ($270 billion), perhaps because the study defined art so broadly—encompassing advertising, broadcasting, motion pictures, publishing, and arts-related merchandizing, as well as the performing and visual arts themselves. This prompted a piece entitled, "Who Knew? Arts Education Fuels the Economy" in the respected *Chronicle of Higher Education*, which noted similar findings from business groups. Among these were the Partnership for 21st-Century Learning, a coalition of corporate and educational leaders and policy-makers, which said, "Education in dance, theater, music, and the visual arts helps instill the curiosity, creativity, imagination, and capacity for evaluation that are perceived as vital to a productive U.S. work force."[67] The Conference Board, an international business research organization, polled employers and school super-intendents, finding "that creative problem-solving and communications are deemed important by both groups for an innovative work force."[68] And IBM, in a report based on face-to-face interviews with more than 1,500 CEOs world-wide, concluded that "creativity trumps other leadership characteristics" in an era of rising complexity and continual change.[69]

But this upbeat rhetoric is not reflected in enrollment statistics or the attitudes of college and university administrators. "Art schools now face a growing challenge," according to a recent report in *Artnet News*, "as application numbers and enrollment figures are falling." The situation is especially dire for programs offering MFAs—the art degree equivalent to a Ph.D. "This year is the worst in memory, perhaps in this millennium," commented one program head, who said that applications recently had fallen by 50 percent.[70] Things are even worse in the humanities, where the number of BA degrees has fallen to its lowest point in 70 years. To put this in perspective, *Inside Higher Ed* recently reported an 8.7 percent decline in humanities enrollments in the *last two years* alone. U.S. and world history declined by 12 percent, with more rarefied areas like archeology and classical studies dropping by 19 percent.[71]

A critical point in the arts and humanities "crisis" talk came when President Obama joked about the foolishness of an art history credential. Channeling the sentiments of Republican leaders, Obama observed that "young people no longer see the trades and skilled manufacturing as a viable career. But I promise you, folks can make a lot more, potentially, with skilled manufacturing or the trades than they might with an art history degree."[72] Naturally, Obama's remarks brought heated responses from academic leaders. Carol Geary Schneider, president of the Association of American Colleges and Universities, weighed in, saying, "In recent years, we've sunk into a 'what's in it for me' approach to learning, making career earnings the litmus test both for college and for different majors," later adding, "The president speaks well in principle about our responsibilities to one another in a democratic society. But he seems to have forgotten that college can build our desire and capacity to make a better world, not just better technologies."[73]

Needless to say, things got a lot worse with the ascension of Donald Trump to the White House. Budget-cutting aside, Trump's assaults on the Department of Education, the Corporation for Public Broadcasting, and the National Endowments for the Arts and Humanities indicate more than the casual biases of his predecessor. In the name of purported populism, the Trump agenda came straight from the right-wing Heritage Foundation, which furnished the White House with a report stating that "actors, artists, and academics are no more deserving of subsidies than their counterparts in other fields," adding that "taxpayers should not be forced to pay for plays, paintings, pageants and scholarly journals."[74]

Fair enough, you might say. But if these agencies only provide welfare for creative types, why has the U.S. Congress supported for them for six decades? Their reasoning has been the same as in nearly every society in the world concerned with the preservation and enhancement of its "national" culture. Other countries spend lavishly in advancing art and literature as essential expressions of their heritage and values, whether this entails the preservation of antiquities or the funding of contemporary artists. The U.S. devotes but 0.02 percent for federal money to the arts—in comparison with Finland (0.47%), Germany (0.36%), France (0.26%), Sweden (0.29%), Canada (0.21%), United Kingdom (0.14%), Australia (0.14%), and Ireland (0.07%).[75]

Historic preservation should be reason enough to sustain arts and humanities funding. According to the venerable American Academy of Arts and Sciences, the arts and humanities are a key force in "achieving long-term national goals for our intellectual and economic well-being."[76] The group argued that such learning is essential in cultivating future generations who are knowledgeable, analytical, and tolerant. The American Academy said that

> the humanities—including the study of languages, literature, history, film, civics, philosophy, religion, and the arts—foster creativity, appreciation of our commonalities and differences, and knowledge of all kinds … [T]hey help us understand what it means to be human and connect us with our global community.

PEN America president Suzanne Nossel saw even more serious consequences in cuts to federal cultural support. "Trump's declaration of war on the arts and humanities must be seen in the context of his repudiation of the American ideals—grounded in the Enlightenment—of self-expression, knowledge, dissent, criticism, and truth," Nossel recently wrote.[77]

> Concepts like the search for truth, the open exchange of ideas, and the esteem for culture may read like empty platitudes etched in the walls of ivy-covered universities. But they are principles that undergird not just a liberal arts education but also the Common Core curriculum taught in the hundreds of thousands of U.S. public schools.[78]

Along with numerous other critics, Nossel sees the President's repudiation of "culture" as a rejection of the very foundations of inquiry and free thinking that underlie American democracy itself.

This chapter has detailed America's contradictory attitudes toward creative education. Most families love the idea of art in schools and usually see youngsters as naturally imaginative. But these days, children are finding less and less creativity in the classroom. Schools pressured by tight budgets and the metrics of standardized testing think they need to prioritize basic skills. Meanwhile, students worried about college or a job after high school often get the message that music or art study won't help their chances. Competition only worsens student anxieties in universities, where arts and humanities options stand at their lowest in decades. Meanwhile, public support for art enrichment programs outside schools continues to dwindle in an era of government austerity.

While America's current moment is worrisome indeed, its backstory is deeply rooted in history. The inherent contradictions between democracy and capitalism certainly have played a big role, along with the often unpredictable ways they have evolved over time. This chapter has looked at ways that contests over knowledge and education have affected creativity, especially in recent history. Discussion in the next chapter turns to twentieth-century debates about what were termed the "culture industries" of mass media and consumer society, as well as their eventual eruption into the "culture wars" of recent decades. Central to this discussion will be the ways that people's knowledge and beliefs are socially influenced by the transmedia worlds in which they find themselves, as well as their relationships to others. Obviously, creativity has played a vital role in this, and will grow only more influential in the years to come.

Notes

1 Pablo Picasso, as quoted in Otto Kroeger and David B. Goldstein, *Creative You: Using Your Personality Type to Thrive* (New York: Atria, 2013), p. xix.

2 "College Enrollment and Work Activity of 2018 High School Graduates," U.S. Bureau of Labor Statistics (Apr. 25, 2019) www.bls.gov/news.release/hsgec.nr0.htm (accessed Jun. 15, 2019).

3 "America's Knowledge Economy: A State-by-State Review," Council of State Governments (2017) www.csg.org/programs/knowledgeeconomy/Elsevier_Report_2015.pdf (accessed Apr. 4, 2018).
4 Jean-Jacques Rousseau, *Emile: Or, On Education* (1763) trans. Allan Bloom (New York: Basic Books, 1979).
5 Northrup Frye, *The Educated Imagination* (Bloomington: Indiana University Press, 1964).
6 Jean Piaget and Barbel Inhelder, *The Psychology of the Child*, trans. Helen Weaver (New York: Basic Books, 1969).
7 Nigel Spivey, "Child Art," *How Art Made the World: A Journey through the Origins of Human Creativity* (New York: Basic Books, 2005) p. 15.
8 John Dewey, *Art and Experience* (1934) (New York: Tarcher Perigree, 2005).
9 Anna Trafton, "Music in the Brain," *MIT News* (Dec. 16, 2015) http://news.mit.edu/2015/neural-population-music-brain-1216 (accessed Sep. 5, 2018).
10 Rossella Valdre, *On Sublimation: A Path to Desire, Theory, and Treatment* (New York: Karnac Books, 2014) p. 32
11 Donald Winnicott, "Transitional Objects and Transitional Phenomena—A Study of the First Not-Me Possession," *International Journal of Psycho-Analysis* 34 (1953): 89–97.
12 Bernard Stiegler, *What Makes Life Worth Living: On Pharmacology*, trans. Daniel Ross (Malden, MA: Polity, 2010), pp. 1–3.
13 *What Makes Life Worth Living.*
14 *What Makes Life Worth Living*, pp. 2–3.
15 Ibid.
16 Benjamin Spock, *The Commonsense Book of Baby and Child Care* (1946) 4th Edition (New York: Sloan and Pearce, 1960).
17 Lea Winerman, "The Mind's Mirror," *American Psychological Association Monitor on Psychology* (Oct. 2005) www.apa.org/monitor/oct05/mirror.aspx (accessed Sep. 4, 2018).
18 Gregory Bateson, "A Theory of Play and Fantasy," in Jerome S. Bruner, Alison Jolly, and Kathy Sylvia, eds., *Play: Its Role in Development and Evolution* (Harmondsworth: Penguin, 1976) pp. 119–29.
19 J.M. Barrie, *Peter Pan* (1902) (New York: Millennium Fulcrum, 1991).
20 Andrew Birkin, *J.M. Barrie and the Lost Boys.* (New Haven: Yale University Press, 1986).
21 Kristen Houghton, "Peter Pan Syndrome," Huffingtonpost.com (Nov. 10, 2011) www.huffingtonpost.com/kristen-houghton/the-peter-pan-syndrome_b_6127308.html (accessed Sep. 5, 2018).
22 "Coloring Enters the 'Peter Pan' Market," AgelessCreativity.com (Nov. 19, 2015) https://agelesscreativity.wordpress.com/2015/11/19/coloring-enters-the-peter-pan-market (accessed Sep 3, 2018).
23 Shelby Ross, "Marketing to Your Consumer's Inner Child," *Marketing Magazine* (Feb. 23, 2016) http://marketingmag.ca/brands/marketing-to-your-consumers-inner-child-column-168707 (accessed Apr. 29, 2018).
24 "Marketing to Children Overview," Campaign for a Commercial-Free Childhood (Apr. 2018) www.commercialfreechildhood.org/resource/marketing-children-overview (accessed Apr. 29, 2018).
25 Henry A. Giroux, *The Mouse that Roared: Disney and the End of Innocence* (Lanham, MA: Rowman & Littlefield, 1999).
26 Benjamin Barber, *Consumed: How Markets Corrupt Children, Infantilize Adults, and Swallow Citizens Whole* (New York: Norton, 2007); Daniel Thomas Cook, *The Commodification of Childhood: The Children's Clothing Market and the Rise of the*

Child Consumer (Durham, NC: Duke University Press, 2004); Peggy Orenstein, *Cinderella Ate My Daughter: Dispatches from the Front Lines of the New Girlie-Girl Culture* (New York: Harper, 2012); Alvin Poussaint, *Captive Audience: Advertising Invades the Classroom* (Northampton, MA: Media Education Foundation, 2009); Shirley Steinberg, *Kinderculture: The Corporate Construction of Childhood* (3rd ed.) (London and New York: Routledge, 2011).

27 Mitchell Resnick and Ken Robinson, *Lifelong Kindergarten: Cultivating Creativity through Projects, Passions, Peers, and Play* (Cambridge, MA: MIT Press, 2017) p. 4.

28 Fletcher Kenway, *Consumer Attitudes and Behaviours: Marketing to the Inner Child*, Esomar World Research (Apr. 2016) www.illuminas.com/wp-content/uploads/2016/04/ESOMAR_paper_S321_07_Fletcher_Kenway.pdf (accessed Sep. 22, 2018).

29 Eric A. Hanushek, "Why Can't U.S. Students Compete with the Rest of the World?" *Newsweek* (Aug. 28, 2011) www.newsweek.com/why-cant-us-students-compete-rest-world-67213 (accessed Sep. 22, 2018).

30 Randy Cohen, "The American Public Says YES to Arts Education!" Americans for the Arts (Mar. 15, 2016) http://blog.americansforthearts.org/2016/03/05/the-american-public-says-yes-to-arts-education (accessed Sep. 5, 2018).

31 "How Do We Fund Our Schools?" PBS.org (Sep. 5, 2008) www.pbs.org/wnet/wherewestand/reports/finance/how-do-we-fund-our-schools/?p=197 (accessed Sep. 5, 2018).

32 Jonathan Kozol, *Savage Inequalities: Children in America's Schools* (1991), reprint edition (New York: Broadway Books, 2012).

33 "Decline of Arts Education in Underserved Populations," Americans for the Arts (2008) www.americansforthearts.org/sites/default/files/pdf/2015/by_program/reports_and_data/research_studies_and_publications/ArtsEd_UnderservedPops_2015.pdf (accessed Sep. 5, 2018).

34 *Arts Education in Public Elementary and Secondary Schools: 1999–2000 and 2009–10* (Apr. 2012) http://nces.ed.gov/pubs2012/2012014rev.pdf (accessed Sep. 5, 2018).

35 Ibid.

36 Valerie Metla, "School Art Programs: Should They Be Saved?" *Law/Street* (May 14, 2015) http://lawstreetmedia.com/issues/education/cutting-art-programs-schools-solution-part-problem (accessed Sep. 5, 2016).

37 Donna Cooper and Maud Lyon, "Schools Are Starving for Arts Education," Philadelphia *Public School Notebook* (Sep. 11, 2015) http://thenotebook.org/latest0/2015/09/11/schools-are-starving-for-arts-education (accessed Jan. 5, 2019).

38 "10 Salient Studies on the Arts in Education," Center for Online Education (2018) www.onlinecolleges.net/10-salient-studies-on-the-arts-in-education (accessed Jan. 5, 2019).

39 *Neuroeducation: Learning, Arts, and the Brain*, Johns Hopkins University (2009) http://steam-notstem.com/wp-content/uploads/2010/11/Neuroeducation.pdf (accessed Jan. 24, 2019).

40 Samantha Smith, "Millennials Less Confident about Nation's Future, but so Were Their Parents, Grandparents When Young," *Pew Research Center* (Feb. 16, 2016) www.pewresearch.org/fact-tank/2016/02/16/millennials-less-confident-about-nations-future-but-so-were-their-parents-grandparents-when-young (accessed Jan. 5, 2019).

41 "UC Irvine Praised for Student Support," *Orange County Register* (Apr. 29, 2016) www.ocregister.com/articles/students-709861-uci-schools.html (accessed Jan. 5, 2019).

42 David Leonard, "California's Upward Mobility Machine," *New York Times* (Sep. 16, 2015) www.nytimes.com/2015/09/17/upshot/californias-university-system-an-upward-mobility-machine.html (accessed Jan. 5, 2019).
43 Stacy Rapcon, "More College Students Are Working While Studying," CNBC.com (Oct. 29, 2015) www.cnbc.com/2015/10/29/more-college-students-are-working-while-studying.html (accessed Jan. 5, 2019).
44 "How Much More Do College Graduates Earn than Non-College Graduates?" Study.com (2018) http://study.com/articles/How_Much_More_Do_College_Graduates_Earn_Than_Non-College_Graduates.html (accessed Jan. 5, 2019).
45 Alfie Kohn, *No Contest: The Case Against Competition (Why We Lose Our Race to Win)* (Boston and New York: Houghton Mifflin, 1986) p. 4.
46 Carole Ames, "Children's Achievement Attributions and Self-Reinforcement: Effects of Self-Concept and Competitive Reward Structure," *Journal of Educational Psychology* 70 (1978).
47 Michael Hiltzik, "When Universities Try to Behave Like Business, Education Suffers," *Los Angeles Times* (Jun. 3, 2016) www.latimes.com/business/hiltzik/la-fi-hiltzik-university-business-20160602-snap-story.html (accessed Jan. 5, 2019).
48 Dan Edmunds, "More Than Half of College Faculty Are Adjuncts: Should You Care?" *Forbes Magazine* (May 28, 2015) www.forbes.com/sites/noodleeducation/2015/05/28/more-than-half-of-college-faculty-are-adjuncts-should-you-care/#6efb36f41600 (accessed Jan. 13, 2019).
49 Ibid.
50 Ibid.
51 David Schultz, "The Rise and Coming Demise of the Corporate University," AAUP (Sep. 15, 2016) www.aaup.org/article/rise-and-coming-demise-corporate-university#.V89ErWVltlo (accessed Sep. 6, 2018).
52 Alai Wong, "Where Are All the High-School Grads Going" *The Atlantic* (Jan. 11, 2016) www.theatlantic.com/education/archive/2016/01/where-are-all-the-high-school-grads-going/423285 (accessed Sep. 5, 2018).
53 Thorstein Veblen, *The Higher Learning in America: A Memorandum on the Conduct of Universities by Business Men* (1917) (Baltimore: Johns Hopkins, 2015) pp. 75, 88.
54 Andrew Rossi, "How American Universities Turned into Corporations," *Time* (May 21, 2014) http://time.com/108311/how-american-universities-are-ripping-off-your-education (accessed Apr. 18, 2018).
55 Paul Willis, *Learning to Labor: How Working Class Kids Get Working Class Jobs* (New York: Columbia University Press, 1977) p. 1.
56 Alan Sekula, *School is a Factory, 1978–80* (Halifax: Nova Scotia College of Art and Design, 1983) n.p.
57 Grace Chen, "Will Community College Tuition Increases Outpace Inflation Rates?" *Community College Review* (Aug. 26, 2016) www.communitycollegereview.com/blog/will-community-college-tuition-increases-outpace-inflation-rates (accessed Apr. 6, 2018).
58 "America's Knowledge Economy."
59 Mike Patton, "The Cost of College: Yesterday, Today, and Tomorrow," *Forbes Magazine* (Nov. 19, 2015) www.forbes.com/sites/mikepatton/2015/11/19/the-cost-of-college-yesterday-today-and-tomorrow/#51708786060c (accessed Jan. 9, 2019).
60 Shaien Nasirpour, "President Trump Has Called Student Debt an 'Anchor' Weighting Down Young Americans." *Bloomberg News* (Feb. 17, 2018) www.bloomberg.com/news/articles/2018-02-17/student-debt-in-america-has-hit-a-new-record (accessed Jan. 9. 2019).

61 Abby Jackson, "Ivy League Admission Letters Just Went Out—Here Are the Acceptance Rates for the Class of 2021," *Business Insider* (Apr. 1, 2018) www.businessinsider.com/ivy-league-harvard-yale-princeton-acceptance-rates-class-of-2021-2018-3 (accessed Jan. 12, 2019).
62 Samantha Olsen, "High School Students Are Stressed Out about College: The Reality of Burning Out before College, *Medical Daily* (Aug. 12, 2015) www.medicaldaily.com/high-school-students-are-stressed-out-about-college-admissions-reality-burning-out-347476 (accessed Apr. 22, 2018).
63 James S. Murphy, "The SAT-Prep Industry Isn't Going Anywhere," *The Atlantic* (Mar. 4, 2014) www.theatlantic.com/education/archive/2014/03/the-sat-prep-industry-isnt-going-anywhere/284430 (accessed Jan. 9, 2019).
64 *Digest of Educational Statistics*, National Center for Educational Statistics (2015) http://nces.ed.gov/programs/digest/d15/tables/dt15_322.10.asp?current=yes (accessed Sep. 5, 2018).
65 Richard Florida, "Cities and the Creative Class" (2003) http://creativeclass.com/rfcgdb/articles/4%20Cities%20and%20the%20Creative%20Class.pdf (accessed Sep. 13, 2018).
66 The Arts and Economic Growth, National Endowment for the Arts (2016) www.arts.gov/news/2016/arts-and-cultural-production-contributed-7042-billion-us-economy-2013 (accessed Sep. 13, 2018).
67 "Creativity and Innovation," Partnership for 21st Century Learning (n.d.) www.battelleforkids.org/networks/p21 (accessed Jun. 15, 2019).
68 Sunil Iyengar and Ayanna Hudson, "Who Knew? Arts Education Fuels the Economy," *Chronicle of Higher Education* (Mar. 10, 2014) www.chronicle.com/article/Who-Knew-Arts-Education-Fuels/145217 (accessed Sep. 13, 2018).
69 Ibid.
70 Brian Boucher, "After a Decade of Growth, MFA Enrollment is Dropping," artnetnews.com (Oct. 18, 2016) https://news.artnet.com/art-world/mfa-programs-enrollment-dropping-707645 (accessed Jan. 16, 2019).
71 Scott Jaschik, "The Shrinking Humanities Major," *Inside Higher Ed* (Mar. 4, 2016). www.insidehighered.com/news/2016/03/14/study-shows-87-decline-humanities-bachelors-degrees-2-years (accessed Jan. 6, 2019).
72 Scott Jaschik, "Obama vs. Art History," *Inside Higher Ed* (Jan. 31, 2014) www.insidehighered.com/news/2014/01/31/obama-becomes-latest-politician-criticize-liberal-arts-discipline (accessed Jan. 5, 2019).
73 Ibid.
74 Susan Nossel, "Donald Trump's Proposed Cuts to the Arts and Humanities is an Attack on Reason," *Business Insider* (Jan. 29, 2017) www.businessinsider.com/trump-funding-cuts-arts-humanities-nea-neh-2017-1 (accessed Jan. 8, 2019).
75 "Comparisons of Arts Funding in Selected Countries," Canada Council for the Arts (2005) http://intranet.americansforthearts.org/sites/default/files/Comparisonsofartsfunding27Oct2005_0.pdf (accessed Jan. 8, 2019).
76 Susan Nossel, "Donald Trump's Assault on the Enlightenment," *Foreign Policy* (Jan. 26, 2017) http://foreignpolicy.com/2017/01/26/donald-trumps-assault-on-the-enlightenment-nea-neh-funding-cuts (accessed Jan. 8, 2019).
77 Ibid.
78 Ibid.

Chapter 8

Industries of Culture

Masterpieces vs. Dream Machines

If you missed the last *Fast and Furious* movie, you were one of the few people who did. Not only do these films keep breaking box office records, but now they feature the world's most highly paid actor, Dwayne "The Rock" Johnson. That car crash movies starring a former pro-wrestler would rise to the top of heap says a lot about the state of popular culture in neoliberal America. More specifically, the movies are a sobering reminder of the way market logic has overtaken artistry in an increasingly globalized technological society. And while Hollywood product is not the only barometer of consumer tastes and values, the extraordinary success of *Fast and Furious* films shows an American public voting with its pocketbook in new—but not completely surprising—ways. The question is: What makes these movies so popular?

This chapter begins with a look at the *Fast and Furious* franchise as a case study in the Hollywood blockbuster phenomenon—in its studied appeal to popular appetites for spectacle and familiarity. While most viewers simply see such films as escapist fun, their very success often depends on a reinforcement of prevalent tropes and fantasies. Hence, themes of freedom and family often are entwined with stereotypical renderings of power and conflict familiarized to seem natural or inevitable. This embedding of values into consumer products reveals how ideology travels in the contemporary landscape. The ability of media to shape people's preferences and beliefs was outlined by Walter Lippmann in his classic 1922 book *Public Opinion*, which made the simple point that for most people the larger world around them is "out of reach, out of sight, and out of mind."[1] Like the shadows in Plato's cave, much of what the public comes to know is mediated by representations of reality rather than actual experiences of it, whether such renderings are delivered by friends or news outlets.

Then as now, such representations often come from third parties with specific agendas—such as attracting large audiences or selling commercial products. Aside from entertainment, even purportedly "objective" stories are always produced from a particular point of view, as cable outlets like Fox News and MSNBC demonstrate. This puts average citizens in a bit of a quandary—and it is one of the trickiest dilemmas of living in a media age—as viewers navigate a terrain in which fact and fiction are often impossible to separate. Of course, most people cling to convictions that they can sort out the truth in media messages—even as habits of mind, unconscious preferences, subtle marketing schemes, or outright deceptions push back against them. It's a cat-and-mouse game consumers and media producers constantly play.

Empirical studies of this began in the 1960s as television networks gained the ability to reach into living rooms throughout the country. From that point onward, it became clear that broadcast journalism heavily influenced what people believed. In studies conducted into the 1990s, the overlap between coverage and public concern ran as high as 90 percent depending on locale.[2] These data gave credence to longstanding theories of mass media as a powerful driver of ideology. Initially seen as deceiving viewers with direct falsehoods, news and entertainment later were understood to work in subtle and less direct ways. Owing to the ubiquity of media in everyday life, the constant repetition of similar messages seemed to reshape opinion over time.

Much of the discussion in this chapter explores the concept of the "culture industry" as originally formulated by intellectuals in the World War II era. At the time, the workings of propaganda were not understood broadly, although the popularity of media could be widely seen. Over the coming decades, culture industry theory gradually would be revised and updated as sociologists and communications scholars began studying audience behavior. The introduction of television and other new technologies would also temper this evolving discourse. Eventually, the influence of news and entertainment would rise to the level of public concern and policy debate. Contention over media industries would even reach into the art world by the 1980s, as the function of culture, broadly writ, became increasingly politicized. The resulting "culture wars" further polarized a U.S. divided by political partisanship—as issues of morality, patriotism, and what it meant to be a "real American" entered the picture. Obviously, these matters have a familiar ring in the current Trump era.

Of course, everyone knows that movies are multi-billion-dollar business—with those working in Hollywood simply calling it "The Industry." As that term suggests, the entertainment business is a monolith owned and controlled by a small number of huge corporations. Decades ago, communications scholar Ben Bagdikian began tracking these patterns in his now-famous book series *The Media Monopoly.* [3] Early on, Bagdikian noted that TV and movie companies were being absorbed by big multinationals like AT&T, Viacom, and Time Warner, with little intrinsic interest in program content or even quality, but under enormous pressure to deliver results to shareholders. This simple economic reality and its demand for predictable box office returns would eliminate any risk-taking ventures, while obliging a startling homogeneity in Hollywood products. Creativity became measured by investment return rather than the merits of ideas—as repetition and familiarity overtook novelty and difference. It's worth mentioning that the multi-billion-dollar advertising industry had been operating this way for decades.

But not everyone accepts the culture industry argument. These days, most people believe they live in a world of unbounded media options, as online streaming, YouTube channels, and new subscription services have expanded the already-bewildering profusion of cable channels and streaming options. Lost in this is the reality that a new cadre of quasi-monopolistic corporations now control most of those offerings as well. Amazon, Apple, Facebook, Google, and Microsoft now comprise the newest face of a media monopoly that *The New York Times* recently dubbed "Tech's Frightful Five."[4] As Farhad Manjoo wrote, "This is the most glaring and underappreciated fact of internet-age capitalism: We are, all of us, in inescapable thrall to the handful of American technology companies that now dominate much of the global economy." While products offered by these companies may seem to offer unlimited consumer choice, a more careful look reveals an ever more sophisticated continuity of brands, themes, and motivating ideals.

Fast and Furious

The Fate of the Furious was the eighth installment (dubbed "F8") in the *Fast and Furious* franchise launched in 2001—with ninth and tenth films planned through 2022. "One of the givens of *Fast and Furious* is that the latest movies will be bigger and more enjoyably ludicrous than the last. The miniskirts will be shorter, the toys zoomier, and stunts more delirious," wrote critic Manohla Dargis in *The New York Times*. [5] Like its predecessors, "F8" depicted the adventures of a cadre of renegade street racers, alternately dodging cops and an ever-changing set of bad guys. The film's diverse protagonists—portrayed by Vin Diesel, Tyrese Gibson, Dwayne Johnson, and Jason Stratham, among others—went up against cyberterrorists and the Russian government as they were tasked by an international security agency with heading off a nuclear war. Spectacular crashes and machine-gun battles filled up most of the chaotic narrative. Like most successful Hollywood franchises, *Fast and Furious* has a clear formula and sticks to it—making it Universal Pictures' biggest enterprise ever in generating $5 billion in revenues. Such multi-picture series have become the norm in Hollywood, as seen in familiar brands like *Batman* (16 films), *Star Trek* (13 films), *Star Wars* (12 films), *Harry Potter* (11 films), *X-Men* (11 films), and *Superman* (eight films)—all dwarfed by the *James Bond* (25 films) and *Avengers* (28 films) series.[6] While it has become commonplace to blame this trend on entertainment industry greed, the reasons behind formula filmmaking are a bit more subtle—resulting from the complex interplay of producing and consuming interests.

But let's start with the formula itself. Many critics saw "F8" as yet another symptom of what has been termed "sequelitis" in the Hollywood. Right now, an astonishing 139 sequels are in some stage of planning or production, with an average of 40 such films released each year.[7] Most commonly, sequels are based on previous movies, TV shows, comic books, or popular works of young adult fiction. In "Why the Hollywood Sequel Machine is Stronger than Ever," Derek Thompson wrote in *The Atlantic* of a growing risk-avoidance in an industry facing competition from streaming video, expanded cable TV channels, and smartphone entertainment. "What does this all mean? Fewer movies, bigger movies, louder movies, and safer movies. Now that studios are making fewer, more expensive films, there is much more riding on each product," Thompson wrote.[8] Hollywood mitigates this risk with safer subjects and more testing, "relying on sequels and adaptations that it knows have a built-in audience, not only at home, but also abroad, where explosions translate easier than wit."[9]

This makes for movies that critics hate, but audiences love. While sequels have accounted for 70–80 percent of the most popular films in recent years, the Academy of Motion Pictures and Sciences has yet to nominate a single one for a "Best Picture" Oscar. "The sequel market thrives on conformity and predictability. The movies are average. But they're average on purpose," Thompson further explained: "They are the product of a Hollywood's exquisitely designed factory of average-ness, which has evolved as this industry has transitioned from a monopoly to a competitive industry that can no longer afford to consistently value art over commerce," Thompson observed. "Hollywood needs to know what its fragile audience wants and when it asks us, we tell them: make something like the last average thing I saw."[10] Of course, companies making everything from fast-food to porn have always said the same thing: that they simply respond to consumer demand. But as any drug dealer will tell you, getting the client hooked in the first place is vital to the business plan. *Fast and the Furious* has been ingenious in this regard, enticing viewers with bombastic pyrotechnics and bare skin served up with narrative machismo and outsider camaraderie. This all-American combo has a couple of inventive twists, not the least of which is an up-to-date diversity in its multicultural cast. Refreshing as this sounds, another logic runs through all *Fast and Furious* films—and it explains part of the franchise's success.

Underpinning every *Fast and Furious* product is the relentless pursuit of *winning at any cost*, as well as the reward of doing so. The opponent may be a criminal gang, or spy network, or other car racers, but the ongoing drama of competitive struggle is ever present—heightened by the anti-authoritarian bent of the *Fast and Furious* crew. Never afraid to break the law for a greater good, the gang travels the world in search of the perfect race, confronting competitors in Azerbaijan, Brazil, Cuba, Dominican Republic, Japan, Spain, and the U.S. At the end of every *Fast and Furious* film there is some kind of victory, reconciliation, or vindication. And then there is "family"—America's go-to value for selling just about anything. In its use of family, *Fast and Furious* pulls out the time-honored us-against-the-world motif, but with an up-to-date "family of choice" slant. As one fan site put it:

> Your family is everything. Whether it be your actual family related by blood, or your best friends that you've met throughout your life. Your family will get you out of those rough spots that we all pretend to be too macho to relate to.[11]

Let's not forget the cars. "From the chrome-plated opening credits to the final vehicular barrel roll, humans take a backseat in this car fetish franchise, which is upfront about what's on offer," wrote critic Andy Wright.

> Numerous scenes are spent in the cabs of flashy, neon-colored racers, tricked out with computer screens, nitrous oxide tanks, extraneous LED lights and ostentatious artwork. But even when viewers aren't sitting in the cab or careening along the road with the drivers, it rarely feels like we've exited the car.[12]

Director Rob Cohen admitted, "The cars really are the co-stars of the movies," photographed to full effect against the muted backgrounds of East Los Angeles urban neighborhoods.[13] And while the actors are portrayed as working-class types, they also are meticulously commodified. As Wright further observed:

> There is something highly calibrated about the men in the film. They are decorated with complicated tattoos, wear chunky necklaces, mesh tank tops and wrist cuffs. The male physique is carefully honed; to see characters with their shirts off is to know time and money has been lavished on their exterior.[14]

Viewed through the lens of everyday creativity, these films depict the highly evolved aesthetic of car culture, replete with the impulses of appropriation, hybridity, and resistance seen in certain forms of contemporary art. Yet, ideologically speaking, the *Fast and Furious* franchise also is a paean to indie materialism, reflecting capitalism's relentless ability to co-opt even the most oppositional impulses. These guys may wear dirty T-shirts, but they will pay anything (or steal if necessary) for a chrome-plated exhaust pipe. Ready to fight the "system" at every turn, they nevertheless have a soft spot when it comes to their cherished commodities. And, of course, the heroes of *Fast and Furious* often pay the ultimate price for rebellion by dying in crashes on screen. Fans of the series know that popular *Fast and Furious* actor Paul Walker died in exactly this way in real life while racing his street-modified Porsche at twice the speed limit.

The Culture Industry

The term "culture industry" dates to the Golden Age of Hollywood, when the studio system was in its heyday and movies had become a staple of American life. By the end of the 1930s, nearly seven in ten in the U.S. were going to the cinema on a weekly basis, where they were thrilled by lavish productions from Paramount, Warner Brothers, and Metro-Goldwyn-Meyer. Movie stars like Cary Grant and Katharine Hepburn were household names, as a rising publicity engine promoted their glamour, prestige, and wealth. Not everyone was taken in by all of this, of course—especially not the small cadre of industry dissidents in Los Angeles during the World War II era. Among these were European social critics Max Horkheimer and Theodor Adorno, who had left the continent as fascism was reaching horrific proportions. Associated with the Frankfurt School of Austria's Goethe University, Horkheimer and Adorno rubbed shoulders with Hollywood leftists like Charlie Chaplin and Fritz Lang—while writing their now-famous essay, "The Culture Industry: The Enlightenment as Mass Deception," published in 1944.[15]

Horkheimer and Adorno likened American movies, magazines, and radio to the propagandistic apparatus of the Third Reich in their ability to shape public opinion and influence personal beliefs. But rather than promoting fascism, the U.S culture industry was an instrument of capitalist ideology—pure and simple—in both structure and messaging. Horkheimer and Adorno's culture industry arguments partly derived from earlier thinkers, notably Karl Marx (alienation and commodity fetishism), Max Weber (bureaucratic rationalization), and Sigmund Freud (desire and drives). As in the Marxist labor/management dichotomy, "The Culture Industry" depicted producers and consumers as wholly separate bodies, with messages flowing one-directionally from Hollywood studios to their passive audiences. Movies were both economic objects and tools of mystification, their manufacture bureaucratically organized. The program's philosophical "mass deception" lay in advancing Enlightenment "reason," while reserving its freedoms and benefits for moneyed elites.[16] Following principles of rationalization, the culture industry maintained a clear goal orientation in both its own purposes and the stories it told. Everything operated in the interest of profit, advantage, or symbolic representations thereof. Notably, Horkheimer and Adorno built psychoanalytic theory into their concept of "culture"—recognizing consumers as desiring subjects, continually seeking pleasure in entertainment or sublimating their drives by working. Movies may have seemed to make audiences happy, yet they always left them hungry for more—their libidinous appetites but temporarily sated.

Influencing Horkheimer and Adorno was the work of fellow Frankfurt School scholar Walter Benjamin, who also had quibbles with Enlightenment "reason." Ironically, it was the invention of photography that began the story. Seen as a quintessentially "modern" phenomenon, when the first photographs came along in the mid-nineteenth century, their machine-made quality made them seem objective and "scientific" by nature. It wasn't until photojournalists began staging battlefield scenes during the Civil War that the "truth" of such images came into question. Observing photography's manipulative aspects in the 1930s, Benjamin drew a distinction between the new medium and such conventional artistic forms as painting and sculpture. In his "The Work of Art in the Age of Mechanical Reproduction," Benjamin argued that original artworks required viewers to be present (in a museum or gallery) and focused in thoughtful "contemplation" of the artwork's "aura."[17] In contrast, photographs (and especially films) overwhelmed viewers with a presumed reality that, in fact, had been cropped, enlarged, or otherwise doctored. This made the painting a source of wisdom, and the photograph a set-up for confusion. To Benjamin, the ubiquity of mechanical images signaled nothing less than a wholescale perceptual shift brought on by the modern era.

In structural terms, Horkheimer and Adorno's "The Culture Industry" saw Hollywood as a monopolistic stronghold, cranking out a standardized line of goods. For the first time in American history, the entire nation was addressed from a centralized source. Iconic movie moguls such as Louis B. Mayer, Jack Warner, and Darryl Zanuck promoted a pantheon of larger-than-life "stars" at the top of the bureaucratic hierarchy—their stature making viewers feel insignificant in comparison. Every aspect of production was meticulously monitored and controlled in the interest of quality standards and consistent messaging. Any "creative" impulses among workers had to conform to the narrow metrics of guaranteed ticket sales. All of this had a commercializing effect on American leisure. As movies became the nation's favorite pastime, people found themselves paying money to enjoy what previously had been "free time." There is no overstating the significance of this monetization of everyday life. "Amusement has become an extension of labor under late capitalism," Horkheimer and Adorno wrote in reference to both the viewing activity and its effect on consciousness.[18] Movies distracted people from their day-to-day lives, allowing them to forget their jobs, problems, or inequities while entering an addictive dream world. But the escape was always temporary, as audiences eventually returned to the less-than-perfect reality of their own existence.

Films also restructured the human imagination. Romantic fantasies always came true, as virtue triumphed over evil. Horkheimer and Adorno explained that:

> All films have become similar in their basic form. They are shaped to reflect facts of reality as closely as possible. Even fantasy films, which claim to not reflect such reality, don't really live up to what they claim to be. No matter how unusual they strive to be, the endings are usually easy to predict because of the existence of prior films which follows the same schemas.[19]

While the studios did this for practical reasons, audiences were powerfully drawn to the new forms of wish-fulfillment—which also supported the goal-oriented imperatives of capitalism. Linking success and happiness to work and consuming, the system offered "role models for people to turn themselves into what the system needs."[20] Real life was not like movies, but the powerful culture industry suppressed this "truth," while diminishing the possibility for a genuine "mass culture." Horkheimer and Adorno saw "authentic" popular culture as an artist enterprise—autonomously driven, critically informed, and capable of dissent. They embraced Kantian views of aesthetics as a special form of perception, capable of depicting the world from a distanced perspective. All of this was negated by the monoculture of the culture industry, an apparatus so pernicious that it was able to neutralize alternative thinking. "What resists can only survive by fitting in," they wrote. "Once its particular brand of deviation from the norm has been noted by the industry, it belongs to it as does the land-reformer to capitalism."[21]

In the decades since "The Culture Industry" first appeared, its premises would be modified and critiqued. Contemporary theorists still find utility in the essay's analysis of cultural commodification and capitalist rationalization, although many see its opposition of producers to audiences as simplistic and overly pessimistic. Adorno himself responded to some of this in his 1969 "Culture Industry Reconsidered," wherein he ceded to audiences a degree of self-consciousness missing in the original argument. "People are not only, as they say, falling for the swindle," Adorno wrote. "They desire a deception which is transparent to them." Recognizing the illusionistic artifice and "knowing fully the purpose for which it is manufactured," viewers nevertheless "force their eyes shut and voice approval, in a kind of self-loathing, for what is meted out to them."[22]

Vast Wastelands

In the 1950s, television dramatically raised the stakes in culture industry debates, as the new medium overtook cinema as America's prime source of entertainment. Movie attendance fell and TV ascended in a textbook example of creative destruction, as discussed in Chapter 3. While broadcast technology inherently accounted for much of this, the new medium also lent itself to novel forms of audience analysis and market forecasting. Viewer preferences would be meticulously "metered" and replicated in nearly identical offerings—privileging preferences of "average" viewers. This only furthered the famous normativity of the 1950s. An outgrowth of the radio business, early TV programming was delivered by local broadcast stations owned by the ABC, CBS, and NBC networks. Between 1950 and 1960, the number of stations rose from 98 to 440 as TV set ownership skyrocketed from 6 million to 60 million.[23] Meanwhile, movie theaters saw their business drop by 50 percent.[24] Because the first home TV receivers were expensive, television initially catered to affluent tastes, with programming limited to "serious" dramatic anthologies (*Kraft Television Theatre, Playhouse 90*, etc.), as well as news and children's shows. But as viewership mushroomed in the 1960s, sponsors pushed for more accessible fare (like game shows and comedies) to capture consumer attention. Non-commercial TV didn't come along until the passage of the 1967 Public Broadcasting Act.

Far more than movies, television brought the nation together as a single audience, with the average American watching three hours of TV daily by 1960. For a post–World War II America with rapidly growing suburbs, TV provided a safe and alluring form of stay-at-home entertainment. The new medium also gave business a pipeline into every household. Producers quickly abandoned any illusions of "quality" as they competed for advertising revenues for this new national market. The result was very much like the assembly-line culture Horkheimer and Adorno had lamented. This was aided by data collection technologies never available to the film industry. First used by radio companies in the 1940s, Nielsen Media Research initially began collecting television viewing statistics through customer diaries and later by monitoring devices ("set meters") installed in viewers' homes.[25] Before the age of home recordings or digital tracking, the "Nielsen Ratings" afforded media producers unprecedented accuracy in discerning the type of programs advertisers likely would sponsor.

The U.S. Census of 1950 showed a nation that was 89.5 percent white and 61 percent married, with 90 percent living as a family with a male head of household.[26] Men also held jobs at twice the rate of women. The Baby Boom had pushed the average age downward to 29. Any form of commerce with national aspirations catered to these demographic norms, often ignoring groups seen as superfluous to maximum profit. This explains the ethnocentric, patriarchal, and heteronormative tone of 1950s culture, as well as its omnipresent materialism and militarism. Nowadays, psychologists use the term "heuristics" to describe how majority "common sense" relies on shorthand rules-of-thumb to maximize communication. This is where stereotyping comes in—when real or imagined characteristics of groups stand in for more complicated ones. Such quickly grasped forms of mental shorthand were (and remain) vitally important in the fast-paced format of 30-minute television, not to mention 30-second TV advertising. For this reason, early programming was awash with clichéd depictions of women and men, families, racial and ethnic groups, "real Americans" and foreigners.

Such tendencies were on full display in top-rated shows of the 1950s such as *Bonanza, Gunsmoke, Wagon Train, The Andy Griffith Show*, and *The Beverly Hillbillies*. The formulaic commonality among the programs is hardly incidental. While cowboys and rural life had disappeared from mainstream America, viewers still yearned for myths of heroism and self-determination in the "common man." As the U.S. was facing student rebellions and a turbulent world abroad, nostalgic tales of justice and virtue lent a form of relief, even if the villain was a hapless millionaire banker. Also popular were assorted "family" dramas and comedies including *The Donna Reed Show, I Love Lucy, Leave it to Beaver*, and *The Adventures of Ozzie and Harriet*. As described by Joseph McFadden, these latter programs

> portrayed the conservative values of an idealized American life. Studiously avoiding sticky social issues like racial discrimination and civil rights, the shows focused mostly on White middle-class families with

> traditional stay-at-home moms and implied that most domestic problems could be solved within a 30-minute time slot, always ending with a strong moral lesson.[27]

Gender stereotypes were common and often overtly sexualized. Even as advertisers measured average behaviors, advertising tended to promote extremes of masculinity and femininity. Rugged men with chiseled features appeared with impossibly proportioned women having flawless skin. This generally made people feel badly about their looks—and willing to buy things to improve them. Like the rest of the employment market, men held most entertainment industry jobs—especially in leadership roles. Men ran the companies, wrote the stories, and directed them. Women appeared on screen, of course—mostly as wives, girlfriends, or temptations. Minorities of just about any kind were largely absent from 1950s television and movies, except when portrayed as servants or enemies. African-Americans, Chicanx-Latinx Americans, and Asian-Americans appeared in such roles as the servile Rochester on *The Jack Benny Show*, the bumbling Sergeant Garcia on *Zorro*, and the inarticulate Hop Sing on Bonanza. Native Americans were brutalized *en masse* on TV western shows, as were Germans and Japanese in World War II dramas and comedies. Homosexuality remained a taboo, although occasionally referenced by innuendo or close friendship. Poverty often was criminalized, as were mental illness and addiction. Disability was rarely seen at all.

A backlash arose in the 1960s, which sent the TV industry reeling. In 1961, Federal Communications Commission (FCC) director Newton Minnow gave his "Vast Wasteland" speech to the National Association of Broadcasters, in which he described

> a procession of game shows, violence, audience participation shows, formula comedies about totally unbelievable families, blood and thunder, mayhem, violence, sadism, murder, western bad men, western good men, private eyes, gangsters, more violence and cartoons. And, endlessly, commercials—many screaming, cajoling, and offending. And, most of all, boredom.[28]

Minnow famously would conclude: "Keep your eyes glued to that set until the station signs off. I can assure you that you will observe a vast wasteland."[29] Initially writing off Minnow's comments as snobbish elitism, the TV industry would soon find itself needing to adapt. Ironically perhaps, the homogenizing medium of television—and all its profit motives—had to follow the concerns of a changing society.

Partly owing to a rising population of young adults, the "counterculture" of the period also empowered a wide swath of those excluded or otherwise disenfranchised from so-called "mainstream" America. The presidencies of John F. Kennedy and Lyndon Johnson set the U.S. on a liberal course for the entire decade. Landmark pieces of legislation included the Equal Pay Act (1963), Equal Opportunity Act (1964), Civil Rights Act (1964), Voting Rights Act (1965), Immigration and Naturalization Act (1965), and Higher Education Act (1965), as well as the establishment of the Peace Corps and the National Endowments of the Arts and Humanities. With some reluctance, Johnson continued U.S. involvement in the Vietnam War, which grew increasingly unpopular as the decade progressed. Televised body counts and gruesome imagery from the Vietnam War played a vital role in turning the American public against the President by 1968, when he declined to seek re-election. Much of the resistance was driven by the draft-eligible Baby Boom generation—which also made up the largest age cohort in the U.S. population.

This growing youth market drove 1970s television and advertising. As viewership for westerns declined, television reached for young adult audiences with *Rowan and Martin's Laugh-In*, (1968–1973), *All in the Family* (1971–1979), and *Saturday Night Live* (1975–present). Supporting TV's new initiatives, advertisers began to shift their focus in what would become a two-sided formula, according to *Advertising Age*. "On one hand, 'Youth Market' covered the marketing of products to children, teen-agers and young adults." But also the concept "meant the selling of 'youthfulness' as a product attribute and referred to the ability of youngsters to influence family or household purchases."[30] Campaigns like "The Pepsi Generation" and Noxzema's "Cover Girl" appeared, playing into rising prejudice against anything linked to maturity or the aging process. This shift hit women especially hard in a culture already idealizing youthful vitality, driving what would become multi-billion-dollar industries for cosmetics, apparel, weight-loss, surgery, and fitness products. While disparaging anyone with wrinkles or gray hair, it also fed fantasies about girlish innocence and dependency. In *The Beauty Myth*, sociologist Naomi Wolf described this as a backlash against women who had "taken control over their own bodies" in the 1960s, but later faced a new form of tyranny as "youthful beauty" became a new form of internalized control.[31] Age discrimination became commonplace in entertainment industry casting, where studies showed careers of actresses declining after age 34 (for men it was 51).[32]

Before long, critics would see a broader intellectual downside to the era's infatuation with youth. As spontaneity and "playfulness" became familiar media motifs, grown-up values of reflection and caution seemed to be suffering, as discussed in Chapter 4. To media scholar Neil Postman, this signaled a rejection of such "adult" characteristics as "the capacity for self-restraint, tolerance for delayed gratification, a sophisticated ability to think conceptually and sequentially, preoccupation with both historical continuity and the future, high valuation of reason and hierarchical order."[33] Initially targeting young audiences and consumers, the industries soon found youthful values complementing other agendas. Television shows had always relied on simple narratives, with ratings data showing that flashy—or even shocking—stories captured attention best. All of this contributed to what Benjamin Barber would term an "infantilization" of the American public. Whether watching movies based on comic books or impulsively shopping at malls, consumers found themselves immersed in three archetypal dualisms: "Easy-over-Hard, Simple-over-Complex, and Fast-over-Slow."[34] As Barber would write, "Affiliated with an ideology of privatization, the marketing of brands, and a homogenization of taste, this ethos of infantilization worked to sustain consumer capitalism, but at the expense of both civility and civilization and a growing risk to capitalism itself."[35]

Questions from the Audience

While the Frankfurt School's culture industry premise made a lot of sense, it didn't give audiences much credit for independent thought. The concept also treated them as an undifferentiated mass without internal variation. Beginning in the 1960s and 1970s, new understandings of mass communication began to modulate "negative" culture industry and "vast wasteland" views, updating their totalizing assessments with more open-ended thinking. Partly this came with rising attention to "culture"—broadly defined as shared patterns of belief, communication, and behavior—and the emergence of media and cultural studies as fields of study. At the same time, shifts were taking place across academic disciplines, which recognized ways that knowledge and even "truth" itself were shaped by context and social interaction. In their 1966 book *The Social Construction of Reality*, Peter Berger and Thomas Luckman contrasted narrow categories of "scientific and theoretical knowledge" with broader "customs, common interpretations, shared routines, habitualizations, the who-is-who and who-does-what in social processes and the division of labor."[36] To Berger and Luckman, these forms of constructed knowledge often traveled through institutions like the media industry.

Adding yet more nuance to the culture industries discussion, Louis Althusser would write in the late 1960s of capitalism operating as a sophisticated system that worked not by coercion but by seductively "calling out" to individuals with promises of a better life.[37] Responding to the elitist implications of culture industry theories, Hans Magnus Enzensberger took Marx's theories of mystification in a different direction. In the 1970s, Enzensberger asserted that capitalism didn't exactly deceive the masses into "false" desires (for products or services), but instead found ways of satisfying "real" desires (for happiness or security, for example) via substitution. Pepsi ads didn't so much convey anything about the drink, per se, as much as they depicted people having fun. Later culture industry theory would adopt these ideas in describing citizens "socialized" into roles of subservience, their yearnings for engagement or fulfillment transformed into consumer desires.[38]

Think about this in relation to today's creativity craze, especially the booming self-help "finding your inner artist" market and the much-hyped creative industries. As in earlier eras of economic hardship and political turbulence, Americans again are yearning for ways to feel better about themselves. It might be true that some forms of creativity actually can help with this. At the same time, the rabid commercialism of creative activities often reduces them to money-making schemes. To combat these regressive tendencies, would-be creatives need to be smart about how and why they make choices. Such critical capacities exist in everyone, although they often get little encouragement and support.

Attention to such critical thinking came as literary theorists began to re-examine the process of reading—and by extension, the varied ways symbolic messages are transmitted and received. Broadly stated, this meant shifting attention from "texts" to audiences. Previous thinking had defined reading as a simple retrieval of messages from written works—beliefs dating to times when elite minorities of "writers" penned works for less educated masses of "readers." The new view saw ordinary readers as attaching their own meanings to texts—much as they did when interpreting newspapers or the Bible. One key work in the emerging "reader-response" theory was Stanley Fish's 1967 *Surprised by Sin: The Role of the Reader in* Paradise Lost.[39] Fish wrote that "meaning is an event, something that happens not on the page, where we are accustomed to look for it, but in the interaction between the flow of print (or sound) and the active mediation of the reader-hearer."[40] Fish said that readers often expressed their own interests in whatever meanings they found—in effect, discovering what they were already looking for.

The specific term "reader response" was coined by Wolfgang Iser in his book *The Act of Reading* (1976). Influenced by phenomenology, Iser argued that ideas from books make sense when readers compare them to what they already know. And because such existing base-knowledge differs from reader to reader, experiences of texts likewise will vary. Iser wrote, "The reader's role is defined by the vantage point by which he joins the written work, and the meeting place they converge. All, as component parts, operate to shape the reader's role as found within the text."[41] These ideas were readily embraced by scholars in many fields concerned with identity and "difference" during the 1970s. Soon Elaine Showalter and Edward Said, among others, would extend Iser's reader-response theory in describing how women and non-Westerners read contrary meanings into works.[42] Later in the decade, Michel de Certeau framed the act of "everyday" reading in defiant terms: "Such readers are travelers," de Certeau wrote, explaining, "They move across lands belonging to someone else, like nomads poaching across fields they didn't write, despoiling the wealth of Egypt to enjoy it themselves."[43]

Complementing reader-response theory were evolving theories of language, notably the structuralism of Ferdinand de Saussure. Asserting that words were more than mere lexicological labels, de Saussure wrote in his *Course in General Linguistics* that such singular "signs" evoked variable meanings, which were influenced by prior knowledge, context, and other aspects of "collective social interaction."[44] For example, even a simple word like "dog" might conjure up different images in the minds of listeners, depending on their past experiences or current thinking. Such mental images similarly attached to different kinds of associations or memories. This new understanding of language gave more substance to theories of social construction, especially as some forms of meaning came to override others. It also gave credence to the variability among interpretations—further undermining the authority of writerly intentions.

In the 1970s and 1980s, poststructuralism took matters of variability further, with Roland Barthes and Jacques Derrida questioning both the stability of established meanings and the capacity of the subject to know itself. Looking beyond literary works, Barthes famously applied these methods in "reading" to everything from professional wrestling to car advertising. In his book *Mythologies*, Barthes showed how prevailing interpretations of media messages can covey political ideology.[45] He said that as dominant meanings were repeated over and over in public discourse, they had a conditioning effect that people often didn't notice. The effect was to "naturalize" certain ways of seeing the world. As Barthes later put it, "The very principle of myth transforms history into nature."[46] Here again, a connection to the new creativity can be made. Much of this book has shown the fallacies of singular or universal views of the creative process. There is no single way of defining "art" or "music." Evidence abounds of how societies around the world have seen creativity in endlessly variable forms, just as each person experiences them differently. Narrowly defining creativity in commercial or utilitarian terms indeed is a function of myth, as Barthes described.

Delving deeper into the reader's mind, the linguistically informed psychoanalysis of Jacques Lacan explored the unconscious, describing the subject itself as an effect of language. Lacan also brought a visual emphasis to aspects of his work, which film theorists of the era would draw upon. In 1973, Laura Mulvey applied Lacanian principles in formulating the desiring "male gaze" through which viewers read themselves into movies and television programs.[47] Mulvey pointed out that TV and movies tended to arrange female characters within the frame to be viewed in certain ways. More than men, women would be shown facing the camera, their appearances often highlighted in various ways. Female characters would be watched by male actors as the scenes unfolded, just as camera operators (invariably male as well) would film them. All of this implicitly invited viewers to join in the objectifying cinematic gaze.

Lacanian theory also examined people's longstanding fondness for stories and tendencies to identify with characters. Western culture historically had favored stories with "resolved" endings, as Aristotle originally discussed. Lacan would note that youngsters developed a fondness for narrative at an early age, with the step-by-step nature of stories generating curiosity and expectation. He linked this desire for narrative closure to primal anxieties and unconscious desires for closeness and security—manifest as viewers imagine themselves in stories while hoping for positive outcomes. To film scholar Teresa de Lauretis, the gaze phenomena of photography, film, and television gave media more potency than written or spoken texts. Somewhat like Benjamin, de Lauretis argued that camera-based stories captured viewers' attention in powerful ways, ensnaring them in a "narrative identification that assures the 'hold of the image,' the anchoring of the subject in the flow of the film's movement."[48]

Later psychoanalytic theory would see this hold as negotiable, with direct identification accepted, altered, rejected, or even ignored. Scholars would revisit the male gaze as well, recognizing that women often saw works differently from men and derived their own meanings in what would be termed the "female gaze." Postcolonial and queer gazes would follow in a similar vein. Before long, poststructuralist theorists added that gendered interpretations of texts were not necessarily fixed, nor were any specific readings guaranteed in advance. Viewers might very well be taken in by a dominant way of seeing things, but they could just as likely choose to see a film or TV show from the "wrong" position. In other words, one's enjoyment of a work need not be tied to one's anatomy, gender, age, social class, ethnicity, nationality, or sexual orientation.

By the 1980s, the new field of "Visual Culture" would take gaze theory even further, as scholars linked it to problems with Enlightenment reason like those that troubled Horkheimer and Adorno. Historians noted that the European fascination with optical rendering predated the invention of the camera, and could be traced to longstanding obsessions with realism in drawing and painting.[49] In the eighteenth and nineteenth centuries, artists and scientists had devised all sorts of devices (telescope, camera obscura, diorama, "magic lantern," etc.) in the pursuit of "realistic" rendering. This encouraged philosophical views of the external world (and experiences thereof) as something that could be studied, organized, differentiated—and, most of all, controlled. Optical single-point perspective focused seeing on the *individual* doing the looking, further reinforcing notions of mastery. Before long, the Western gaze was linked to power of many kinds: the imperialist's gaze toward subjugated populations, the plantation owner's gaze toward servants, the police officer's gaze toward suspects, and the voyeuristic gaze of men toward women. Even among "free" people, the gaze was not held equally, as the poor or indentured were expected to "lower their gaze" in the presence of gentry.

Enlightenment visuality also supported capitalism, of course. This linkage of vision to ownership already had been evident in paintings of land, estates, and the décor of wealthy properties. Also, "seeing" had been equated with forms of possession dating to published engravings or other images of items for sale. This derived from the display of goods in marketplaces or shop windows, which initially served as a primitive form of advertising. Well before the invention of photography, the visual commodification of desire was attached to the mere "sight" of goods. Film historian Anne Friedberg wrote of the broad-based intensification of spectatorship in nineteenth-century society, reinforced in commercial displays and architecture, the rise of printing and visual imagery, and even the layout of urban streets and avenues—all generating what she termed a "frenzy of the visible."[50] This set in place a powerful paradigm linking vision to many values: success, wealth, and togetherness. In general terms, theories of visuality suggested that images and moving pictures (especially when commercialized) conformed to viewers' desires and ways of getting them satisfied—for better or worse. Contemporary scholars in many disciplines would examine this history: deconstructing ideologies within media texts, exposing their programs of economic exploitation (Marxism), patriarchy (feminism), elitism (cultural studies), binary reason (deconstruction), imperialism (postcolonialism), and heteronormativity (queer theory).

Culture Wars: Old and New

Media always have lured audiences with sex and violence—two ubiquitous human impulses frowned upon by polite society, but offering endless fascination at the same time. In his famous theory of "repressed" drives, Freud argued that social conventions oblige people to behave properly and to push their primitive instincts aside.[51] But erotic and aggressive drives nevertheless keep percolating in the unconscious, often popping up unexpectedly in jokes and innuendo. While Freud's theory of a generally regulated ("sublimated") society seemed an adequate explanation for early twentieth-century bourgeois piety, Michel Foucault would argue against this "repressive hypothesis" in accounting for ongoing demand for sex in media.[52] In the 1970s, Foucault wrote that modernity's dogmatic puritanism pushed sexual thinking into all manner of talk, writing, and imagery. In keeping with his other writings on social control, Foucault further saw the "official" discourse on sex as a means of policing people's desires while directing them in normative ways. It wasn't only that sex-talk was proscribed, but also that certain attitudes to sex were specified: heterosexuality, traditional marriage, the objectification of women, etc.

Unsurprisingly, pornography featured prominently in culture industry debates. Contrasting longstanding feminist views of pornography as the quintessential expression of male power, the 1980s saw new "sex-positive" views in which women exercised agency in sex and its representations.[53] Asserting that voyeurism didn't have a fixed identity, sex-positive theory argued that anyone might be drawn to porn—although not necessarily that *same* porn. Queer theory in the 1990s would further critique notions of unified spectatorship, pointing out that gay and lesbian viewers sometimes chose to read heterosexual narratives against the grain. This also was a time when independent filmmakers and artists were actively contesting mainstream media on many levels. Scores of regional and national organizations emerged such as Women-Make-Movies and the Frameline Festival to support independent production, along with groups devoted to media by African-American, Chicanx/Latinx, and Asian-American producers.

Then there is media violence. Throughout human history, violence has been a narrative driver in historical, fictional, and religious literature—just as it serves to capture attention in contemporary media. But how this works is not always as obvious as one might think. Many people see fighting as a "natural" (albeit unfortunate) part of human behavior, owing to animalistic survival instincts of the kind described by Charles Darwin. This long-disproven view portrays people inevitably caught in competitive struggles that require aggressive behavior, escalating to violence when other means fail.[54] Men especially are sanctioned to fight in this way, encouraged in a culture where depictions of violent masculinity, armed combat, and military conflict remain staples of popular entertainment, not to mention daily news coverage. But there is disagreement over how much media depictions actually *cause* violent behavior. While the mental health field sees a strong conditioning effect behind media violence, communications scholars note that most consumers of such images never act violently. This has led to a more careful approach to media violence, viewing it as a "risk" (like drinking beer) to which some people are more vulnerable than others. Also, not all media violence works the same way. Movies like *Fate of the Furious*, for example, almost never show negative consequences of crashes or fights—letting viewers focus instead on the visceral excitement of full-screen explosions with booming sound effects.

But this hasn't diminished common-sense beliefs that media cause aggressive behavior. Year after year, attempts to prove the "effects" of movie and TV violence continue, invariably yielding few clear results, even among youngsters.[55] Similar beliefs that computer games provoke aggression also remain unsubstantiated. This is *not to say* that violent media are harmless. Nothing that pervades contemporary society so completely can leave viewers unaffected. Communications analysts have long noted how news in particular distorts people's view of their surroundings, as it continually reports on murder and mayhem at rates far exceeding statistical reality.[56] For example, parents remain alarmed over the "epidemic" of violent school attacks when less than a half dozen such incidents have occurred each year for two decades.[57] Such disparities come from what communications expert George Gerbner termed the "mean world syndrome"—appearing in viewers constantly exposed to images of threat and harm.[58] To Gerbner, the real danger of media violence lies in the fearful mindset it instills by distorting views of certain kinds of people, groups, and nations—as well as the ways it distracts attention from more prevalent forms of harm such as poverty, illness, or environmental pollution.

Despite these controversies, for most of its history the U.S. government has taken a hands-off approach to sex and violence depicted in publishing, movies, and television. Free-market attitudes have cast media consumption as a matter of personal choice, while constitutional free-speech provisions also have given media producers broad legal rights. The Hollywood entertainment industry shrewdly was able to head off pious legislators in implementing its own Motion Picture Production Code (in effect from 1930 to 1968) prohibiting offensive content, and later developing movie and TV ratings systems (the familiar, G, PG,

PG-13, R, and NC-17) that put responsibility in the hands of viewers.[59] A few famous court actions were brought against movie producers, art museums, and video-game makers—nearly all of which would be defeated on First Amendment grounds. In 1973, the U.S. Supreme Court decided in favor of a mail-order pornography dealer in its landmark *Miller v. California* decision.[60] The case redefined obscenity as "lacking serious literary, artistic, political, or scientific value"—criteria so broad that clever defense lawyers invariably have prevailed.

Then came the "Culture Wars." In the 1980s, President Ronald Reagan was swept into office on a campaign of Christian values, conservative economics, and get-tough social policies. The actor-turned-politician won over hardline conservatives and moderate "Reagan Democrats" with his amiable speaking style and common-sense approach to public policy. Perhaps more than any prior President, Reagan understood the power of media and culture, using his office as a massive public relations engine. Catering to the "Moral Majority," Reagan assailed liberal education policies, as he launched "wars" against abortion, drugs, communism, and pornography. First Lady Nancy Reagan promoted abstinence to America's youth in her infamous "Just Say No" campaign, as Republican legislators looked for ways to get on the Reagan bandwagon, often by tracking down examples of liberal excess or perceived wrongdoing.

In the late 1980s, conservatives found a perfect target in the National Endowment for the Arts (NEA), a taxpayer-supported agency awarding small grants to museums, schools, and individuals.[61] Controversies first erupted from an NEA-sponsored showing of a 1987 image by artist Andres Serrano of a crucifix submerged in orange fluid. Entitled "Piss Christ," the work was excoriated as a "dishonor to the Lord" by then-campaigning North Carolina Senator Jesse Helms.[62] Soon legislators began combing through records of the NEA's 100,000 grants. They only found four "morally repugnant" projects, but exploited them to maximum potential—bringing copies of Robert Mapplethorpe's homoerotic photography and "Dread" Scott Tyler's desecration of the American flag to the entire Congress. Bowing to political pressure, the NEA began cancelling grants in a clear pattern of discrimination against feminists, LGBT artists, and people of color. This overall chilling effect led to diminished funding for the agency, from which it never fully recovered.

But hardliners pressed onward with messianic zeal, with Reagan advisor Patrick Buchanan writing his widely circulated editorial, "In the War for America's Culture, Is the 'Right' Side Losing?"[63] As Buchanan put it, "There is a religious war going on in this country. It's a culture war, as critical to the kind of nation we shall be as was the Cold War itself, for this is a war for the soul of America." As the expression caught on, the "culture war" would become a catchphrase for rising polarization over a broad swath of issues: artistic expression, media censorship, education reform, gun control, homosexuality, immigration, privacy, and religious freedom. Sociologist James Davison Hunter's 1991 *Culture Wars: The Struggle to Define America* was one of many books on the subject to appear at the time.[64] Hunter saw the conflict emerging from opposing philosophies he defined as "orthodox" (traditional, idealist, and universal) and "progressive" (contemporary, realist, and diverse). This ongoing conflict between orthodox and progressive views will inform much of the debate over singular versus plural forms of creativity, discussed in the next chapter.

The culture wars retreated to the political fringes following the inauguration of moderate Democrat Bill Clinton. But they came back with a vengeance in Clinton's 1998 impeachment over his affair with White House intern Monica Lewinsky. Journalism scholars later would observe the media firestorm coinciding perfectly with rise of cable news and its relentless search for sensation, shock, and scandal. With many others, reporter Telly Davidson would conclude in his book *Culture War* that "Monicagate" and impeachment were a crescendo "to the entire decade of '90s culture warring."[65] In a later anthology, Laurent Berlant and Lisa A. Duggan would see deeper patterns in the Clinton debacle:

> What is this "scandal" a case of? The case is not only about persons, the presidency, celebrity culture, or law. It is also a moment of stunning confusion in norms of sexuality; of fantasies of national intimacy—what constitutes "ordinary sex" and "ordinary marriage," let alone the relation between law and morality, law and justice.[66]

The culture wars were intensified by 9/11 in both new and old ways, as President George W. Bush escalated his rhetoric into a culturally loaded set of oppositions: good/evil, West/East, reason/unreason, and Christian/Muslim. Lacking credible intelligence, Bush military leaders hastily targeted Iraq as the source of the attacks, igniting a wave of anti-Muslim hate crimes on U.S. soil. In the heightened atmosphere of worry and impending war, anyone disagreeing with Bush's view quickly became labeled unpatriotic or subversive. Fearing for the U.S. economy, the President repeatedly urged the nation to "go shopping more" and maintain normal affairs: "Get down to Disney World. Take your families and enjoy life," he intoned.[67] Fox News soon launched an ongoing "Culture Wars" segment hosted by Bill O'Reilly, which would express conservative outrage over reproductive freedom, "homosexual activism," and other aspects of a rising secularism.[68] Hundreds of articles, periodicals, and books would appear on the resurgent culture wars phenomenon, including *Culture Wars: A Monthly Magazine* and a 1,200-page *Culture Wars: An Encyclopedia of Issues, Viewpoints, and Voices.* [69]

Of course, no one could have predicted the election of Donald Trump and the new culture war he would launch. *National Review* editor Rich Lowery has written recently, "The nation's foremost cultural warrior is President Trump."[70] But Lowery saw a new twist in the current conflagration. "There is no way Trump could be a credible combatant in the culture war as it existed for the past 40 years," he wrote, adding that Trump "reoriented the main lines of battle away from issues related to religion and sexual morality." Instead, the nation found in Trump an aggressive economization of culture in the name of business pragmatism, as discussed in Chapter 3. Previous hot-button culture war issues on both domestic and international fronts have been rendered as fiscal conflicts—for example, that taxes cripple growth, immigrants steal jobs, social programs waste money, environmentalism burdens business, and trade accords are giveaways.

Without doubt, the culture wars and culture industries now have reached a point of convergence—although this is not without precedent. "The Frankfurt School Knew Trump Was Coming," was the headline from a recent issue of *The New Yorker* discussing the prescience of the culture industry's early critics.[71] Horkheimer and Adorno had witnessed first-hand the tactics of European fascists, and they knew how effective media propaganda could be in igniting populist fervor. Anxious populations are easily swayed, as Horkheimer and Adorno would discuss, noting the dangers when reality and fiction blurred. "Lies have long legs," they wrote, especially in the "conversion of all questions of truth into questions of power" as leaders "attack the very heart of the distinction between true and false."[72] In comments that could apply to Trump, Adorno would later observe:

> Today there is a growing resemblance between the business mentality and sober critical judgment, between vulgar materialism and the other kind, so that it is at times difficult properly to distinguish subject and object. To identify culture solely with lies is more fateful than ever, now that the former is really becoming totally absorbed by the latter, and eagerly invites such identification in order to compromise every opposing thought.[73]

Given such historical resonance, it's no surprise that the decades-old concept of the culture industries now is seeing a revival. While reader-response theory gave audiences more credit than Frankfurt School scholars had done, the workings of discursive power in media remain undiminished. Overt political propaganda and popular franchises like *Fast and Furious* continue to exert a pull on audiences, while more subtle forms of ideological persuasion embed themselves in such commonplace activities as one-click shopping and social networking.

This chapter has looked at the culture industry as it was first seen in monolithic terms and later as a dialogical phenomenon. Both views merit ongoing attention in understanding the operations of media and ideology. Many forces operate simultaneously in individuals, groups, and societies—with deep (and often unconscious) psychological tendencies intersecting with increasingly subtle messages and more vaporous institutional forms. Difference, desire, and power are a few of the constants in this unending conversation. Another is the broad-based atrophy of creative thinking within a population continually bombarded by messages from advertisers, news outlets, and the entertainment industry. The problem isn't simply that certain ideas or products are being exaggerated or mischaracterized. After all, any middle-schooler will tell you that TV and the internet can't be relied upon as sources of the truth. The bigger issue lies in the overwhelming volume of information presented, the disorientation and confusion this causes, and the resentment people come to feel about the onslaught itself. In an age in which even a child suspects the veracity of media content, is it any wonder that adults are expressing feelings of unnameable resentment and anxiety?

Of course, this line of argument doesn't sit well with apologists for the market economy and the media infrastructure on which it increasingly relies. In keeping with mainstream American beliefs in liberty and free expression, counterarguments invariably propose that consumer "choice" and citizen "rights" demand ever-expanding arrays of products and services. For a time, scholars in the field of cultural studies joined this chorus in asserting that people express their "symbolic creativity" by shopping, selecting music, getting dressed in the morning, and deciding what games to play on their iPhones.[74] And while there is no denying that choice and selection remain important aspects of creative behavior (and art-making itself), this counterargument often disregards the culture industry case altogether. The answer to this standoff is that both factors are always in play. Culture continues to shape a person's beliefs and preferences, even as the individual pushes back with a degree of autonomy. One might even say that creativity can arise from the tension between these two forces. But this presumes a degree of balance or fairness in the exchange, informed, perhaps, by a common awareness of the stakes in the game.

This chapter has explored the question of mass-produced entertainment, beginning with a look at Hollywood movies. While many complain about formula filmmaking and a seemingly endless array of remakes and sequels, it seems that audiences often prefer familiar stories. These conflicting realities highlight a longstanding debate about popular culture and everyday creativity. Decades ago, social critics argued that money-driven Hollywood "culture industries" were brainwashing audiences with capitalistic values. But later studies by linguists and media scholars argued that matters were not so simple, inasmuch as viewers interpreted what they saw in varied and often contradictory ways. These debates intensified as TV extended the reach of programming (and advertising) into living rooms everywhere. They have become even more confusing in a digital age when the line between illusion and reality seems even more of a guessing game.

What happens when the game becomes harder to see? How does this critical "choice" effectively operate when the choice itself becomes obscured or rendered invisible? Or, more bluntly, what happens when the very "creativity" of self-definition, leisure pursuits, work, and schooling becomes subservient to the marketplace? Critics of the new creativity argue that the broad-based anxiety and dissatisfaction pervading American culture is directly tied to this simultaneous advocacy and regulation of creative impulses. People become frustrated when told that survival in a sluggish economy is simply a matter of personal ingenuity, that the key to success lies in finding inventive solutions, and that life would improve if they only could become more "creative." Like beauty and love, the new creativity is held up as something that everyone wants but that nobody can have in sufficient measure. Is it any wonder that many Americans are feeling confused and upset? Discussion in the next chapter follows the trail of culture into the new "creative industries," which promise both continuities and departures from the culture industry debate.

Notes

1 Walter Lippmann, *Public Opinion* (New York: Macmillan, 1922) p. 29.
2 Maxwell McCombs, *Setting the Agenda: The Mass Media and Public Opinion* (2nd ed.) (London: Polity, 2014) p. 31.
3 Ben H. Bagdikian, *The Media Monopoly* (5th ed.) (New York: Beacon, 1997).
4 Farhad Manjoo, "Tech's Frightful Five: They've Got Us," *New York Times* (May 10, 2017) www.nytimes.com/2017/05/10/technology/techs-frightful-five-theyve-got-us.html (accessed May 10, 2018).
5 Manohla Dargis, "Review: Family Values and Hot Rides in 'Fate of the Furious,'" *New York Times* (Apr. 12, 2017) www.nytimes.com/2017/04/12/movies/fate-of-the-furious-review-fast-and-furious-8.html (accessed Jun. 6, 2019).
6 "Movie Franchises," The Numbers.com (2018) www.the-numbers.com/movies/franchises (accessed May 1, 2018).
7 Simon Brew, "139 Movie Sequels Currently in the Works," msn.com (Jun. 18, 2015) www.msn.com/en-gb/entertainment/movies/news/139-movie-sequels-currently-in-the-works/ar-AAbORny (accessed May 1, 2018).
8 Derek Thompson, "Why the Hollywood Sequel Machine is Stronger than Ever," *The Atlantic* (May 14, 2014) www.theatlantic.com/business/archive/2014/03/the-hollywood-sequel-machine-is-stronger-than-ever/359726 (accessed May 1, 2018); Derek Thompson, The Reason Why Hollywood Makes So Many Boring Superhero Movies, *The Atlantic* (May 13, 2014) www.theatlantic.com/entertainment/archive/2014/05/hollywoods-real-superhero-problem/370785 (accessed May 1, 2018).
9 "The Reason Why Hollywood Makes So Many Boring Superhero Movies."
10 Ibid.
11 Mikael, "Lessons Learned from The Fast and the Furious," returnofkings.com (2018) www.returnofkings.com/11423/lessons-learned-from-the-fast-and-the-furious (accessed May 1, 2018).
12 Andy Wright, "How the 'Fast and Furious' Set Designers Made SoCal the Focus," atlasobscura.com (Oct. 25, 2016) www.atlasobscura.com/articles/how-the-fast-and-furious-set-designers-made-socal-the-focus (accessed May 2, 2018).
13 Ibid.
14 Ibid.
15 Max Horkheimer and Theodor Adorno, "The Culture Industry: The Enlightenment as Mass Deception," in Gunzelin Schmid Noerr, ed., *Dialectic of Enlightenment: Philosophical Fragments* (1947), Edmund Jephcott, trans. (Stanford: Stanford University Press, 2002) pp. 94–136.
16 Theodor Adorno and Max Horkheimer, "Preface to the 1969 Edition," *Dialectic of Enlightenment: Philosophical Fragments*, p. xii.

17 Walter Benjamin, "The Work of Art in an Age of Mechanical Reproduction," in Hannah Arendt, ed., *Illuminations* (London: Fontana, 1968) pp. 214–8.
18 "The Culture Industry," p. 107.
19 "The Culture Industry," p. 110.
20 "The Culture Industry," p. 112.
21 "The Culture Industry," p. 106.
22 Theodor Adorno, "Culture Industry Reconsidered" (1969) *New German Critique* 6 (Fall 1975) www.icce.rug.nl/~soundscapes/DATABASES/SWA/Culture_industry_reconsidered.shtml (accessed Apr. 30, 2018).
23 "Television," *World Book Encyclopedia* (Chicago: World Book, 2003) p. 119.
24 "The American film Industry in the Early 1950s," encyclopedia.com (2017) www.encyclopedia.com/arts/culture-magazines/american-film-industry-early-1950s (accessed May 2, 2018);
25 "TV Ratings," Nielsen.com (2018) www.nielsen.com/us/en/solutions/measurement/television.html (accessed May 13, 2018).
26 "1950 Overview," U.S. Census Bureau (2018) www.census.gov/history/www/through_the_decades/overview/1950.html (accessed May 3, 2018).
27 Joseph McFadden, "The Relationship Between Television and Culture," in Understanding Media and culture (Minneapolis: University of Minnesota, 2010) http://open.lib.umn.edu/mediaandculture/chapter/9-2-the-relationship-between-television-and-culture (accessed May 3, 2018).
28 Newton Minnow, "The 'Vast Wasteland' of Television Speech" (1961) www.terramedia.co.uk/reference/documents/vast_wasteland.htm (accessed May 2, 2018).
29 Ibid.
30 "Youth Market," AdAge (Sep. 15, 2003) http://adage.com/article/adage-encyclopedia/youth-market/98946 (accessed May 2, 2018).
31 Naomi Wolf, *The Beauty Myth: How Images of Beauty Are Used Against* Women (New York: HarperCollins, 1991) p. 17.
32 Lisa Wade, "Female Movie Starts Peak at Age 34, but Men See Success till the End," SociologicalImages.com (Feb. 5, 2014) https://thesocietypages.org/socimages/2014/02/05/female-movie-stars-peak-at-age-34-but-men-see-success-till-the-end (accessed May 4, 2018).
33 Neil Postman, *The Disappearance of Childhood* (New York: Vintage, 1982) pp. 98–9.
34 Benjamin Barber, *Consumed: How Markets Corrupt Children, Infantilize Adults, and Swallow Citizens Whole* (London and New York: Norton, 2007) p. 83.
35 *Consumed*, pp. 3–4.
36 Peter L. Berger and Thomas Luckman, *The Social Construction of Reality: A Treatise on the Sociology of Knowledge* (New York: Anchor, 1966) p. 65.
37 Louis Althusser, *Lenin and Philosophy and Other Essays*, trans. Ben Brewster (New York: Monthly Review, 2001).
38 Henry A. Giroux, "Neoliberalism, Democracy and the University as a Public Sphere," *TruthOut* (Apr. 22, 2014) www.truth-out.org/opinion/item/23156-henry-a-giroux-neoliberalism-democracy-and-the-university-as-a-public-sphere (accessed Jul. 4, 2018).
39 Stanley Fish: *Surprised by Sin: The Role of the Reader in* Paradise Lost (Cambridge, MA; Harvard University Press, 1967).
40 *Surprised by Sin*, p. 4.
41 Wolfgang Iser, *The Act of Reading: A Theory of Aesthetic Response* (Baltimore: Johns Hopkins University Press, 1978) p. 36.

42 Elaine Showalter, *A Literature of Their Own* (Princeton: Princeton University Press, 1998); Edward Said, *The World, The Text, and The Critic* (Cambridge, MA: Harvard University Press, 1983).
43 Michel de Certeau, "Reading as Poaching," in *The Practice of Everyday Life* (Berkeley: University of California Press, 1984) p. 175.
44 Ferdinand de Saussure, *Course in General Linguistics* (1916) ed. Charles Bally and Albert Sechehaye, trans. Charles Bally (New York: Philosophy Library, 1959).
45 Roland Barthes, *Mythologies*, trans. Annette Lavers (Paris: Editions du Soleil, 1957) p. 115.
46 *Mythologies*, p. 129.
47 Laura Mulvey, "Visual Pleasure and Narrative Cinema," *Screen* 16, no. 2 (1975): 6–18.
48 Teresa de Lauretis, *Alice Doesn't: Feminism, Semiotics, Cinema* (Bloomington: Indiana University Press, 1984), p. 144.
49 Robert S. Nelson, "Descartes' Cow and Other Domestications of the Visual," in Robert S. Nelson, ed., *Visuality Before and Beyond the Renaissance* (Cambridge: Cambridge University Press, 2000) p. 4.
50 Anne Friedberg, "The Mobilized and Virtual Gaze in Modernity: Flâneur/Flâneuse," in Nicholas Mirzoeff, ed., *The Visual Culture Reader* (London and New York: Routledge, 1998) pp. 252–62.
51 Sigmund Freud, *The Ego and the Id* (1923) (New York: CreateSpace, 2010).
52 Michel Foucault, *The History of Sexuality: Volume 2: The Use of Pleasure*, trans. Robert Hurley (1984) (London: Penguin, 1992) pp. 15–36.
53 See Carole Vance, ed., *Pleasure and Danger: Exploring Female Sexuality* (London and New York: Routledge, 1984).
54 Robert Augros and George Stanciu, *The New Biology: Discovering the Wisdom of Nature* (Boulder: Shambhala, 1987).
55 Jonathan Freedman, *Media Violence and Its Effects on Aggression: Assessing the Scientific Evidence* (Toronto: University of Toronto Press, 2002).
56 Barry Glassner, *The Culture of Fear: Why Americans Are Afraid of the Wrong Things* (New York: Basic Books, 2000).
57 *The Culture of Fear*, p. 55.
58 George Gerbner, *The Mean World Syndrome: Media Violence and the Cultivation of Fear* (DVD) (Northampton, MA: Media Education Foundation, 2010).
59 Bob Mondello, "Remembering Hollywood's Hays Code, 40 Years On," National Public Radio (Aug. 8, 2008) www.npr.org/templates/story/story.php?storyId=93301189 (accessed May 15, 2018).
60 "Miller v. California," Oyez.com (1973) www.oyez.org/cases/1971/70-73 (accessed May 7, 2018).
61 "National Endowment for the Arts: Controversies in Free Speech," National Coalition Against Censorship (n.d.) http://ncac.org/resource/national-endowment-for-the-arts-controversies-in-free-speech (accessed May 7, 2018).
62 Elizabeth Kastor, "Funding Art that Offends," *Washington Post* (Jun. 7, 1989) www.washingtonpost.com/archive/lifestyle/1989/06/07/funding-art-that-offends/a8b0755f-fab9-4f7f-a8ef-2ccad7048fe2/?utm_term=.7802c2c9f55b (accessed May 14, 2018).
63 Patrick Buchanan, "In the War for America's Culture, Is the 'Right' Side Losing?" *Richmond News Leader* (Jun. 24, 1989). Republished in Richard Bolton, ed. *Culture Wars: Documents from the Recent Controversies in the Arts* (New York: Free Press, 1992) pp. 31–2.
64 James Davison Hunter, *Culture Wars: The Struggle to Define America* (New York: Basic Books, 1991) pp. 109–12.

65 Telly Davidson, *Culture War: How the '90s Made Us Who We are Today* (Jefferson, NC: McFarland 2016) p. 245.
66 Lauren Berlant and Lisa A. Duggan, "Introduction," *Our Monica, Ourselves: The Clinton Affair and the National Interest* (New York: NYU Press, 2001) p. 4.
67 Justin Fox, "A Look Back at Bush's Economic Missteps: Telling Us to Go Shopping," *Time Magazine*, (Jan. 19, 2009) http://content.time.com/time/specials/packages/article/0,28804,1872229_1872230_1872236,00.html (accessed Jun. 7, 2019).
68 "Culture Wars: President Bush and Secularism," (Mar. 4, 2004) www.foxnews.com/story/2004/03/05/culture-wars-president-bush-and-secularism.html (accessed May 8, 2018).
69 *Culture Wars: A Monthly Magazine* (1995–2006) www.culturewars.com/CultureWars/index/indexlist.htm (accessed May 8, 2018); Roger Chapman and James Ciment, *Culture Wars: An Encyclopedia of Issues, Viewpoints, and Voices* (London and New York: Routledge, 2009).
70 Rich Lowery, "Donald Trump's New Culture War," *National Review* (Jan. 24, 2018) www.nationalreview.com/article/444156/donald-trump-culture-war (accessed May 8, 2018).
71 Alex Ross, "The Frankfurt School Knew Trump Was Coming," *The New Yorker* (Dec. 5, 2016) www.newyorker.com/culture/cultural-comment/the-frankfurt-school-knew-trump-was-coming (accessed May 8, 2018).
72 Theodor Adorno, *Minima Moralia: Reflections on a Damaged Life* (1951) trans. E.F. N. Jephcott (London: Verso, 2005) sect. 71.
73 *Minima Moralia*, sect. 22.
74 Paul Willis, *Common Culture: Symbolic Creativity in the Cultures of the Young* (Boulder: Westview, 1990).

Chapter 9

Creative Economies

"Big Magic" or Empty Promises?

Most Americans believe in magic. It might seem hard to fathom, but it's really true. Magic is one of those primal instincts people just can't shake—with even die-hard skeptics having a few irrational beliefs. Magical thinking is hardwired in the human psyche, psychologists say, causing people to believe that luck is possible and objects can have power.[1] Add to this America's idealism and its famously reckless optimism, and it's clear that a certain mysticism infuses U.S. culture. America's magical thinking never wavered in the recent tight economy. And studies bear this out. *Scientific American* reports that seven in ten still believe in miracles, god, and heaven, with over 40 percent also accepting ghosts and extrasensory perception.[2] Harris Polls show 29 percent buying into astrology, 26 percent believing in witches, and 21 percent convinced the government is hiding aliens.[3] Famed magicians Penn & Teller have said that audiences crave the escape of make-believe and the "wondrous possibilities" beyond the here-and-now.[4] In this way, magical thinking works a lot like religion.

Creativity long has been viewed as magic, especially in Western culture. For most people, creativity is hard to predict or even explain, miraculously appearing in some people but not in others. Societies throughout time have treated artists as supernatural beings or visionaries. The ancient Greeks regarded artists and poets as "seers," whose inspiration came from magical "muses." Later on, such views stoked public obsessions with artistic genius, which would lead to a booming market for artistic masterpieces. These days, creativity gets credit for what only can be described as magic: a pathway to happiness, self-knowledge, and empowerment; a wellspring of innovation and new ideas; a cure for an ailing economy. Everywhere, creativity's mystical powers are promoted and sold in courses, craft items, toys, websites, and even advertising firms bearing the name.

"Imagination" is where magic and creativity intersect, broadly writ—and the vagueness of imagination hardly is incidental. Certainly, it takes some imagination to believe that creativity can do all that it promises. In this sense, the very idea of "creative imagination" is both the method and the message of the sales pitch. Discussion in this chapter begins with a look at popular writing on magical creativity, as seen in the recent outpouring of books with titles like *The Magic of Creativity, Writing Magic, Drawing is Magic, Life Changing Magic,* and *Be the Magic.* What does this craze say about America's current mood—the country's hopes and aspirations, as well as its doubts and worries? To answer this question, discussion next turns to the new "creative industries" and the magical qualities their proponents assert. This is where ideology enters the picture, especially as the aspirations of the creative industries come face to face with the lived realities of those inside them.

Skeptics argue that at best the creative economy has been oversold and its downsides minimized. As concepts of artistry are generalized and rendered in businesslike terms, humanistic values get pushed aside by a creeping functionalism, with any critiques likewise minimized. This fits within a broader "economization of culture" not unlike that of the culture industries discussed in the last chapter—but with a few important distinctions. Emphases on money-making and efficiency remain paramount in the new environment, with creativity hyped as a saleable commodity. But new is a disingenuous downplaying of sameness and conformity, even as ideas and products change little. The mantras of the new creative economy are self-expression, originality, and being "different"—promoted so much so that nobody feels they possess them.

"Creativity and imagination have always been intimate parts of capitalism," wrote Max Haiven in his book *Crises of the Imagination, Crises of Power.* [5] Haiven is among a growing list of critics linking creativity to earlier forms of exploitation, which confused people into accepting capitalistic values while taking advantage of them. A current example is the pauperization of many young creatives, who willingly accept low-wage, part-time jobs in artsy companies seen as offering "meaningful" work, but without the benefits or security of regular jobs. Another is the way consumers increasing believe they are "expressing themselves" in purchasing seemingly varied products aimed at their personal preferences, when oftentimes the items are mere variations of existing goods. In both cases, creativity is a new kind of bait.

This chapter uses the lens of "magic" to pose the ethical and political questions about the recent creativity hype. What kind of people find opportunity in the new creative economy? What sorts of promises are made to them? Who wins and who loses when business enters the picture? And what about entire nations promoting themselves as creative havens or "magical kingdoms"? The creative industries often concentrate geographically in newly gentrified urban zones, encouraged by city planners seeking to revitalize their communities. Unfortunately, the development of "creative zones" often pushes out poor populations and makes their living conditions worse. And none of this is simply an American phenomenon. Great Britain, Germany, and Australia discovered the creative economy before the U.S., advancing the magical qualities of their countries with nationalistic zeal though ad campaigns, branding, and tourist promotions. Finally, the internet has played a huge role in advancing the creative industries and the positive feelings they promise. The much-celebrated profusion of personal expression seen in Facebook, Instagram, and YouTube postings may empower many to produce culture, but it also means that online "artists" never get paid for their work—even as they build a massive profit machine for internet giants.

Magical People

"The central question upon which all creative living hinges is: Do you have the courage to bring forth the treasures hidden within you?" With this entreaty, author Elizabeth Gilbert introduced her recent bestseller *Big Magic: Creative Living Beyond Fear*, which offered an artistic cure for an anxious America.[6] Speaking directly to widespread feelings of disaffection and powerlessness, *Big Magic* romanticized artistry in Gilbert's signature blend of sentiment and cliché—packaging familiar motifs (human creativity, divine creativity, etc.) with a self-help twist about creating one's "self" in new and better ways. While it's easy to write off *Big Magic* as yet another feel-good advice book (which it surely is), I think it's time to take Gilbert's approach to creativity seriously and ponder why such ideas now get so much traction.

Publicity doesn't hurt. Reviewers effused over *Big Magic* as a "book-length meditation on inspiration" (*Newsday*) to "unlock your inner artist" (*Woman's Day*) and "dream a life without limits" (*Publishers' Weekly*).[7] This message also resonated in quarters promoting creativity as an innovation engine and economic tonic. While no one would dispute the good of a little artistic dabbling, at what point does such boosterism border on delusion? Or, put another way, when does fantasy disguise reality? Might America's fondness for make-believe explain the nation's political confusion and disaffection? Do fairytale versions of life infantilize a citizenry that so desperately hopes for happy endings? Certainly, the fantasy version of reality offered by certain politicians would fail any thoughtful analysis. Nevertheless, many leaders treat constituents like children, with entire governments encouraging populations to set worries aside and simply "Be Creative."

In *Magical Thinking and the Decline of America*, historian Richard L. Rapson took a long look at the nation's romantic idealism. "Probably in no other society of the world can one write the script for one's life as completely as in the United States. This fact has made the nation the 'promised land' for much of the world over the past two centuries," Rapson wrote. "The flight into endless self-improvement and innocent optimism has a long lineage in our past."[8] Perhaps anticipating "Make America Great Again" sloganeering, Rapson pointed to the disconnection between America's self-image as an "exceptional" driver of human history and the growing evidence of the nation's falling fortunes. This has led to a growing "flight from knowledge and reality into faith and fantasy," according to Rapson, largely resulting from "an American public increasingly in thrall to the fairytales told by the mass media."[9]

It also promotes a "cultural fixation on the individual, the personal, the biographical, the confessional, and, all too often, the narcissistic," and hence the rise of new "magic words" like "self-awareness," "personal growth" and other aphorisms entreating everyone to "be all that you can be."[10]

Individualism lies at the heart of American idealism, dating to the country's Enlightenment Era origins, when the independent citizen was invented as a counterpoint to deific and royal authority. Necessary as individualism was (and remains), no one could have predicted how its value could be magnified and distorted in neoliberal times. The initial affirmation of personal identity, which encouraged people to vote and participate in society, soon morphed into "striving to get ahead" and "winning at any cost." Eventually, the "self" would become an American obsession of theological proportions. "The purpose of nearly all the current gospels is to put believers 'in touch' with *themselves*," Rapson further explained.[11] This new brand of secular "faith" also comports well with the religiosity many Americans still profess, especially evangelical strains that promise economic gain to the faithful.

But let's get back to *Big Magic*. Written in memoir style, the book is reminiscent of Gilbert's breakthrough bestseller *Eat, Pray, Love*, in which the author recounted exiting from a dreary marriage to travel the world—savoring food in Italy ("Eat"), discovering spirituality in India ("Pray"), and finding a hot Brazilian boyfriend ("Love").[12] While many readers found the book inspiring, others thought it smacked of class privilege. *Big Magic* offered more of the same, while also zeroing in on feelings of insecurity common in the first decades of the twenty-first century. Many people want to be "creative" (especially with all of the recent hype), but also feel they just don't measure up. To Gilbert, this all boils down to *fear*. "You're afraid you have no talent. You're afraid you'll be rejected or criticized. You're afraid your work will be taken seriously. You're afraid of being a one hit wonder," she wrote.[13] The solution is all about spreading your wings, flying free, and letting your inner artist emerge. "You don't need anyone's permission to live a creative life," she concluded.

Gilbert wasn't kidding about the "magic" in all of this. The book was peppered with quirky anecdotes about strange coincidences and unexpected flashes of insight. For example, she wrote:

> When I refer to magic here, I mean it literally. Like, in the Hogwarts sense. I'm referring to the supernatural, the mystical, the inexplicable, the surreal, the divine, the transcendent, the otherworldly. Because the truth is, I believe that creativity is a form of enchantment.[14]

But when she got down to the nitty-gritty of how to actualize this process, her story ditched the creative economy altogether. After debunking myths of aesthetic genius in the name of everyday ("genuine") creativity, Gilbert dove into a starving-artist narrative. "Here's what I did during my 20s, rather than going to a school for writing," she began:

> I got a job as a waitress at a diner. Later, I became a bartender as well. I've also worked as an au pair, a private tutor, a cook, a teacher, a flea marketer, and a bookstore clerk. I lived in cheap apartments, had no car, and wore thrift shop clothes. I would work every shift, save all my money, and then go off traveling for a while to learn things.[15]

"Here's a trick: stop complaining," Gilbert advised: "First of all, it's annoying. Every artist complains, so it's a dead and boring topic." Contradictions notwithstanding, this nasty prescription both romanticizes artistic poverty and lays all failure on the individual. Never mind that structural conditions create lousy situations like adjunct faculty working for pennies, writers who never get published, and media crews hired one project at a time. Rather than questioning this negative pattern, Gilbert's exhortation to "stop complaining" did just what tight labor markets have always done—by putting all blame on the worker. The message is that with just a little more persistence, hard work, or courage, the wondrous "magic" will arrive. In other words, *Big Magic* told readers to embrace the mystifying operations of the creative economy in the belief that enchantment will provide a reward for their extreme effort, sacrifice, and transcendental faith. Forget that corporate boards are reaping fortunes while underpaying employees, or that things could be done differently. In the highly individualized world of magical creativity, workers never see themselves as a unified pool, and, in fact, often never see each other at all. Control becomes internalized as work shifts to the psyche itself. Meanwhile, would-be creatives are told to worship the system and do anything they can to gain admission.

Historically, Marxists explained this contradiction as a simple fact of capitalism—as the structural advantage of employers played out in the internalized self-defeatism of the laboring class. Caught up in the here-and-now of getting work and doing a good job, employees couldn't see their inherent disadvantage and hence unwittingly supported the system. Today, that blindness has moved well beyond the workplace itself, now reinforced in the institutional apparatus of school, media, consumerism, and social interaction, not to mention the continuing privatization of formerly public accommodations provided by government. The persuasive surround of culture now has become economized in what Michel Foucault termed a *dispositive*—defined as a "thoroughly heterogeneous ensemble consisting of discourses, institutions, architectural forms, regulatory decisions, laws, administrative measures … the system of relations that can be established between the elements."[16]

In *Creativity and Other Fundamentalisms*, Pascal Gielen dissected the deep transformation of public consciousness occurring as "creativity generates a new 'faith' not unlike prior faith in art."[17] Increasingly isolated from each other as "atomized individuals," the new creative subjects often find themselves "lacking the traditional grounding that institutions used to offer."[18] This new fundamentalism is enhanced in the fast-paced and ever-changing world of commercial creativity, where workers find themselves without a voice or a sense of job continuity. Gielen explained that "with no clear idea of the future, creative individuals are bouncing from one wave to another." Online connectivity, so often seen as a blessing, only makes anxieties about the future worse. As Gielen observed:

> Anticipation is central to high speed networked society, both in investments and in keeping up with peers, trends, new products. This produces a lack of depth, continuity, history, memory, stability, and solid social connection. Networks destroy permanence and the essence of actor-network theory, which is communication.[19]

This leaves many Americans frustrated and angry about a problem they cannot name—often lashing out against anything or anyone they can blame: government, immigrants, each other, or even themselves. It also provides a perfect set-up for superhero leadership figures proffering quick fixes, or magical cures like those promised by creativity. The postmodern blurring of fact and fiction doesn't help matters either, as critic Dana Polan once pointed out. In a post-truth age of "fake news," many people find themselves in a quandary over what they can believe. Unlike the industrial age, "Here ideology may have as a function less to reproduce productive relations than to disenfranchise workers by offering no interpellative space," Polan wrote.[20] This environment supports "dominant power by encouraging a serialized sense of the social totality as something *one can never understand* and that always eludes one's grasp."

Fairy-Tale Success

One of the craziest books about the new creativity was *Fairy-Tale Success: A Guide to Entrepreneurial Magic.* Without a shred of irony, authors Adrienne Arieff and Beverly West framed their how-to business guide as a juvenile girl-power parable: "Cinderella is a self-determined and innovative entrepreneur, a born CEO with a killer marketing sense, and a great sense of herself … an icon for today's entrepreneurial age," they wrote.[21] Here again, one sees the familiar motif of self-knowledge as a prerequisite: "Only when Cinderella truly knows who she is and what she stands for does she begin to come up with an idea and construct an effective plan for launching herself in the world and finding her fairytale success."[22] Also in evidence are the themes of self-sacrifice and meaningful work as seen in *Big Magic*: "Do what makes you happy, not what makes you rich. Don't get distracted or discouraged by what is classified as a 'glamorous' or 'acceptable' career—what's in your heart is what's important!" Arief and West effused.[23]

All of this might be comical if *Fairy-Tale Success* didn't conform so perfectly to the claims and underlying problems with the creative economy. Among other things, the book exudes the *ex nihilo* principle of wealth and success conjured "out of nothing," while romanticizing the artsy underclass on whose efforts the story hinges. The rags-to-riches Cinderella allegory combines all of the elements described thus far in this book: desire, commodification, mystification, and infantilization—but with a more targeted address to the younger, often female, populations driving the creative economy. But to *Fairy-Tale Success*, the artist's personal "grit" is all that matters. "With just a little innovation and imagination, you can make magic with the resources you have lying around," the book concluded.[24] Of course, not everyone agrees. "One trouble with the Cinderella myth is that it completely externalizes the success factor," commented Mike Roy in a series entitled "The 12 Artist Myths."[25] This externalization typically comes in projections of a magical savior or event appearing from nowhere. As Roy continued:

> Somehow we find it so easy to cling to the idea that, even though we may be locked in a tower, someone or something will eventually come to our rescue. We believe that *somebody* will discover us and lead us out of our poverty-like state.[26]

But let's get specific—and return for a moment to what launched the creativity craze, now well into its second decade. Economist Richard Florida still gets much of credit for popularizing the idea, owing to the popularity of his 2002 bestselling book, *The Rise of the Creative Class: And How It's Transforming Work, Leisure, Community, and Everyday Life.* [27] Florida's case was relatively simple, as it drew on the values long-embedded in concepts of art and creativity. He argued that mainstream business and political leaders had been mistaken in crediting America's success to the country's determination, resources, and strength. Instead, Florida said, "America's growth miracle turns on one key factor: Its openness to new ideas, which has allowed it to mobilize and harness the creative energies of its people."[28] But new ideas don't just fall from the sky, Florida added. They come from entrepreneurs who devise products, write software, and start new businesses.

And who are these innovative people? "There's a whole new class of workers in the U.S. that's 38 million strong: the creative class," Florida would explain.[29] "The Super Creative Core of this new class includes scientists and engineers, university professors, poets and novelists, artists, entertainers, actors, designers and architects, as well as the thought-leadership of modern society."[30] To this list Florida would add creative figures in business and finance, law, healthcare, and any other fields that entail "problem-solving that involves a great deal of independent judgment."[31] Florida estimated this previously overlooked group of workers accounted for 30 percent of U.S. employees, attributing its initial growth to post–World War II research spending, expansion in educational opportunities, and, to some extent, the counterculture's openness to diversity and fresh ideas. Furthermore, Florida located the epicenters of this rising creativity in the enclaves of artists, academics, bohemians, and other "no collar" workers drawn to large cities.

Florida's "creative class" gave a name and an upbeat premise to a number of already-unfolding phenomena. For decades, the U.S. economy had been shifting, first from industry to services, and later from a service to an information economy—as media, communications, and technology firms accounted for ever larger shares of exported goods. In this new context, all sorts of previously non-economic items became valued for their profit-making potential (online profiles, consumer data, etc.). At the same time, the nature of money became increasing abstract in an era of "financialized capital." This was seen in the rise of credit cards, borrowing, and the new kinds of unregulated securities (not connected to hard assets) that eventually caused the debt bubble resulting in the 2008 recession. In these terms, the economic "crisis" of the new millennium was tied to both money and the imaginary instruments that had come to stand in for it. Seen this way, it's hardly surprising that Florida and his followers would seize upon creativity as the Rosetta stone of a new era of American prosperity. But it also illustrates how "creative" accounting and financing might have been part of the problem in the first place.

Not surprisingly, the creative economy also changed the ways people thought about art itself. In "The Death of the Artist—and the Birth of the Creative Entrepreneur," William Deresiewicz wrote in *The Atlantic* about how the *new* creative economy transformed *old* ideas into money-making schemes. First among these were the magical attributes of artistry itself. "'He's an artist,' we'll say in tones of reverence about an actor or musician or director. 'A true artist,' we'll solemnly proclaim our favorite singer or photographer, meaning someone who appears to dwell on a higher plane."[32] Many people don't know that ideas of mystical artistic genius only came along with the Romanticism of the 1800s. Before that painters and sculptors were classed with tailors and carpenters—"grouped with other artisans, somewhere in the middle or lower middle, below the merchants, let alone the aristocracy." These earlier artisans could be esteemed, of course, as "master craftsmen" (the Dutch masters, for example) — but this usually was attributed to skill or technique rather than mystical vision.

The creative economy waters down both genius and master craft, while nevertheless drawing on their positive associations. It says that "anyone" can become a creative entrepreneur with a little artsy cleverness and flexibility. (As *Big Magic* put this, "When artists are burdened by the label of 'genius,' they lose their ability to take themselves lightly, or to create freely."[33]) This flexibility also tells young creatives that it is okay to float from job to job and adapt to different work demands. Unfortunately, this view also downplays an incremental de-skilling taking place in the creative economy. As Deresiewicz explained, old-school mastery conformed to the 10,000-hour rule, where artist/artisans "trained intensively in one discipline, one tradition, one set of tools, working to develop one artistic identity as a writer, painter, or choreographer."[34] In the past, almost no successful artist was known in more than one genre.

"One of the most conspicuous things about today's young creators is their tendency to construct a multiplicity of artistic identities. You're a musician *and* a photographer *and* a poet; a storyteller *and* a dancer *and* a designer," Deresiewicz said. Technique and expertise are less necessary. "The point is versatility. Like any good business, you try to diversify."[35] It goes without saying that much of this is driven by money, which in itself is another huge difference between the market-driven imperatives of the new creative economy and the generally non-materialist character of traditional artistry. "Entrepreneurialism is being sold to us as an opportunity. It is, by and large, a necessity. Everybody understands by now that nobody can count on a job," Deresiewicz concluded.[36] And the sales pitch indeed has become ingenious—matching, as it does, near-ubiquitous feelings of economic anxiety with the creative spark in everyone.

Creative entrepreneurialism also advances under the guise of upward mobility, often pitched to young people while they are in school. Colleges and universities find themselves under administrative pressure to produce employable graduates in a tight economy, as discussed in Chapter 7. Students and their families want this too. The results are seen in the overall expansion of "practical" programs of study in such areas as business and engineering, as well as the growth in career preparation and marketing courses in art schools. This has a conditioning effect on young people, many of whom are already edgy about their future prospects. In addition to changing study options for all students, this tells art majors to set aside "impractical" directions like conceptualism or social activism. It even can turn students against each other. As one critic recently explained, "Not only does business studies become a key component of the 'art school' provision, but fierce competition among individual students figures more prominently as the pedagogic strategy."[37]

Ironically, left-leaning intellectuals helped set the stage for the creativity craze. Marx himself often romanticized the craft abilities of preindustrial carpenters, coppersmiths, potters, and shoemakers, whose artistry was eliminated by the factory assembly line. A similar nostalgia for working-class creativity fed the early cultural studies movement of 1950s and 1960s—valorizing, as it did, forms of popular expression excluded from "high culture." Cultural studies focused a lot on young people, too—a population that would soon grow into an enormous market. As its numbers increased, the boomer generation would gobble up music with "working-class" sensibilities, starting with the Beatles and Rolling Stones, and later in bands like the Hollies and Sex Pistols. American folk and hip hop music of the era similarly appealed to subcultural sensibilities, while updating country and western and rhythm and blues forms. In 1970, John Lennon's well-known song, "Working Class Hero" became an anthem for this idea.

"Existential rebellion has become the more-or-less official style of Information Age capitalism," stated Thomas Frank in *Commodify Your Dissent.* "However the basic impulses of the countercultural idea may have disturbed a nation lost in Cold War darkness, they are today in fundamental agreement with the basic tenets of Information Age business theory."[38] Frank's words echo growing concerns about the market's ability to co-opt its opposition. This has less to do with any capitalistic plot than it does the new creative economy's inherent profit motive—as well as the focus on evolutionary "change" common to both business and artistry. In either case, it is clear that the old culture industry theory opposition of capitalism to creativity no longer holds.

The New Spirit of Capitalism by sociologists Luc Boltanski and Eve Chiapello has become a key explanatory work on this topic.[39] Looking at the market's affective dimensions, the book said the system wouldn't work if people truly hated it. Moreover, business is inherently dynamic, whether this means updating products, finding new ways of making them, or managing criticism of the system itself. "Capitalism will face increasing difficulties, if it does not restore some grounds for hope to those whose engagement is required for the function of the system as a whole," Boltanski and Chiapello wrote.[40] Hence, Depression Era unhappiness led to New Deal income redistribution, and post–World War II standardization triggered later diversification. Central to this thesis is the market's built-in need to appropriate *new external elements*, and thereby refresh itself. This occurs most effectively when capitalism "discovers routes to its survival in critiques of it."[41]

Entrepreneurial creativity is a perfect example of Boltanski and Chiapello's "new spirit." They noted that activism of the 1960s and 1970s often was driven by the "artistic critiques" from writers and musicians at odds with bureaucracy, conformity, and a rigid establishment. In the shared zeitgeist of the counterculture, capital's new spirit likewise adopted values of flexibility, individualism, and antiauthoritarianism—of exactly the kind seen in *Big Magic* and elsewhere in new creativity discourse. Today, this same spirit appears in the dressed-down personas of "cool capitalists" such as Facebook's Mark Zuckerberg or Google's Sundar Pichai. Not surprisingly, both CEOs are huge creativity fans, with Zuckerberg chanting to college students, "It's not about creating companies, it's about creating change," and Pichai imploring them "to pursue their passions, take risks, and be creative."[42]

Lost in this billionaire rhetoric is the downside of creative entrepreneurship. Passion and risk are fine for those who can afford a few losses. But for others it can translate into disappointment and poverty. So intense is the boosterism for the new creativity that considerations of anything else get pushed aside. Searching online, one finds endless stories about rising creative industries within regional and national economies—with scarcely a word about low wages, job insecurity, or fractional employment. Part of the problem is the vagueness of creativity as a conceptual category—into which any enthusiasm or dismay can be projected. Both the U.S. Department of Commerce (DOC) and the National Endowment of the Arts (NEA) count as "creative" such diverse activities as publishing, advertising, tourism, movies, TV, computer games, art education, design, architecture, fashion, jewelry-making, photo-processing, and the sale of any "creative" anything—as well as what people usually think of as the arts themselves. This is how creativity gets credited with $729.6 billion in economic impact, or 4.2 percent of the U.S. economy.[43]

But by any accounting, the true job situation is *abysmal* in creative fields—with federal statistics showing 90 percent of workers finding only part-time jobs, and full-time employment continuing to slide downhill. Creatives are self-employed at 3.5 times the rate of the general population, with most subsisting on short-term, project-by-project income.[44] And the situation is worsening. "Although arts and cultural production has increased, employment has decreased," according the most recent DOC/NEA reports showing a 17 percent drop in full-time jobs in the last 15 years.[45] Income doesn't look good either, even for those with full-time work. The U.S. Bureau of Labor Statistics shows creatives making $35,540 per year, which falls 15 percent below the national average of $41,000.[46] Tight job markets are bad for workers in lots of ways, as employment analysts have long recognized. The Labor Department now cautions job-seekers, "Because competition for these jobs is often intense, workers may have a hard time finding enough creative work to make a living," adding that "creativity takes a lot of trying" and "to succeed, you must be able to handle stress and frustration, accept criticism and failure, and learn to persevere."[47] To improve their chances, applicants are encouraged to get experience through "hands-on learning: paid or unpaid" as interns, assistants, or simply doing the work as a "hobby." Echoing a refrain heard from many creatives, the Labor Department said, "In some jobs or certain projects, artistic freedom may be limited. You have to remember you're creating for other people, not yourself."

Why would anyone put themselves through this? Conditioning is part of the answer. The powerful appeal of creative work links to values that many would-be creatives bring to the transaction. The educated, art-oriented class of young people often draws from middle-income groups, distinguished from the working class by expectations of "meaningful employment." As explained by Angela McRobbie, "To some extent middle-class status nowadays rests upon the idea that work is something to which one has a passionate attachment."[48] As McRobbie argued in her ironically titled book *Be Creative*, "Middle-classification processes come to be linked directly to self-entrepreneurship as an ideal. This is not, however, upward mobility; instead, it is an ideological effect, giving young people, especially young women, the feeling of being middle-class and aspirational."[49] This helps to explain unfortunate tendencies among creatives to forgo such niceties as full-time pay checks, health insurance, or retirements plans, which a prior generation of industrial workers fought so hard to get. In exchange for meaningful work and self-management, the creative class "bypasses mainstream employment with its trade unions and its tranches of welfare and protection."[50]

Besides the lack of money and benefits is the vulnerable state that comes from being hungry for work—often obliging job-seekers to accept whatever tasks or terms an employer chooses. Social scientists use the term *precarity* to describe this employment uncertainty, with many seeing a new class emerging—not quite living in poverty, but nevertheless subject to the exploitation typical of economic disadvantage. For some time, intellectuals like Pierre Bourdieu, Michael Hardt, and Tony Negri have described:

> a class-in-the-making, internally divided into angry and bitter factions consisting of a multitude of insecure people living bits-and-pieces lives, in and out of short-term jobs, without a narrative of occupational development, including millions of frustrated educated youth who do not like what they see before them.[51]

While this situation is especially dire in the creative industries, it reaches across nearly all employment sectors, as businesses either struggle to survive or simply get greedy. No one is especially happy about this. But the growing gap between rich and poor leads some to suspect the motives of companies gaining the most—many of which are in creative fields.

Aside from employment, critics point the economization of creative life itself—manifest in attitudes of creatives toward their jobs, where their work takes place, how they allocate time, and why they choose the field. Artistic workers today rarely see themselves as magical visionaries, although the creative industries still capitalize on such ideals—especially in discourses promoting obscurity and poverty as essential to the artistic psyche. This conjoins with what Marion von Osten has termed the "contingent subjectivity" of today's artistic-types in which any "failures in the free market are reinterpreted as personal challenges" or recast as "learning opportunities."[52] In another reversal, "the artists' studio or loft formerly seen as a symbol for the convergence of labor and leisure in everyday life" has been transformed into a panoptic work space the creative can never leave. Even the term "loft" is now more associated with upscale gentrification than anything "artistic."[53] Among other things, this means that "[t]ime can no longer be clearly assigned according to dual parameters like work and leisure, production reproduction, employment and unemployment," as Gerald Raunig observed in his book *Factories of Knowledge, Industries of Creativity*. [54] To be specific:

> Forms of striating time become frayed: for instance, a time of poorly paid or unpaid internship, a time of looking for work without pressure from the employment office, a time of preparing for new projects, the time for unpaid practices of self-organization, the time for paperwork, the time for

> an electronic correspondence, a time for brief regeneration, a time of training and continuing education, a time of unpaid sick leave.[55]

All of this changes the meaning of creative artistry. "The slogan that 'everyone is creative' represents a broad cultural shift in our society," stated Haiven in *Crises of Imagination*. "Artists are being held up not as poverty-stricken malcontents, but as triumphant 'pioneers of the new economy.' Today when the idea of a good, steady, lifelong job seems impossible, corporate propaganda encourages us all to see ourselves as artistic souls."[56] This is perhaps the most worrying aspect of the new creative economy: its generalizing tendencies. Initially broadening the category of "artist" to include anything creative, the movement now expands "creativity" to include everyone. And while this might seem a wonderfully democratic idea, the hitch lies in just how the new creativity defines itself. As Haiven concluded, "This fundamental shift in what we could call the 'economy of creativity' drastically alters what sorts of creativity we think of as valuable and it orients humanity's creative energies toward earning ever greater profit for a few."[57]

Fantasy Islands

Magical beings live in enchanted places. For millennia, people have envisioned imaginary worlds in which their hopes and aspirations play out unencumbered by immediate realities. This has been the impetus for utopian visions of "a better place" as seen in Atlantis, Erewhon, the Garden of Eden, Shangri-La, and Walden—as well as Thomas More's 1516 *Utopia* that gave the idea its name. Cities feature prominently in this as sites of assembly and containment, often set apart from the world around them. Entire mythologies attach to such ancient cities as Athens, Babylon, Beijing, Cairo, and, Rome, much as they still do to places like Berlin, London, Mumbai, Paris, and Tokyo. Famed urban studies scholar Lewis Mumford wrote:

> The city, as one finds it in history, is the point of maximum concentration for the power and cultural community. It is the place where the diffuse rays of many separate themes of life fall into focus … Here, in the city the goods of civilization are multiplied and manifolded.[58]

To put some current numbers behind this, according to most recent Census Bureau data, half of the U.S. population lives in the nation's 48 largest urban areas, the top three of which are New York City (18 million), Los Angeles (12 million), and Chicago (8 million).[59]

In this context, it is no surprise that creative economy discussions have focused heavily on metropolitan areas—often to the exclusion of anyplace else. So-called "creative enclaves" and "creative zones" are said to drive the world's new "creative cities" and regions. In what quickly would become a point of controversy, Richard Florida initially described the creative class as a quintessentially urban phenomenon in its aggregation of highly educated, artistic, and tech-savvy younger people in what he termed "creative clusters" of the kind seen in Boston and northern California's Bay Area. In doing so, Florida drew on the thinking of progressive urban theorist Jane Jacobs who, in the 1960s, opposed top-down planning approaches in favor of more organic approaches.[60] Notably, Jacobs saw potential in the "human capital" of depressed zones often slated for destruction ("slum clearance"), much as Florida would also see the potential of rundown bohemian neighborhoods. The big difference lay in Jacobs' attention to disadvantaged minorities and Florida's advocacy of inherently privileged groups—a point he later would concede.[61]

But let's think about how cities have changed. Throughout most of the world, the modern city was an outcome of industrialization, as factories and railways began to concentrate formerly agrarian populations in places where employment was growing. Economic necessity drove workers to cities, many of whom would accept low wages. As immigrants and the poor collected in urban enclaves, their neighborhoods soon became sources of worry for better-off middle- and upper-class groups—even as conditions in slums and ghettos deteriorated. Uprisings by underpaid workers began in the early twentieth century, often with bloody reprisals from factory owners. By mid-century, the racial and ethnic character of urban poverty would result in riots. In the 1960s, the federal government stepped in with a swath of urban renewal, welfare, housing programs aimed at the poor—only to have many of them discontinued in the 1980s. This also was the moment when American industrial jobs began to shrink due to globalization and technological advances in manufacturing. Before long, jobs began shifting to such non-manufacturing "service economy" sectors as retail, health care, hospitality, education—and, significantly, information technology.

By the 1990s, cities across the U.S. were still struggling with this shift. Many of them looked to tourism and their own histories in an effort to capitalize on "culture" in reinventing downtown areas. All manner of historical districts appeared as former factories were converted into mini-malls, restaurant courts, and lofts catering to an upscale clientele drawn to "local character" and authenticity. Initially, city planners embraced this influx of new money into sagging manufacturing neighborhoods via the phenomenon of "gentrification." But it didn't take long for downsides to become apparent, especially as existing residents found themselves surrounded by boutique cafés and gourmet restaurants. Rents skyrocketed and evictions took place as gentrification pushed lower-income people elsewhere—often to worse places. Meanwhile, lots of people made fortunes, especially savvy real-estate investors with the foresight to "buy low" and "sell high" when neighborhoods improved. Sociologist Sharon Zukin was one of many critics of what she termed "spiky reurbanization" (favoring some, but not others). "Who benefits from the city's revitalization?

Does anyone have a right to be protected from displacement? These stakes make it important to determine how the city's authenticity is produced, interpreted, and deployed."[62]

Appearing in 2002, Florida's *The Rise of the Creative Class* didn't so much invent this process as much as it gave the phenomenon a name and specificity. To Florida, rising "creative centers" were economic dynamos:

> Not only do they have high concentrations of creative class people, they have high concentrations of creative economic outcomes, in the form of innovations and high tech industry growth. They also show strong signs of overall regional vitality such as increases in regional employment population.[63]

Balking at theories that the internet had diminished the importance of "place," Florida countered that new creative centers relied on the synergies of physical proximity and depended on urban lifestyle amenities to pull creatives to one place. Along with many others caught up in this enthusiasm, he predicted a trickle down of wealth and opportunities benefiting everyone in creative cities.

A decade later, Florida was singing a different tune, as evidence grew that the creative class mostly benefited itself. While working-class people in creative zones sometimes saw modest income increases, such gains were wiped out by higher rents and consumer prices. As Florida eventually conceded, creative-class money "flows disproportionately to the more highly-skilled knowledge, professional, and creative workers. On close inspection, talent clustering provides little in the way of trickle-down benefits."[64] Worse still, the growing affluence of creative pockets in places such as New York City and San Francisco contributed to broader income inequality in those cities. Speaking of America's leading creative cities, journalist Joel Kotkin would add that "the gap between rich and poor is the widest it's been in a decade."[65]

Then there is diversity—purportedly a noble value within the creative class and a major reason creatives flocked to urban zones. In his original formulation of "Technology, Talent, and Tolerance," Florida effused over the presence of LGBT people in the arts and Asian immigrants in the tech industry. But as time progressed, it became clear that the real demographic driving the movement was overwhelmingly white, single, and heterosexual. Noting this demographic narrowness, Kotkin additionally observed, "The places that attract the 'creative class' are also the ones with the fewest families and children."[66] In this vein, much of the discourse on the creative economy also remained silent about suburban and rural life in general. In 2015, researchers conducted the first-ever statistical analysis of diversity in a creative city. Appearing in the journal *Applied Geography*, "Neighborhood Diversity and the Creative Class in Chicago" found "no clear pattern of increasing or decreasing diversity" in ethnic/racial and sexual orientation terms, concluding that "associations between measures of diversity and creative class employment appear to be relatively weak."[67]

Always the nimble scholar, Florida did further back-peddling. His next book, *The New Urban Crisis* was all about the very *old* problems of inequality and segregation, heightened in many cities with bustling creative economies. Taking the issue head on, Florida wrote that "economic segregation is greater in bigger, denser metros with large concentrations of high tech industries, college graduates and members of the creative class."[68] Naturally, Florida was not going discount his original thesis completely—instead tempering his initial claims by saying that equality itself was uneven among creative cities, resulting in a "new age of winner-take-all urbanism, in which the talented and the advantaged cluster and colonize a small, select group of superstar cities, leaving everybody and everywhere else behind."[69] Answering other gripes, Florida conceded "deeply disturbing connections between race and the creative economy," explaining, "Whites, (specifically, non-Hispanic whites), hold almost three-quarters of creative-class jobs (73.8 percent) while making up two-thirds of the population" while "Blacks hold just 8.5 percent of creative-class jobs, despite making up 12 percent of the population."[70]

Enchanted Kingdoms

> The Department of International Magical Cooperation is a division of the Ministry of Magic. Like the British Foreign & Commonwealth Office, the duties of this department are to work with magical governments of other countries, set standards for trade, create regulations, interpret international magical law, and work with the department of magical games and sports, and be present at the International Confederation Wizards.[71]

In this detailing of J.K. Rowling's Harry Potter saga, fans depict a magical "country" existing inside a non-magical one. Harry Potter aficionados are well acclimated to the notion of a wizard "shadow regime" running parallel to the bureaucracy of the "muggle" world. While clearly a fantasy (and a subversive one at that), the global phenomenon of "Pottermania" also is embraced by the very *real British government* as a strategy for economic renewal and national rebranding. Motifs of Rowling's fiction appear on the "Visit Britain" tourism website, for example, deploying motifs of myth and magic to mine England's cultural past while appealing to younger demographics.

The Harry Potter phenomenon is but one example of creative industrialization now occurring around the world, as governments promote their nation's unique characteristics through entertainment, advertising, cultural tourism, and sporting spectacles like the Olympic Games. Britain is a classic example of this, launching its "Cool Britannia" campaign nearly two decades ago. Since then, nations on every continent have joined in—with Luxembourg, Bermuda, and Singapore now at the top of the Global Creativity Index.[72] (Interestingly, despite its oversize media industries, the U.S. ranks 37th on this list.) Historians long have noted the importance of fantasy in the construction of national identity and ambition—often expressed in mythological or religious forms. Such thinking can be traced to the ancient Greeks, whose pantheistic belief systems included a bewildering array of deities, their powers of enchantment, and often quirky proclivities. While early Christianity took a different path, it still hung on to the supernatural, leaving in place templates for magical belief and superstitions still lingering today. Popular notions of creativity, creation myths, and *ex nihilo* creativity directly or indirectly tap into such mystical "beliefs," while encouraging conceits that entire populations are magically endowed.

Magical thinking also played an historic role in colonialist ambition, as superstition colored theological doctrine in nations throughout Europe. Voyagers to "new worlds" like America often imagined themselves on spiritual quests or divine missions as they ventured into unknown territories. The very newness of "discovery" lent itself to all manner of magical projection, even as this wonder could degenerate into anxiety and fear. Notions of divine purpose often justified the subjugation of indigenous populations seen as primitive, threatening, or otherwise "godless." European settlers of the U.S. colonies would draw on hero and conquest stories in developing what Richard Slotkin called "the mythology of the American frontier."[73] And while most American settlers had consistent views of "God," their thinking could vary when it came to the "wild" or demonic natives. Keep in mind that "witch-burning" had been a staple of Western culture since Roman times, later to be sanctioned by Pope Alexander IV in the persecution of heretics, sorcerers, and presumed Satanists.

Undercurrents of magical thinking persisted in America's view of itself as an unstoppable force in the world. In the early twentieth century, sociologist Max Weber explained how mysticism played into capitalist ideology through abstract beliefs in moral reward. Weber wrote that "magical and religious forces, and the ethical ideas of duty based upon them, have been among the most important formative influences on conduct."[74] While superstitious beliefs in the "natural" authority of monarchs and deities had long since given way to "reasoned" approaches to governance, magic retained its grip on the idea of "nation" nevertheless. Weber saw magical thinking underpinning of the process of "rationalization" he would critique so virulently in *The Protestant Ethic and the Spirit of Capitalism*. In contrast to other socialists of his day, Weber argued that faith and belief were what really drove market ideology (as opposed to profit and greed) at a time when "lust for gain" was seen as immoral and anti-social.[75] Individualistic and goal-oriented themes common to religion (heaven) and economy (profit) served to reconcile these contradictions, much as they animate free-market boosterism today.

There is no discounting the role of media in all of this—both in historical and contemporary aspects. Writing in *Imagined Communities,* Benedict Anderson described modern nations as mental constructs held together by group belief rather than geographical boundaries. Initially generated through newspapers and books, solidarity within national populations only intensified as mass media became ubiquitous social phenomena. Add to this the enormous popularity of fantasy literature, movies, television, consumer products, and theme parks, and the staying power of magical thinking becomes clearer, often undergirded by the infantilizing allure of fantasy worlds. Nearly everyone has a fondness for one such place, be it Camelot (Chrétien de Troyes), Earthsea (Ursula Le Guin), Middle Earth (J.R.R. Tolkien), Narnia (C.S. Lewis), Neverland (J.M. Barrie), Oz (L. Frank Baum), Panem (Suzanne Collins), Terabithia (Katherine Paterson), or Wonderland (Lewis Carrol).

Imagineering

The principle of travel into imaginary worlds got very real in the tourist marketing of creative economies. Just think about ads for the "enchanted" deserts of the American Southwest, the "fairytale" castles of Europe, and the "mystical" ruins of South America—not to mention pitches evoking the "mysterious" Orient and the "dark magic" of Africa. At its core, tourism operates at the intersection of curiosity, vulnerability, and social power. Long associated with the pleasurable release from work, tourism also resonates with familial or romantic togetherness, combined with themes of escape and discovery. Most histories of Western tourism date it to the travels of European aristocrats in the seventeenth and eighteenth centuries, as they would voyage forth into foreign lands for self-improvement or sometimes simply growing up. Ever present were the twin specters of safety and risk, along with a fascination with difference and "otherness" in various forms. All of this reflected back upon a preoccupation with the self and its ability to know the world. As tourism became a middle-class activity in the nineteenth and twentieth centuries, greater emphasis accrued to the idea of travel as a respite from work, with relaxation, recuperation, rest, and pleasure gaining importance. But it wasn't until the post–World War II era that tourism became a driver of the global economy. Travel agencies, packaged tours, and resort hotels boomed in the 1960s as the marketing of national "destinations" became more pronounced.

Creativity always has played a central role in tourism, beginning with its emphasis on museums, architecture, cuisine, and local handicraft. Tourists often would see themselves as "creative" in their choice of itineraries and experiences of novelty, discovery, and other cultures. Beginning in the 1990s, tourism scholars began discussing the growing role of what was termed the "experience economy" in which travel became a component of broader demands for immersive, staged experiences in entertainment. This fed the development of computer game culture, but also the growing popularity of theme parks emerging from film and TV franchises. Disney remains most famous as a pioneer in this area. Launched as an animated film studio in the late 1920s, the company entered the television age with a series entitled

"Disneyland"—which it shrewdly realized could become an amusement park brand. Inspired by its 1950s TV series *The Magical World of Disney*, the company built a place where viewers could physically inhabit the media fantasy, replete with staff "Imagineers" to facilitate. In the following decades, Universal Studios and Warner Brothers, among others, would launch parks with attractions based on franchises such as *Batman, Indiana Jones, Iron Man, Shrek*, and *Star Wars*—all catering to this experience-driven impulse. The three companies now have theme parks around the world: Disney (Anaheim, Hong Kong, Lake Buena Vista-FL, Paris, Shanghai, Tokyo), Warner (Abu Dhabi, Los Angeles, Madrid, Queensland), Universal (Los Angeles, Orlando, Osaka, Singapore).

Expansive as it might seem, the globalization of the U.S. theme park industry is dwarfed by the rise of "cultural tourism" in the international creative economy, as nations around the world have ramped up efforts to market themselves as unique brands. This new cultural tourism differs from the past by emphasizing the immersive experience of cities and regions rather than familiar landmarks or particular historical sites. Writing in *Annals of Tourism Research*, Greg Richards has described a "creative turn" in what travelers want, as tourists increasingly seek participation in local cultures, rather than mere observation. "Production and consumption factors are therefore increasingly entwined," Richards observed.[76] UNESCO has noted this as a global phenomenon: "Creative tourism is travel directed toward an engaged and authentic experience, and it provides a connection with those who reside in this place and create this living culture."[77] In turn, countries began to encourage regions to develop new offerings such as dance workshops in the Caribbean, crafting holidays in Ireland, silk weaving in Japan, and a wide range of festivals in which locals mixed with visitors. According to the international Creative Tourism Network, this new approach offered such other benefits as diversification of tourist offerings "without any investment," sustainability as it relies on "authenticity and creativity" as main resources, and a positive effect on the "self-confidence of the locals." The movement also was seen to offer a material stake in tourism to previously resentful locals.[78]

Skeptics argue that creative tourism does little to change the unequal relationship between visitors and locals. In this view, creativity is simply the latest face of an incremental colonization of those seen as folksy, quaint, or otherwise "exotic"—with the creative hunger for authenticity carrying invasive implications. More strident critics further worry that the commodification of ever more detailed aspects of local culture has a damaging effect in turning everyday life into an array of objects for sale. As Richards further observed, while the "colonization of the lifeworld" has already been observed in cultural tourism, within "creative tourism it takes on a new dimension because it tends to involve more elements of everyday life and the intangible, embedded culture of the host community."[79] Driven by the imperatives of national governments hungry for economic revival, creative tourism risks becoming yet another rationalization and economization of a previously off-limits field.

Creativity is part of what has been termed the "magic tourism multiplier" effect—referring to the secondary expenditures visitors make in addition to lodging, transportation, and destination costs.[80] It's the things people buy or do as they causally shop, wander around, or spontaneously spend in ways impossible to track in tourism accounting. And it's exactly the kind of abstract economic benefit that proponents of creative tourism can hype without the nuisance of verifiable statistics. At the other end of this extreme is the direct selling of magic itself as its own special object in the experience economy. Never far from the cutting edge of tourism, the theme park industry has kept pace with this trend as the Wizarding World of Harry Potter parks have begun operating around the world. The creative centerpiece of the Wizarding World experience became its $49.95 Interactive Wands, "allowing guests to directly influence their environment and 'cast spells' to produce effects in both Diagon Alley and Hogsmeade."[81] Parents I know tell me that the add-on wands are far more than incidental purchases these days, driving kids crazy with want. No longer satisfied with merely seeing the magic or going on rides, visitors now can "become" Harry, Hermione, or Ron.

Tomorrowland

"Any sufficiently advanced technology is indistinguishable from magic." This famous "law" from science fiction writer Arthur C. Clark is often recited by fans of the genre, as well as people in the "real" fields of science and technology. As the mind behind *2001: A Space Odyssey* (1968), Clark has been honored by the United Nations as well as university research institutes bearing his name. His law derives philosophically from the ancient belief that technology was an essentially creative endeavor, with the Greek word *tékknē* meaning "art, skill, craft, technique or principle."[82] The word's magical connotations come from the inexplicability of the creative process, which often give technology the otherworldly associations that drive science fiction. This thinking is seen today in Apple's "Magic Mouse" and companies like Magic Software Enterprises. As Richard Stivers explained in his book *Technology as Magic*, "Our expectations for technology have become magical and our use of it is increasingly irrational," adding that "magic began historically as the attempt to influence nature, which was experienced as sacred."[83]

Technology is seen as an enormous boon to creativity today, manifest in software like Adobe Creative Cloud (Photoshop, Illustrator, Lightroom, and Premier) and Apple GarageBand, further enabled by sites such as Facebook, Instagram, Spotify, SoundCloud, Twitter, and YouTube. In the twenty-first century, digital technology has turned the once-specialized crafts of photography, music production, and video-making into ubiquitous activities that even a ten-year-old can do. It has radically decentralized forms of creative expression once restricted to industrial studios and high-end firms—and, in so doing, also blurred the line between "professional" and "amateur." Add to this the endless copying of material made possible by digital rendering, and the ongoing paradigm shift in creativity becomes even more apparent. Is it any wonder so many believe "We're All Artists Now"?[84]

The interplay between technology and creativity has caused a certain sloppiness in the creative economy conversation, as the concepts are conflated, combined, or confused with each other. "Technology" is just as vague as creativity, and, as such, prone to all manner of projection or ideological inflection. Pessimistic views common in the post–World War II era cast technology as a malignant force, capable of assuming a life of its own, and responsible for such human threats as nuclear weaponry, rampant consumerism, and environmental ruin. Optimists saw technology as a tool for progress and good in its abilities to save lives, enhance productivity, and improve communication. Between these two views is the structural notion of technology as merely a "tool," subject to positive or destructive uses (as seen in a hammer or automobile). In this sense, technology works in the same open-ended way as creativity.

The creative industry conversation of the first decade of the twenty-first century saw none of this nuance in its delirious utopianism. In what only can be described as magical thinking, proponents only chose to look at positive outcomes in their projections of revolutionary growth. They also failed to see that technology companies often cared little about art or creativity as anything more than a means to an end. This blind optimism and instrumentalism drove a broad reconceptualization of creativity, changing personal expression itself into an object that would be sold. Ironically, this new approach to profit-making was driven by the democratic implications of home-grown, everyday creativity, as seen in Adobe "Yes You Can Have it All" promotional materials:

> Creative Cloud gives you all the apps you need for any creative field you want to explore. You can seriously make almost anything. Create and share an animated video in two minutes. Or make a feature film with the same editing tools they're using in Hollywood. Make a website for your band. Or your business. Discover what a creative genius you are.[85]

So you're now a "genius" to Adobe, not simply an artist. Apple long has incorporated artistic accomplishment into its marketing. Who else could take a common object like a mobile phone and turn it into a "creative" device, as seen in stunning "Shot on an iPhone" photography billboards. Of course, this consumer flattery isn't much different from what companies have always done when suggesting that products can make a person good-looking, popular, or happy. But creativity is a relatively new angle in catering to narcissism and ever more potent appeals to the individualized ("Think Different") self. Critical theorists describe this as a form of subject construction, referring to ways that media not only send messages to consumers, but also work on the ways people see themselves. Both Adobe and Apple encourage aspirations to artistic accomplishment (enabled by certain products), fully recognizing that consumers may never realize their goal. Like being beautiful or handsome, no one can ever be creative enough.

Let's not forget "networking" in all of this—as both metaphor and reality in the internet age. In her book *Magic and Loss*, cultural critic Virginia Heffernan reviewed the strange evolution of the internet from a technology one once "went on" to an existential condition one can "never escape." Likening the effect to a form of involuntary enchantment, Heffernan wrote of her gradual disaffection with literature in the face of online culture:

> Something happened to the novels—Hillary Mantel, a reissue of John Updike—that I ordered in hard copy from Amazon. The spell that had been cast over me by inked letters on white pulp was broken. Or more accurately: a new spell had been cast, on a separate part of my brain.[86]

While not completely discounting the wonders of this new enchantment, Heffernan lamented the commercializing feel of online life and its degradation of artistic sensibility:

> Cruise through the gargantuan sites—YouTube, Amazon, Yahoo!—and it's as though modernism never existed. Twentieth-century print design never existed. European and Japanese design never existed. The Web's aesthetic might be called late-stage Atlantic City or early-stage Mall of America.[87]

Online culture transforms the self in other ways, too. Initially seen as a way of "connecting" and being together, some of the hype about social networking has died down. One can have hundreds of Facebook "friends" if desired, but interaction with them differs greatly from face-to-face relationships. For many users of social networking sites, this has to do with the scripted character of the self, as well as the controlled nature of the interactive exchange. *New York Times* writer Amanda Hess traced this to the online figure of the "avatar"—a term commonly associated with computer games, but also more generally applicable to all virtual representations of the self. The term itself was appropriated from Indian Hindu culture in the nineteenth century by British literati intrigued with its magical signification of a "god descending to the earth in mortal form."[88] Vishnu appeared as a fish, tortoise, and half-lion and half-man, for example. Adopted by U.S. science fiction writers and the game industry in the 1980s and 1990s, the avatar became a synonym for the digital self. Key in this was one's ability to create whatever real or imagined identity one wanted.

The concept of the avatar applies to the many, varied ways people represent themselves in the internet age. "Our avatars represent a self-image fractured across dozens of sites and text bubbles and email chains," Hess wrote.

> We present ourselves differently on Twitter and Tumbler and Slack depending on the norms built into each space. On Facebook, I'm

> posed by a professional photographer ... on SnapChat, I'm burrowing into my office chair ... on Candy Crush, I'm a cartoon otter man.[89]

If the avatar once was a projection of religious values rendered in earthly forms, it now functions more as a mask than anything else. On social networks, avatars allow people to put forth the best versions of themselves. In games, it allows for the acting out of primitive fantasies. But in every use, the avatar is a reminder of the self. All of this gives users a sense of autonomy—from the online construction of the self to the solitary experience of sitting at one's computer. For all of the much-touted "social" aspects of online life, the lived experience of the internet is actually a solitary one, not unlike reading a book. This atomization of internet life rarely is cast as such, especially by the creative industries. Instead, much is made of the pleasures of working on one's own. What is in reality an unorganized, self-supervised, and often exploited workforce is cast in this way as free, empowered, and entrepreneurial.

Is it any wonder that people speak of the new economy as magical—in mysteriously creating social connectedness, entrepreneurial opportunities, and "something" out of "nothing"? Of course, notions of capitalism-as-magic have an historical basis, dating all the way to Adam Smith's famous invisible hand metaphor. But consider this. It is estimated that right now a mere 10 percent of the world's population is without some form of internet access, usually in remote regions or the world poorest countries. Why else would Facebook be launching the latest generation of high-tech drone gliders to ping signals down to the poorest people on earth? Maybe social conscience plays some role. But more likely the company sees an untapped resource in the developing world—or what investment brokers call "emerging" or "frontier markets." According to *Harvard Business Review*, this is where Western multinationals expect to find 70 percent of their future growth—selling things like laundry products and refrigerators, as well as building roads and infrastructure.[90] This is the face of twenty-first-century colonialism.

This chapter has framed the creative industries in terms of magic to highlight the degree to which irrationality, speculation, and wishful thinking often color the industry's upbeat discourse. As one of humankind's most primal impulses, the universal appeal of creativity has proven hard to resist. But as discussion has shown, creativity's benefits often come with strings attached in the fast-moving world of capitalistic enterprise. At their core, the creative industries appeal to fundamental desires for personal autonomy and security at a time when many people feel these very things are under assault. All of this is taking place as the operations of economics and politics become ever more abstract and hard to grasp. The next chapter will delve further into this process, examining in more detail the ways that creativity combines with other factors to shape people's views of themselves, each other, and the ways they come together in collective endeavors. Like that famous moment in the *Wizard of Oz*, it's time to draw back the Wizard's curtain.

Notes

1 Natalie Wolchover, "Why Everyone Believes in Magic (Even You)," *LiveScience* (Apr. 12, 2012) www.livescience.com/19665-belief-magic.html (accessed Mar. 29, 2019).
2 Sander van der Linden, "How Come Some People Believe in the Paranormal?" *Scientific American* (Sep. 1. 2015) www.scientificamerican.com/article/how-come-some-people-believe-in-the-paranormal (accessed Mar. 19, 2019).
3 Courtney Coren, "Harris Poll: Belief in God Falls," Newsmax.com (Dec. 23, 2013) www.newsmax.com/US/harris-poll-religion-god/2013/12/23/id/543450 (accessed Mar. 19, 2019).
4 Joseph Stromberg, "Teller Speaks on the Enduring Appeal of Magic," *Smithsonian Magazine* (Feb. 22, 2014) www.smithsonianmag.com/arts-culture/teller-speaks-on-the-enduring-appeal-of-magic-97842264 (accessed Mar. 26, 2019).
5 Max Haiven, *Crises of the Imagination, Crises of Power Capitalism, Creativity, and the Commons* (London and New York: Zed Books, 2014) p. 12
6 Elizabeth Gilbert, *Big Magic: Creative Living Beyond Fear* (New York: Riverhead, 2015) p. 8.
7 *Big Magic*, front matter, n.p.
8 Richard L, Rapson, *Magical Thinking and the Decline of America* (New York: Xlibris, 2007) p.11.
9 *Magical Thinking and the Decline of America*, p. 21.
10 *Magical Thinking and the Decline of America*, p. 20.
11 *Magical Thinking and the Decline of America*, p. 131.
12 Elizabeth Gilbert, *Eat, Pray, Love: One Woman's Search for Everything Across Italy, India, and Indonesia* (New York; Riverhead, 2007).
13 *Big Magic*, p. 15.
14 *Big Magic*, p. 34.
15 *Big Magic*, pp. 100–11.
16 Michel Foucault, *Power/Knowledge: Selected Writings and Essays*, ed. Colin Gordon (New York: Vintage, 1980) p. 194; see also Michel Foucault, *The Birth of Biopolitics: Lectures at the Collège de France 1978/79*, ed. Michel Senellart, trans. Graham Burchell (Basingstoke: Palgrave, 2008).
17 Pascal Gielen, *Creativity and Other Fundamentalisms* (Mondriaan Fund: Amsterdam, 2013) p. 18.
18 *Creativity and Other Fundamentalisms*, p. 36.
19 *Creativity and Other Fundamentalisms*, p. 25.
20 Dana Polan, "Postmodernism and Cultural Analysis Today," *Postmodernism and Its Discontents: Theories, Practices*, ed. E. Ann Kaplan (London and New York: Verso, 1988) p. 53.
21 Adrienne Arieff and Beverly West, *Fairy-Tale Success: A Guide to Entrepreneurial Magic* (Avon, MA: AdamsMedia, 2014) p. 12.
22 *Fairy-Tale Success*, p. 27.
23 *Fairy-Tale Success*, p. 40.
24 *Fairy-Tale Success*, p. 75.
25 Mike Roy, "The Cinderella Artist Myth" (Sep. 14, 2014) http://artistmyth.com/cinderella-artist-myth (accessed Mar. 22, 2019).
26 Ibid.
27 Richard Florida, *The Rise of the Creative Class: And How It's Transforming Work, Leisure, Community, and Everyday Life* (New York: Basic Books, 2002).
28 Richard Florida, "America's Looming Creativity Crisis," *Harvard Business Review* (Oct. 2004) https://hbr.org/2004/10/americas-looming-creativity-crisis (accessed Mar. 28, 2019).

29 Ibid.
30 *The Rise of the Creative Class*, p. 69.
31 "America's Looming Creativity Crisis."
32 William Deresiewicz, "The Death of the Artist—and the Birth of the Creative Entrepreneur, *The Atlantic* (Jan/Feb. 2015) www.theatlantic.com/magazine/archive/2015/01/the-death-of-the-artist-and-the-birth-of-the-creative-entrepreneur/383497 (accessed Jun. 7, 2019).
33 *Big Magic*, p. 69.
34 "The Death of the Artist."
35 Ibid.
36 Ibid.
37 Angela McRobbie, *Be Creative: Making a Living in the New Creative Industries* (London and Malden, MA: Polity, 2016) p. 75.
38 Thomas Frank, "Why Johnny Can't Dissent," in Thomas Frank and Matt Weiland, eds., *Commodify Your Dissent: Salvos from the Baffler* (New York: Norton, 1997) p, 34.
39 Luc Boltanski and Ève Chiapello, *The New Spirit of Capitalism*, trans. Gregory Elliot (London: Verso, 2005).
40 *The New Spirit of Capitalism*, p. 213.
41 *The New Spirit of Capitalism*, p. 27.
42 Nina Zipping, "Mark Zuckerberg: Entrepreneurship Is About Creating Change, Not Just Creating Companies, *Entrepreneur Magazine* (Jun. 24, 2016) www.entrepreneur.com/article/278128 (accessed Mar. 22, 2019); "Experiment, Take Risks, Be Creative; Sundar Pichai's Advice for Students," *Business Standard* (Jan. 6, 2017) www.business-standard.com/article/current-affairs/experiment-take-risks-be-creative-sundar-pichai-s-advice-for-students-117010501253_1.html (accessed Mar. 22, 2019).
43 "The Arts and Economic Growth 2017," National Endowment for the Arts and U.S. Department of Commerce, (2017) www.arts.gov/sites/default/files/NEAartsandeconomy_April2017_FINAL.pdf (accessed Mar. 22, 2019).
44 "The Creative Workforce," National Assembly of State Arts Agencies (2017) https://nasaa-arts.org/nasaa_research/facts-figures-on-the-creative-economy (accessed Jun. 16, 2019).
45 Ibid.
46 "Career Outlook: Careers for Creative People," U.S. Bureau of Labor Statistics (2015) www.bls.gov/careeroutlook/2015/article/creative-careers.htm (accessed Jun.16, 2019).
47 "Career Outlook: Careers for Creative People."
48 Angela McRobbie, *Be Creative: Making a Living in the New Creative Industries* (London and Malden, MA: Polity, 2016) p. 46.
49 *Be Creative*, p. 12.
50 *Be Creative*, p. 11.
51 Matthew Johnson, ed., *Precariat: Labour, Work, Politics* (London and New York: Routledge, 2015) p. 1.
52 Marion von Osten, "Unpredictable Outcomes/Unpredictable Outcasts," in Gerald Raunig, Gene Ray, and Ulf Wuggenig. eds., *Critique of Creativity: Precarity, Subjectivity and Resistance in the "Creative Industries"* (London: MayFly, 2011) p. 139.
53 Ibid.
54 Gerald Raunig, *Factories of Knowledge, Industries of Creativity*, trans. Aileen Derieg (Los Angeles: Semiotext(e), 2013) p. 102.
55 *Factories of Knowledge, Industries of Creativity*, p. 103.

56 Max Haiven, *Crises of Imagination, Crises of Power: Capitalism, Creativity, and the Commons* (London and New York: Zed Books, 2014) p. 201.
57 *Crises of Imagination*, p. 205.
58 Lewis Mumford, *The Culture of Cities* (1938) (New York: Harvest/HBJ, 1970) p. 3.
59 Nate Berg, "U.S. Urban Population Is Up … But What Does 'Urban' Really Mean?" *The Atlantic* (Mar. 26, 2012) www.citylab.com/equity/2012/03/us-urban-population-what-does-urban-really-mean/1589 (accessed Mar. 24, 2019).
60 Jane Jacobs, *The Death and Life of Great American Cities* (New York: Random House, 1961).
61 Joel Kotkin, "Richard Florida Concedes the Limits of the Creative Class," *Daily Beast* (Mar. 20, 2013) www.thedailybeast.com/richard-florida-concedes-the-limits-of-the-creative-class (accessed Jun. 16, 2019).
62 Sharon Zukin, *The Naked City: The Death and Life of Authentic Urban* (London and New York: Oxford University Press, 2009) p. xii.
63 *The Rise of the Creative Class*, p. 219.
64 "Richard Florida Concedes."
65 Ibid.
66 Ibid.
67 Bradley Bereitschaft and Rex Cammack, "Neighborhood Diversity and the Creative Class in Chicago," *Applied Geography* 63 (Sep. 2015) www.sciencedirect.com/science/article/pii/S0143622815001642 (accessed Mar. 24, 2019).
68 Richard Florida, *The New Urban Crisis: How Our Cities Are Increasing Inequality, Deepening Segregation, and Failing the Middle Class—And What We Can Do About It* (New York: Basic Books, 2017) p. 115.
69 *The New Urban Crisis*, p. ii.
70 *The New Urban Crisis*, p. 115.
71 "Department of International Magical Cooperation," Harry Potter Wiki (n.d.) http://harrypotter.wikia.com/wiki/Department_of_International_Magical_Cooperation (accessed Mar. 25, 2019).
72 "Global Creativity Class," *Global Creativity Index 2015*, Martine Prosperity Institute (2016) http://martinprosperity.org/media/Global Creativity-Index-2015.pdf (accessed Mar. 25, 2019).
73 Richard Slotkin, *Regeneration Through Violence: The Mythology of the American Frontier, 1600 1860* (Norman: University of Oklahoma Press, 2000).
74 Sam Westminster, ed., *The Essential Weber: A Reader* (London and New York: Routledge, 2003) p. 109.
75 R.H. Tawney, "Forward," in Max Weber, *The Protestant Ethic and the Spirit of Capitalism*, trans. Talcott Parsons (1930) (Oxford: Oxford University Press, 2010) p. I(e).
76 Greg Richards, "Creativity and Tourism: The State of the Art," *Annals of Tourism Research* 38, no. 4 (2011): 1230.
77 UNESCO. (2006). "Towards Sustainable Strategies for Creative Tourism: Discussion Report of the Planning Meeting for 2008 International Conference on Creative Tourism," Santa Fe, New Mexico, UNESCO (Oct. 2006) http://unesdoc.unesco.org/images/0015/001598/159811e.pdf (accessed Mar. 26, 2019).
78 "The Creative Tourism's Assets," Creative Tourism Network (2012) www.creativetourismnetwork.org/about (accessed Mar. 26, 2019).
79 "Creativity and Tourism," p. 1244.
80 Norbert Vanhove, "The Magic Tourism Multiplier," in *The Economics of Tourism Destinations* (London and New York: Routledge, 2010) p. 239.

81 "Interactive Wands & Spell-Casting in the Wizarding World—Complete Guide," *Orlando Informer* (n.d.) https://orlandoinformer.com/universal/interactive-wands (accessed Mar. 26, 2019).
82 "Techne," Oxford Dictionaries (2017) https://en.oxforddictionaries.com/definition/techne (accessed Mar. 27, 2019).
83 Richard Stivers, *Technology as Magic: The Triumph of the Irrational* (New York: Continuum, 2001) pp. 1–2.
84 Laura M. Holson, "We're All Artists Now," *New York Times* (Sep. 4, 2015) www.nytimes.com/2015/09/06/opinion/were-all-artists-now.html (accessed Mar. 3, 2019); Sam Louie, "We're All Artists," *Psychology Today* (Apr. 29, 2017) www.psychologytoday.com/us/blog/minority-report/201704/were-all-artists (accessed Mar. 3, 2019).
85 "Yes You Can Have it All," Adobe Corporation (n.d.) www.adobe.com/creativecloud/buy/students.html (accessed Mar. 24, 2019).
86 Virginia Heffernan, *Magic and Loss: The Internet as Art* (New York: Simon & Schuster, 2016). p. 4.
87 *Magic and Loss*, p. 24.
88 Amanda Hess, "What Do Our Online Avatars Reveal About Us?" *New York Times* (May 10, 2016) www.nytimes.com/2016/05/15/magazine/what-do-our-online-avatars-reveal-about-us.html?mcubz=1&_r=0 (accessed Mar. 27, 2019).
89 Ibid.
90 Matthew Ryring, Mark W. Johnson, and Hari Nair, "New Business Models in Emerging Markets," *Harvard Business Review* (Jan.–Feb. 2011) https://hbr.org/2011/01/new-business-models-in-emerging-markets (accessed Mar. 29, 2019).

Part IV

Creative Societies

While generalities sometimes are sloppy, they also can yield critical analytical dimensions—especially when constant Tweets, texts, and "breaking news" keep one focused on immediate details. National or societal views give perspective, while often forming the basis for public policy and federal politics. This section takes this wider view, looking at creative economies around the world, population attitudes toward creativity, and the prospects for digging the U.S. out of its anxious funk. Discussion begins with arguments for creativity as a collaborative enterprise rather than the exclusive product of individual minds. Initially advanced by psychologists and sociologists, concepts of collective consciousness and shared imagination increasingly find their way into contemporary discourses of communications, economics, media studies, neuroscience, philosophy, and politics.

Decades ago, Fredric Jameson argued that many in the U.S had lost the ability to imagine the future. He said that many were seeing "progress" as an empty promise and "utopia" an idealistic fantasy.[1] Jameson concluded that the postmodern age had ushered in an intensified "aestheticization" of public culture, which suppressed people's capacities to envision a better life, replacing it with empty consumerism and nihilist confusion. At the same time, he held out a certain possibility in the collective sensibility that Raymond Williams called the "structure of feeling," if this subjective domain could be activated democratically. He saw critical consciousness and even dystopian creative works as vehicles for this. Keeping this in mind, *Anxious Creativity* concludes with discussions of ways art and media can both reflect and inspire common aspirations for a more generous and caring world, even if these visions remain more as ideals or "creative" horizons than present reality.

"Creativity isn't your baby. If anything, you're its baby."[2] With these words, author Elizabeth Gilbert neatly summarized the generative effect of creative activity on makers themselves. Chapter 10, "Becoming Creative: The One and the Many," reviews how artists and writers speak about the sense of discovery they encounter when working on a painting or essay. Partly, this has to do with the open-endedness also common in other forms of research. But it also pertains to the relational quality of works made with other people in mind. This notion of "becoming" has been explored by thinkers including Judith Butler, William Connolly, Gilles Deleuze, James Leach, Friedrich Nietzsche, Gilbert Simondon, Alfred North Whitehead—all of whom pondered how the "self" unfolds over time. Contemporary neuroscience has explored both these processes of change and the interconnections between people's minds, confirming earlier theories of "actor networks" (Bruno Latour) or a dynamic "social brain" (Steven Sloman, Philip Fernbach) quite in contrast to conventional views of a static individual identity. Put another way, creativity derives largely from meanings people have in common.

In his book *Doing Things Together*, sociologist Howard Becker wrote of his experience as a jazz musician to illustrate the process of collaborative creativity.[3] If players don't share certain core understandings, they can never play a song or improvise together. Chapter 11, "Distributed Creativity: Toward a Sharing Economy," addresses how this fundamental premise explains many kinds of collaborative effort—in creative expression, scientific research, business, and so on. It also contradicts beliefs that individualism and competition are necessary components of creative endeavor. Business groups saw this in the 1950s practice of "brainstorming" developed by Alex Osborn.[4] These days, a broad movement is gaining momentum to encourage people to work together in the interest of innovation, rather than secreting their work in "silos" of separate (and often duplicate) efforts. Leading scholars and policymakers around the world now encourage the concepts of "distributed creativity" (Petre Glăveanu) and a "sharing economy" (Joseph Stiglitz) as a way of moving forward. Others discussed include Teresa Amabile, Claire Bishop, Mikhail Bakhtin, Chris Csíkszentmihályi, Grant Kester, and Jean-Luc Nancy.

How do Americans see the future? This is the theme of Chapter 12, "Imaginary Worlds: Utopia and Virtuality." Many scholars now argue that the current anxious climate has blunted public imagination to the possibility of a better society. The chapter reviews thinking by scholars Alain Badiou, Cornelius Castoriadis, Terry Flew, Paulo Freire, Henry A. Giroux, Fredric Jameson, Jacques Lacan, Ursula Le Guin, and José Esteban Muñoz, all of whom in different ways commented on what C. Wright Mills termed the "sociological imagination." Mills was concerned about people's tendencies to get caught up in the present moment and forget that a different world might be possible. Creative works can hinder or help in keeping utopian thinking alive, even though people's visions of the future may differ. This requires an acknowledgment of "multitudinous creativity" (Giuseppe Cocco, Barbara Szaniecki) as a "complex adaptive system" (Murray Gell-Mann), as well as the recognition that history alone does not determine what might happen next. As Muñoz put it, "We must strive, in the face of the here and now's totalizing rendering of reality, to think and feel a *then and there*."[5]

Notes

1 Fredric Jameson, "Progress versus Utopia; Or, Can We Imagine the Future?" *Science Fiction Studies* 9, no. 2 (Jul. 1982): 147–58.
2 Elizabeth Gilbert, *Big Magic: Creative Living Beyond Fear* (New York: Riverhead, 2015). p. 233.
3 Howard Becker, *Doing Things Together: Selected Papers* (Evanston: Northwestern University Press, 1986).
4 Alex Osborn, *Your Creative* Power (1948) (New York: Myers, 2007).
5 José Esteban Muñoz, *Cruising Utopia: The Then and There of Queer Futurity* (New York and London: NYU Press, 2009) p. 1.

Chapter 10

Becoming Creative

The One and the Many

One more thing about Elizabeth Gilbert's *Big Magic*. For all of its shortcomings, the book makes one *very* valuable point about becoming more creative. Despite her clichéd mumbo-jumbo about finding one's "inner artist," Gilbert is quite direct about the fact that creative ideas are not one's own. While romanticizing artistry as "magic," Gilbert also generously credits her friends and colleagues with passing the enchantment among each other. In seeing this connection, she provides a key insight about the relational quality of creativity. So-called "self-expression" never occurs in a vacuum. The creative process is nearly entirely learned. People imitate what they see around them. And most artworks are made with a viewer in mind. For this reason, creative acts are essentially interactive and subject to external influence—rather than simply the singular "vision" of an isolated maker. People are social creatures, no matter what anyone says.

This chapter takes a close look at the creative person: the way makers see themselves and are regarded by others, how myths and conventions have tempered these views, and what can be done to recognize both exceptional accomplishment and the creativity residing in everyone. Returning briefly to the populist themes of *Big Magic*, discussion turns next to the mythic artist-genius and its role in today's creativity craze. High achievement always has driven American Dreams of boundless opportunity and freedom to succeed. But at what cost? As these celebrated values conjoin with their bedfellows of competition and greed, America's creativity is starting to wither, along with impulses for experiment and innovation, as well as commitments to diversity and alternative voice. Not only does competition seem a detriment to artistic thinking, but concomitant worries about failure now generate a creeping conformity.

Is that all there is? Fortunately, longstanding lines of philosophical thought have taken more expansive and inclusive views of artistry. This chapter will conclude with some of this thinking—and analyze why it gets buried in the current creativity craze. From the ancient Greeks through figures such as Hegel and Nietzsche, philosophers have pondered the *connections* of individuals to society. These questions grew even more pointed as human strife reached global proportions in the twentieth century. Theories of subjectivity and language would critique divisive impulses of all kinds, while pointing to the role of culture in constructing the "self." George Herbert Mead, Gilbert Simondon, Gilles Deleuze, and Bernard Stiegler each made subsequent contributions in reconciling the one with the many, or at least explaining their interactions. Recent advances in cognitive science have given this metaphysics a more practical grounding. Beginning with views of the mind as a computer for the body, scientists now see people's brains connected to each other (and to physical objects) through networks. In this view, creativity and "knowledge" reside not in any one person's consciousness, but instead in shared or distributed forms of cognition.

Magic or Real?

So I'm going to take Elizabeth Gilbert's *Big Magic* on its own terms, and ask you to bear with me. As discussed earlier, Gilbert romanticizes creativity as a mystical force much as seen throughout human history (as alchemy, spirituality, etc). At one point, Gilbert even suggests that creative ideas exist in the world as spiritual objects one can stumble upon or pass to another person. She tells a story about a complicated novel plot she had been pondering, which Gilbert accidentally transmitted to author Ann Patchett in a friendly kiss. Silly as this sounds, it makes an important point in suggesting that creativity is not strictly an aspect of "self." Even as Gilbert wraps her views of creativity in mystical rhetoric, she also makes a break from traditional views of individualized talent and inspiration. Perhaps inadvertently, *Big Magic* radicalizes creativity as a relational phenomenon *between* rather than *within* people.

What does it mean to say that creative work is not one's own? In Gilbert's universe, the question is all explained as *Big Magic*—attributed to muse-like figures and posited more or less as a matter of faith. Now, Gilbert is hardly known as a deep thinker. And she's unlikely to muck up her flowing prose with deconstructionist jargon. Instead, she personifies tricky concepts, speaking about them like people. Gilbert recounts long chats she has with her friend "Creativity," and describes "Genius" as a visitor who drops in to see her unexpectedly. All of which is to say that she externalizes qualities typically treated as elemental components of artistic core being. It's a quirky kind of humility to be sure, but it's also refreshing in world in which the self is so often portrayed as a person's most valued possession. As Gilbert writes, "Your creative work is not your baby. If anything, you are its baby."[1]

Gilbert's quip points to the political and ethical stakes of "becoming creative." Besides taking issue with common assumptions about innate talent, the notion of passing creative ideas from one person to another also counters premises of individual artistry itself, not to mention the competitive impulses that encourage people to secrete themselves away from others, guard their insights, or otherwise avoid cooperating or sharing. This throws into relief the ways artistic "greatness" has been hyped and commodified, while the creative activities of most people are trivialized. Certainly, Gilbert's thinking promotes its own brand of new-age mysticism. But her message of empowerment has a bit of merit, especially as it pertains to those whose creative activities are written off as "amateur," "domestic," and generally unimportant.

Of course, the concept of artistic genius is an artifact of history, emerging, as it did, with the Enlightenment Era's celebration of the autonomous, reasoning subject. This lonely figure emerged from newly forged ideals of individualism and "unique" talent, as discussed in Chapter 5. Virtuosity was construed as a natural or divine gift, often arriving without training and requiring little encouragement. It bears mentioning that this fascination with personal mastery also paralleled the rise of capitalism in Europe and the birth of copyright law. Inspiration may have come from above, but publishing houses, museums, and art dealers gave it prodigious earthly value, as creative traits of "originality" and uniqueness became selling points in a culture increasingly intrigued with novelty. By the time the modern era rolled around, artistic vision also became synonymous with conjurings of "the new."

But let's go back to *before* the solitary genius showed up. For most of that history, creators were more likely seen as deft imitators—and celebrated as such—in millennia of art-making striving to reproduce the wonders of nature and its presumably God-given perfection. Think of Michelangelo's meticulous renderings of the human form and Da Vinci's efforts to replicate perspective, all in the interest of "copying" nature with ever more exactitude. Shakespeare's brilliant plays contained few "original" ideas, frequently borrowing characters and plots from myths or other plays. Elizabethan society recognized imitation as a vital human impulse, much as earlier seen in the ancient Greek concept of *mimesis*. Looking back even further, anthropologists date humanity's "Creative Explosion" to Paleolithic times (10,000–40,000 years ago) when early hominins first sought to replicate their world in carvings and cave paintings. Tradition also figured prominently in pre-genius times. No one thought creative people were "born" with special skills. Craft abilities passed from generation to generation, with apprentices learning to emulate "masters" in painting and sculpture, much as was done in carpentry or bricklaying. In fact, being an artist wasn't regarded much differently from any other trade, with what we now call creative ability simply seen as one of many skills someone could acquire.

Beginning in the nineteenth century, industrialization took the artistry out of many jobs. Machines replaced hand tools in the fabrication of mass-produced goods, as work was mechanized in the factory assembly line.

Artists and writers responded to this rising dehumanization with the Romantic Movement, which placed a heightened emphasis on the subjective experience of the individual. This only intensified society's growing hunger for genius and masterpiece. The artist became an existential role model for audiences seeking transcendence or escape from the drudgeries of working life. While capitalism also played its part in commodifying creative products, deeper yearnings drove the period's culture. The Romantic Era celebrated emotionalism, originality, nature, and nostalgia—in reaction to rationality, mass production, artificiality, and the present. Romantics quite literally encouraged a disassociation from the here-and-now with a retreat into the secluded and presumably "authentic" realm of personal reflection. As John Keats wrote, "Think of my Pleasure in Solitude, in comparison to my commerce with the world."[2]

Romantic notions of solitude and its pleasures comported well with foundational American values of privacy. The U.S. Constitution's Fourth Amendment of 1789 guaranteeing the "right of the people to be secure in their persons, houses, papers, and effects, against unreasonable searches and seizures" was a reaction against intrusive British rule and the quartering of soldiers in people's homes.[3] Hence, privacy laws came to reinforce individualism as a "right" to be separate or protected from others. In later years, privacy accrued moral inflections linked to religious and sexual inclinations. As Romantic Era sentiments grew in the nineteenth century, Samuel Warren and Louis Brandeis wrote in their often-cited *Harvard Law Review* essay "The Right to Privacy" of an overarching entitlement to personal thoughts, sentiments, and emotions. Summarizing this "inviolate principle," Warren and Brandeis said that everyone deserved the "right to one's own personality."[4]

While the Romantic Era eventually would lose ground to rising modern sensibilities, values of aesthetic originality would remain intact. In one way or another, the image of the artist as privileged visionary persisted in Fauvism, Cubism, Futurism, Dada, Surrealism, and Abstract and Neo-Expressionism. Some of these movements directly sought to counter the canonizing, commodifying, and exclusionary practices of a perceived "high culture" and its entitled audiences. But in every instance, the overarching logic of the art marketplace would take the day, turning even the most radical and subversive practices into new products for museums and collectors, much as Boltanski and Chiapello would discuss.[5] As Hegel famously said about antagonisms, opposing forces often start to look alike—even if the one side shapes the outcome. In this case, radical artists unwittingly reinforced the very celebrity system they often opposed. While some of this can be written off as an inevitability of life within capitalism, the answer may lie deeper in personal and social terms.

Definitions and purposes of art have changed a lot over time—often reflecting other social values. To get around this, historian Janet Wolff has suggested use of the term "production" instead of "art" to restore creativity to its place alongside other kinds of work. As Wolff wrote in *The Social Production of Art*: "Insofar as people, including artists, are socially and historically located, and members of particular social groups, their thought, including their artistic ideas, is ideological."[6] Writing in the early 1980s, Wolff was part of a then-rising "social turn" in art history and practice in which creativity was seen as a "constructed" enterprise, rather than a timeless or universal pursuit. This marked a big departure from reigning norms of art appreciation and connoisseurship, which saw art existing outside (and above) everything else.[7] The social turn in art history largely focused on the relationship of creativity to the material realities around it—both good and bad ones. For many art makers of the period, this meant working "outside the studio" in non-traditional formats (installation, performance, and media forms) or in a broader engagement with audiences, especially those not typically thought of as consumers of art.

This push-back against idealized creativity gained further impetus in the 1980s and 1990s from within the art world's own "institutional critique" and postmodern theory movements. In the former, looking at art in societal terms led artists and critics to contemplate the roles of museums, galleries, art patrons, and corporate interests in propping up values of genius and signature style, often for their own benefit or commercial gain. On the theoretical side, art critics were partaking of the broader embrace of structuralist linguistics, which questioned the stability of meaning—and, by extension, the grounds of received knowledge. Postmodernism took direct aim at the idea of originality as well as the modernist infatuation with the new—arguing that both were relative concepts, varying from viewer to viewer and subject to cultural suasion. Some postmodernists even resurrected the philosophical notion of mimesis in arguing the primacy of the "copy" as an end in itself. In a well-known example of this, artist Sherrie Levine displayed photo-reproductions of famous images by Walker Evans and Edward Weston as "her" work in this new spirit of appropriation.

Postmodernism raised important questions about people's abilities to understand themselves and the world around them. Immersed in a sea of images and representations, is it ever possible to tell truth from fiction? In this sense, it's hardly incidental that much of the postmodern conversation centered on aesthetics and media. The resulting intellectual shake-up sent artists looking for new ways to explain creative activity. During the 1980s, "critical theory came into general art discourse in the United States and soon became part of the everyday currency of the art world," one anthology explained.[8] Before long, "the dissemination of 'theory'—that is, writings in history, culture, and other forms of representation from a variety of disciplines and intellectual perspectives—was also for the first time systematically applied to the study of art," drawing from such fields as "continental philosophy, structuralism, feminism, psychoanalysis, literary studies, new historicism, and post colonialism, among others."[9]

This outpouring of theory threw conventions of artistic merit into crisis. As critic Thierry de Duve wrote, "Creativity soon lost its aura," explaining:

> Philosophically speaking, the times were very suspicious of anything more or less resembling the old psychology of the faculties, and creativity, which is a neo-romantic amalgam of the Kantian faculties of sensibility and imagination, became old hat. It had everything against itself: being universal, it could only be "bourgeois"; being

> transcendental, it could only be "metaphysical": being universal, it could only be "ideological."[10]

For better or worse, postmodernism's emphasis on the contingent self soon would be met by the rising politics of difference, which appeared to rescue creative originality by assigning it to specific identities. This suggested that aesthetics may not be the same for everyone, but that individuals nevertheless could carve out their own creative experiences, histories, and ambitions.

Self(ish) Expression

"The cult of individualism is something every American sees but few even try to understand," wrote Aaron Barlow in *The Cult of Individualism: A History of an Enduring American Myth*. "What we rarely consider is that individualism, as practiced in America, is as much a cultural phenomenon as a human one."[11] The nation's relentless preoccupation with individual knowledge, expression, and gain has led many to discount or forget the shared aspects of existing together—deriving from an often ideologically driven opposition to the collective dimensions of life itself. Indeed, these days it's hard to imagine the idea of "being" in the world as anything other than a "self" because singular personhood has been so systematically drilled into people via upbringing, school, work, and entertainment. Yet, for a century or more, academics in many fields have argued there is nothing "natural" or predetermined about people's self-perceptions as isolated beings. Anthropologists, biologists, psychologists, sociologists, and, more recently, even neuroscientists conclude that humans are social creatures above all else, as Aristotle famously set forth when he wrote that "society is something that precedes the individual."[12]

This is borne out in cultural history. In the Europe of the Middle Ages, one's identity was a matter of family, faith, social class, and community rather than a unique aspect of one's self. People saw themselves defined by their relationships with others, the groups to which they belonged, and the histories from which they emerged. Often this was tempered by an overarching religiosity that, while simplistic, also gave people a sense of membership or totality (as sinners, believers, "God's children," etc.). It is often argued that what is now seen as individualism arose most vigorously as "reason" and rationality began to secularize societies, especially in the Western world from the seventeenth century onward. Early sociologist Émile Durkheim proposed that such rationalism would feed people's latent egoism as the influence of theological restraint diminished. These individualist factors conjoined with particular potency in the social experiment known as the United States.

No one needs reminding of the centrality of individualism in American history, as the nation came to see itself as a unique case in human history—from the War of Independence to the current era of American exceptionalism. Idealized notions of "self-knowledge" and "self-expression" have deep roots in the American past, dating to the founding of the republic itself and its grounding in newly established views of the individual. From the beginning, the American self and its acquisitive inclinations were premised on ownership, as discussed in Chapter 3. Underpinning Adam Smith's "invisible hand" thesis was the belief that self-interest made life better for everyone, yielding quality improvements and higher living standards.[13] This might have made sense to America's wealthy founders, but things didn't always go well for everyone else. After all, freedom to thrive also meant freedom to fail in the cut-throat game of upward mobility. Before long, competitive pressures revealed the dark side of America's utopian experiment. Within a few decades of the nation's founding, Alexis de Tocqueville would observe in *Democracy in America* a U.S. embroiled in a "daily battle" in the "vast lottery" that life had become. By the mid-1800s, urbanization and mechanization only intensified inequality and class division in an America of "winners" and "losers." As de Tocqueville further explained, "What our ancestors in the Middle Ages called servile greed, the American calls noble and estimable ambition."[14]

Few people know that the expression "rugged individualism" was coined by Herbert Hoover in the Depression Era of the 1930s. Not unlike Donald Trump's exhortation to "Make America Great Again," Hoover's familiar "Up-By-Your-Bootstraps" philosophy told citizens they could make it on their own, without any meddling by government.[15] In "The Future of American Individualism," Hoover would write: "The spirit need never die for lack of something for it to achieve. There will always be a frontier to conquer or hold as long as men think, plan and dare."[16] Such attitudes held sway until the 1960s, when Christopher Lasch, among others in the counterculture, would criticize America's fetish with the self in *The Culture of Narcissism*, as discussed in Chapter 2. Looked beyond partisan politics, Lasch said the problem was more about "long-term social changes that have created a scarcity of jobs, devalued the wisdom of the ages, and brought all forms of authority (including the authority of experience) into disrepute."[17] Sound familiar?

Today America is the most self-centered nation in the world. And it's also the wealthiest. The U.S. generates the world's highest gross domestic product ($18.5 trillion), exceeding the combined output of the 28-nation Europe Union ($16 trillion), and third place China ($11 trillion).[18] But in a trend that worsens with each passing year, America also leads the world in wealth inequality. America's top 10 percent now earn nine times the amount of the bottom 90 percent.[19] This divide is double what it was half a century ago. Numbers are even more startling among the super-rich. The top 1 percent takes home 38 times what the lower 90 percent receive. Put in terms of actual pay, this means that the top 1 percent earns an average of $1.3 million per person, as compared with $16,000 per year for the bottom half of Americans.[20] The future doesn't look too good either, with millennials and others born later having a 50 percent likelihood of making less than their parents.

In one of the most comprehensive studies of individualism worldwide, sociologist Geert Hofstede and a team of researchers compiled national rankings of ties between people and their social networks (friends, families, groups). In the "individualism" scoring, the U.S. came out in first place (score of 91), followed by nations like Australia (90), United Kingdom (90), Netherlands (80), and New Zealand (79). Lower scores occurred in more collectivist countries, including Brazil (38), Mexico (30), Hong Kong (25), and China (20).[21] These findings are nothing new. Similar studies beginning in the 1950s by psychologists and anthropologists such as Raymond Cattell, Michael Harris, and Michael Minkov all yielded the same results.

Creativity falls into a similar pattern. In perhaps the only recognized assessment of international creative output, the Martin Prosperity Institute's Global Creativity Index (GCI) reports numbers strikingly similar to those for individualism. As above, the top countries are listed as Australia (.97), U.S (.95), New Zealand (.94), Netherlands (.89), and United Kingdom (.88). These contrast with Brazil (.60), South Africa (.56), Mexico (.41), and China (.42).[22] In many instances, high creativity apparently correlates with high inequality—as seen in Great Britain and the U.S. National culture and economic orientation obviously play roles in this. But the intersection of self-interest, industrial creativity, and economic injustice is hard to ignore.

Critics haven't minced words about these contradictions, pointing out that "resourceful" and "flexible" creatives are often underpaid and underemployed. Worse still, the industry paints these exploited workers as role models. As McRobbie has observed, "Artists are being held up not as poverty-stricken malcontents, but as triumphant 'pioneers of the new economy.' Today when the idea of a good, steady, lifelong job seems impossible, corporate propaganda encourages us all to see ourselves as artistic souls."[23] And, of course, as U.S. income disparities reach ever more obscene proportions, the earnings of celebrity artists like Jeff Koons (whose sculptures can fetch $50 million) and Christopher Wool (paintings going for $30 million) now compare favorably with annual incomes of movie "creatives" such as Matt Damon ($55 million), Tom Cruise ($53 million), and Jennifer Lawrence ($46 million). [24]

Superstar artistry is one reason most Americans don't see themselves as creative. Highly visible artistry puts pressure on people to find whatever creativity they can in their daily lives. Imaginative faculties get pushed inward or displaced onto other kinds of activities—which, admittedly, some experts say is a good thing. After all, everyone can't be famous. But more often than not, this means that personal creativity comes down to buying things to adorn one's life. Unfortunately, the purchase of a new pair of jeans provides but a short respite as the competitive aspect of consuming takes over—famously satirized in the 1920s comic strip *Keeping up with the Joneses*, later rendered as a Zach Galifianakis film in 2016.[25] In other instances, this is seen in the culture of self-improvement: reading a book, getting in shape, or maybe taking an art class to nurture that inner artist. This confirms in a general way how America has become more and more a place of atomized subjects—isolated from one another in private domains of consumption and media use. As diversity is the official norm of American society, people are told they can all be "special" like the proverbial snowflake—even as the self is commodified in ever more intimate ways.

The Art Competition

It's no big surprise that creativity has become a horse race. Nobody can be "creative" enough. People in the U.S. compete in just about everything: parenting, sports, intelligence, school, appearance, work, spending, and even relationships—as "America First" has become a national mantra. But when it comes to artistry, those who study competition are of two minds. "The relationship between competition and creativity is not a simple and direct one," creativity expert Mark A. Runco has explained: "Sometimes competition stimulates creativity, and sometimes it does not."[26] Keep in mind that competition is both an internal state (doing one's best) and a relational quality (doing better than others). This is why healthy desires for continual improvement can diminish with fears of failure. Or, as Runco put it, creativity can suffer when extrinsic pressures outstretch intrinsic benefits.

Roland Barthes saw competition in just these terms when he wrote of sports as a social ritual. Obviously, sporting events convey narratives of winning and losing, self versus other, or even good against evil. But it's the unknown outcome that gives any game its real drama. Philosophically speaking, "What wins the race is a certain concept of man and the world," Barthes wrote: "The concept is that man is proven by his actions, and man's actions are aimed, not at the domination of other men, but at the domination of things."[27] By providing symbolic control of the outside world, sports help people come to terms with their feelings of uncertainty and powerlessness. Specifically, Barthes noted that in games like basketball, football, hockey, and soccer, the "goal" is an empty space waiting to be filled. The final score relieves the absence. Hence, sports function much like other belief systems (religion, democracy, capitalism, etc.) as vehicles for psychic yearnings. This also is why no single sporting outcome ever dampens the love of the game. The unending process of contest is the actual attraction.

Competition also is a lifelong event. Newborns are "scored" for heartbeat, breathing, and reflexes, launching a sequence of measurement and comparison to follow them ever after. Height, weight, and health get measured by physicians, along with developmental milestones, language, and intelligence monitoring. And with school come grading, report cards, and standardized testing, not to mention the ratcheting pressures of social life. Most people can remember a childhood art contest or an audition for a school play—not unlike sports team try-outs. Families generally see this as a healthy early exposure to the "life lessons" of winning and losing. But what such competition really teaches is a naturalization of external assessment—thereby instilling a hunger for validation by means of superiority or winning. Unfortunately, one doesn't need to be statistician to grasp that everyone can't be a winner, that most people don't land at the top, and that this mindset leaves most people feeling badly about themselves.

For young people, such early assessments can play a critical role in whether or not they see themselves as creative. Educators commonly observe that children move away from painting and performing as they get older, as discussed in Chapter 7. "Playtime" ends sometime during middle school, society seems to say. In an often-quoted line about this, education expert Sir Ken Robertson has said, "We don't grow into creativity. We grow out of it."[28] Parental pressure and school testing are big factors. In one of the largest studies of youthful creativity done to date, researchers looking at 300,000 children recently found significantly declining abilities "to produce unique and unusual ideas," according to study director Kyung Hee Kim of William and Mary College. Kim said that competitive pressures also affect attitude and mood, making children "less humorous, less imaginative and less able to elaborate on ideas."[29] These days, childhood has become over-scheduled and over-supervised, leaving little room for imagination and unstructured play, Kim added, laying much of the blame on the No Child Left Behind Act, enacted by Congress in 2001 to tie federal funding to standardized test results. How can creativity survive in such an environment? Kim said, "If this trend continues then students who look different or nonconformist will suffer, because they are not accepted."

No less a publication than *Forbes Magazine* recently chimed in with a feature entitled "How Kids Lose Their Creativity as they Age."[30] Citing IQ and Torrance Tests of Creative Thinking conducted since the 1950s, *Forbes* noted scores consistently dropping in American children in recent decades. Terming current public education a "factory system," the article said that schools are crushing artistry in ways that will damage business in the long run, concluding that "creativity is the only factor to differentiate ourselves from robots." To fix this problem, *Forbes* recommended more creative problem-solving and less rote memorization, a heightened focus on critical inquiry over "as is" learning, and the encouragement of diverse thinking over conformity. Schools should reassure students that "mistakes are not evil, and should not be feared." After all, most people learn from trial and error, and failure can be a portal to change.

Nice as this might sound, the reality is that mistakes and failure don't sit well with American education and business. Anything promising a predictable outcome tends to win over the hard-to-define and often-messy process of artistry. This led schooling expert Alfie Kohn to say that competition promotes sameness and repetition more than anything else. "Creativity is anticonformist at its core; it is nothing if not a process of idiosyncratic thinking and risk-taking. Competition inhibits this process."[31]

Going even further, Kohn wrote that "The simplest was to understand why competition generally does not produce excellence is to realize that *trying to do well and trying to beat others are two different things.*"[32] In other words, competition not only confuses a pair of opposing attitudes, but it also favors some kinds of people over others. "In a win/lose framework, success comes to those whose temperaments are best suited for competition. This is not at all the same thing as artistic talent, and may well pull in the opposite direction," Kohn concluded.[33]

This helps explain why competition is hurting creativity in the business world. Companies find themselves caught in a double bind of wanting innovation, but also needing to avoid failure. Such pressure is seen in every giddy announcement of a new phone or car model—instantly met by reports of market success or failure, often regardless of quality or aesthetic merit. This also has led to the risk-avoidance evident in the familiar sequel phenomenon of the movie and TV industry. For over a decade, economists have warned of an American "innovation crisis" resulting from the short-term profit-seeking underlying such behavior.[34] Aside from the corporate bottom line, this also generates a generalized public perception of the marketplace as the final arbiter of all value. Here again, it is worth remembering the relationship of competition to perceptions of scarcity. At a time when everyone is being urged to be creative, some inevitably will be seen as more so than others—as pressures build to acquire as much creativity as possible.

That said, recent evidence suggests that *some* competition can motivate creative people to produce novel, untested ideas. But by the same token, *too much* competition can have the opposite effect, according to researcher Daniel P. Gross of the Harvard Business School. In a paper entitled "Creativity Under Fire: The Effects of Competition on Creative Production," Gross discussed findings from a study of graphic designers competing in multiple rounds of 122 contests.[35] He noticed two telling phenomena. If the tournament had only a few participants, designers felt free to try new ideas and take risks. But in large competitions they reverted to well-established formulas. High innovation also appeared in lower-stakes rounds of contests, but similarly vanished as the prize got closer. Gross explained that competitive "incentives have nuanced, multidimensional effects on creativity," explaining that "some degree of competition is required to motivate high-performing agents to explore originality."[36] Extrapolating from these findings, Gross concluded that competition could be good for those working in small groups or in firms where people saw themselves as peers. On the other hand, very large competitions (like those for government contracts or large foundation grants) drive applicants toward low-risk, conformist thinking.

The question is how to handle this. Can anything be done to temper America's love/hate relationship with winning? Intellectuals struggling with this problem often have gotten stuck on the seemingly all-or-nothing opposition between cooperation and competition. Unfortunately, such thinking tends to polarize people into mutually exclusive camps. In what follows, I want to suggest that, as with *Big Magic*, the answers may not be simple. Maybe a little creative thinking itself can help. But this can happen only if artistry first can be rescued from the singularly individualized and utilitarian corners into which it's being pushed. Regarding creativity as both an individual quality *and* a shared resource can open its meaning for everyone—maybe even changing the way people think about business as well.

Becoming Creative

Consider this. Nobody wants to squelch creativity: not families, educators, or business leaders. Everyone knows that being more creative is a good thing, especially nowadays. But people get drawn into stifling mindsets by worries about the future. This anxiety sends kids the message that creativity is great but less important than practical skills. It tells well-meaning teachers that their jobs depend on numbers. And it pushes corporations to avoid the risks necessary for innovation. What can be done? Can anything temper the quandary of both individuals and companies in their conflicted relationships with creativity? This isn't exactly a new problem. But the dilemma keeps dogging the current creative economy debate nevertheless. This partly comes from messy understandings of creativity itself, even as most people believe they know what it is. Confusion allows creativity to be overdetermined by extrinsic factors like money, self-interest, and competition. Important as these values may seem, they can diminish understandings of artistry as the complex and multifaceted topic it is. Salvaging creativity will entail keeping its definition open and broadly inclusive to accommodate its remarkable varieties and potentials. Fortunately, plenty of models for doing this can be found in history and around the globe. The key point is that simple, short-term ways of looking at creativity foreclose its promises of novelty and experiment.

This book has argued that norms surrounding creativity have been narrowing in the new millennium, as the creative economy popularizes artistry while constricting its definition. Nowadays "everyone" can be an artist, but only in terms of individual interest or marketable pursuit. Rather than a broad-spectrum human endeavor, creativity gets assigned a specific range of tasks. Societies often have used art this way. Today's rationalized and utilitarian view of creativity isn't much different from its deployment as propaganda or nationalistic symbol. The big problem with the new creativity is its suppression of difference and diversity, foreclosure of alternatives and dissent, and an overall reduction of the very novelty and inventiveness that creativity supposedly embodies. The reasons for this aren't very complicated. In anxious times, people crave the certainty of what appear to be clear answers, with little tolerance for ambiguity or complexity. Add economic stress to the mix and materialism can infect even the most idealistic projects. Worries over gain and loss then can enter the picture, as they have in the new creative economy.

Regrettably, this materialism can obscure deeper values—like yearnings for connection, security, and a future safe from harm.

In all of this, anticipation is perhaps the key feature of societal anxiety, tempered by the frustrating reality that no one really knows what lies ahead. Martin Heidegger wrote about this worried state as a fundamental condition of existence, or what he termed "being-in-the-world" (*Dasein*) in his 1929 *Being and Time*. Uncertainty about the future generates an existential anxiety, often lingering in the unconscious even as people forge ahead. When you think about it, this really is what drives many to work hard in school, strive to earn a living, or plan for retirement. It's not so much the money itself as the benefits money brings. To Heidegger, creativity is one of the ways people respond to this anxiety by expressing themselves, making things, and making themselves feel productive. Staunchly rejecting the reduction of art to either an aesthetic ideal or a saleable commodity, Heidegger held out the promise of creativity as an articulation of "being." But this notion of being was not to be confused with individuality as typically conceived. Heidegger said that humans never are truly isolated or cut off from each other. They always exist "within a world" defined by immersion in society and history.[37]

Heidegger's work was a major influence on the theory of knowledge known as "social construction"—the idea that people's thinking is shaped by their upbringing, education, and culture. Extending this idea to interpersonal relationships, George Herbert Mead's principle of "symbolic interactionism" similarly said that people's sense of identity came from those around them.[38] Common to both concepts was the belief that selfhood was not a static entity, but an ongoing event rich with creative potential, as people discovered more and more about themselves. Even as individuals shaped the world around them, they, in turn, were shaped by it—often with little conscious awareness. These theories of subjectivity informed what sociologists would term the "presentation of the self" in everyday life.[39] People almost always act or "present" themselves socially with others in mind. Hence, a reciprocity between self and society occurs as individuals contemplate their effects on others. Speakers think about listeners, for example, and artists consider audiences. This also implies an ongoing set of choices. One can choose to conform to expectations and norms, or do just the opposite, in an open-ended set of possible actions. In this sense, the "self" is always a work in progress.

This open-ended intersection of self and society got further attention in the work of philosopher Gilbert Simondon. Writing in the 1960s, Simondon coined the term "individuation" to describe how people are conditioned to see themselves as singular beings. Contrasting beliefs that a separate "self" defines each human being, Simondon said that the perception of separateness is an "effect" of individuation's influence rather than its origin. Put another way, *individuation* causes *individualization*. Seeing this as an interactive and social process, Simondon added that people were continually generating this effect on each other and in groups through a process he termed "transindividuation."

Simondon's work influenced contemporary thinkers including Gilles Deleuze, Elizabeth Grosz, Bruno Latour, and Bernard Stiegler. Of these, Stiegler made the most use of Simondon's work in discussing the effects of communication and creative expression. Stiegler wrote that "creativity is what produces meaning from significations shared by those who co-individuate themselves through a process of transindividuation," explaining further that in "transindividuation, a psychic individual co-individuates itself with one or several other psychic individuals in such a way that their co-individuation leads them to agreement on the signification of an artefact—word, thing, practice, social convention, ritual, goal, etcetera."[40]

Put in plain English, these thinkers point out that life as a solitary "individual" is inherently stressful because it cuts a person off from the kinds of support others can provide. Adding competition to the mix only pushes people further apart and can worsen anxieties. These attitudes deprive solitary individuals of any full sense of themselves in the world—understandings that only can come from interacting with others and feeling part of a group or community. None of these are *natural* occurrences. Both Mead and Simondon argued that people *are taught* to see themselves as competitive creatures by a cultural system set up to benefit from their isolation, as anxious people buy things to feel better or accept lower pay because they feel powerless. In contrast, transindividuation offers a way of becoming connected to others and supported by them.

Transindividuation helps explain how creative expression can bring about social change. When people look at an artwork together, they share at least some aspects of what they believe the work means, even though they may hold slightly different views. This puts people partially in the same orbit, opening them to other shared ideas and interpersonal influences. Add to this any novelty they might glean from the artist's intentions. In both processes, the viewer's thinking enters a state of potential change. This counters widely held beliefs in an essential self—not only regarding the Enlightenment subject, but also aspects of identity commonly thought to be immutable like gender. Judith Butler's famous principle of "performativity" described subjectivity as enacted on a daily basis. This self may be informed by history, biology, and culture, but it is continually being performed nevertheless, knowingly or unknowingly. In writing about this, Butler noted that people often reproduce received notions of self in this process. But in emphasizing identity-as-process, she held out the promise for creativity in "subversive performative acts."[41]

Change also can occur in makers. Many artists see the creative process as a means of self-discovery, given ways that experiment and discovery can change a person's outlook. Anthropologist James Leach described this as a process of "becoming" as opposed to simply "being." Working on the Rai Coast of Papua, New Guinea, Leach found the concept of simply crafting creative "objects" quite foreign to inhabitants. Instead, they quite literally believed that making things for others "created a new person" in the artist.

Extending this idea, Leach wrote that people in contemporary Western nations often unknowingly reinvent themselves in everyday pursuits like "gardening, exchange, and spiritual activities."[42] Going even further, he argued that this "creative becoming" can supersede conventional notions of social belonging—even those as seemingly primal as kinship. Leach's insight helps explain how people sometimes "outgrow" their families or friends when moving to new places or jobs—in effect, "becoming" new people.

In philosophical terms, the notion of becoming has a history dating to the ancient Greeks. Heraclitus said that life is a continual process of change, during which people may (or may not) strive toward higher purposes in interacting with others. Heraclitus also saw human development as an inherently creative process, writing that "becoming is not a moral but an aesthetic phenomenon."[43] These ideas influenced Friedrich Nietzsche, who also wrote of life as a journey "to become who one is" through ongoing self-discovery.[44] To Nietzsche, such becoming was complicated by the vast differences between people's aspirations. Referencing Heraclitus in this, Nietzsche drew a distinction between an unexamined *knowledge* of things and the thoughtful *interpretation* of such knowledge, writing that, "The world may be knowable, but this alone has no meaning behind it."[45] What mattered was what people did with what they learned, which varied over time and from place to place. Nietzsche would term this variability "perspectivism," a concept that became an important precursor in the later development of deconstruction and difference theories.

Mathematician and philosopher Alfred North Whitehead linked creativity to becoming in one of the founding texts in what would be termed "process philosophy." Wrestling with concepts of time and mortality in his 1929 book *Process and Reality*, Whitehead spoke of existence as an ongoing sequence of adaptation and change. People not only became themselves, but they did so in relationship with others. "The creative advance of the world is the becoming," Whitehead wrote: "It belongs to the nature of a 'being' that it is a potential for every 'becoming.' This is the 'principle of relativity.'"[46] Gilles Deleuze later applied this open-ended view of becoming directly to creativity, noting that intention often drives artistic expression. He argued that creative becoming can represent a break from everyday "common sense," social control, or even history itself, as typically construed. Citing Nietzsche, Deleuze wrote that "becoming isn't a part of history; history amounts only to the set of preconditions, however recent, that one leaves behind in order to 'become,' that is, to create something new."[47] Put another way, creativity was for Deleuze a means to exceed the "actual" of mundane experience through the "virtual" potentials of becoming. He even concluded that history derives from creative becoming, rather than the other way around.

In his recent book *A World of Becoming*, political scientist William Connolly further pondered the topic of becoming over time. Connolly said that becoming implies a transition from one state of being into another, with people finding themselves caught between two attitudes: *inevitability* and *possibility*. These dual modes of anticipation remain continually in play as people consider simultaneous and often hard-to-predict scenarios, as well as the different speeds at which they unfold. As Connolly put it, "We participate in a world of becoming in a universe set on multiple zones of temporality, with each force field periodically encountering others as outside forces, and the whole world open to a certain degree."[48] Creativity enters these openings of the world, especially when people lean toward possibility rather than letting events unfold on their own. "Does the flow of time express inherent development?" he asked. This would imply that becoming simply happens as a consequence of time itself rather than human agency, hence "subtracting an element of real creativity from the world of becoming."[49] Instead, Connolly said that creativity inheres in a spirit of purpose-driven becoming when it is actively pursued.

Shared Creativity

Cognitive science long has seen creativity as a collective phenomenon—especially when kids make things together. In the 1920s, psychologist Lev Vygotsky argued that learning and self-expression emerge from early childhood interactions of self and society. Vygotsky used the term "appropriation" in describing the ways youngsters acquire skills (reading, drawing, etc.) from others around them and later internalize them as their own. Once this happens, children begin to use the skills for their own purposes. In this way, self-expression begins with the modification of skills or knowledge rather than their externally given purposes. Vygotsky wrote in his *The Psychology of Art* of the impossibility of seeing creators and audiences as separate entities, asserting that it is not individual brainpower that defines human beings, but their ongoing relationships with each other.[50] This gives knowledge and creativity an evolving collaborative quality.

More recently, cognitive scientists have specified creativity's interactive aspects further. In this contemporary view, knowledge is not seen strictly as something people carry around them, since individuals have limited capacities to "know" and remember. Rather than retaining massive amounts of data in their heads, people tend to develop generalized concepts that they apply to information picked up from objects, other people, and what perception tells them. "We live in a world in which knowledge is all around us," wrote Steven Sloman and Philip Fernbach in their book *The Knowledge Illusion*. "It is in the things we make, in our bodies and workspaces, and in other people. We live in a community of knowledge."[51] The catch is that people usually remain unaware of their dependence on the world of information around them. Sloman and Fernbach explained that "The nature of thought is to seamlessly draw on knowledge wherever it can be found," but that most people "live under the knowledge illusion because we fail to draw an accurate line between what is inside and outside our heads. And we fail because there is no sharp line."[52]

The knowledge illusion is partly a function of evolution. As the world got bigger and more complex, the human mind shifted more and more to what it does best: discerning patterns within the large amounts of knowledge it takes in. While the brain can hold only limited data (about half a gigabyte), it has enormous abilities to recognize patterns and draw inferences from that information (exceeding any supercomputer).[53] This is somewhat like Nietzsche's distinction between objective knowledge and subjective interpretation. The problem noted by Sloman and Fernbach is that many people today draw *both* knowledge and interpretation from those around them. This is how facts get distorted or otherwise colored by belief, ideology, and other forms of collective thought—even as people remain unaware that they are being influenced.

The knowledge illusion isn't always a bad thing. It simply means that people usually don't realize how much the external world influences them—even telling them how to do simple things. It isn't only other people exerting this influence. Everyday objects and technologies do it all the time. When you walk into a new store, your mind reads the space and merchandise in a continual process of recognition and orientation. The products tell you what they are by their appearance, placement, and packaging. And if you buy an item, the object itself usually tells you how to use it, turn it on, or put it together. Otherwise, there is signage or imagery as well. And if all else fails, one can consult instructions, books, or the internet for help. Most people picking up a guitar for the first time have a vague idea of how it works. But when they hold the instrument, the guitar itself indicates how to operate its strings, frets, and tuning keys. Of course, most people remain unaware of the knowledge illusion because it is so conditioned into them.

This idea of knowledge residing in both people and inanimate objects has a precedent in the realm of critical theory. In the 1980s, researchers at the French Center for the Sociology of Innovation observed that social activity could not be explained by group behavior alone. Instead, they proposed that people, objects, and ideas all interacted in "networks" of both human and non-human "actors." Most associated with Michel Callon and Bruno Latour, the new "actor-network theory" was seen as a way of recognizing the overlooked influence of the material world on language and meaning.[54] This structural view put inanimate objects, technologies, and animals on equal ground with people in describing the factors going into the generation of new ideas and creative products. In doing so, actor-network theory opened further thinking about corporeal experience and "transhumanist" considerations of other species.

The guitar analogy also shows how the body holds and conveys knowledge. Playing an instrument is an example of what cognitive scientists now call "embodied intelligence"—as the guitar becomes an extension of the body in a continual feedback loop with the brain. When an artist sketches with a pencil, the writing implement functions as a prosthetic for actual touch as it glides across the paper. Automobiles, home appliances, and even knives and forks serve much the same purpose in mediating between the body and the material world. Here again, the ubiquity of these items makes them fade into the background of consciousness, enabling people to think they get through the day without any help. Eyeglasses are a typical form of bodily extension that nobody considers a "prosthetic" enhancement for a disability. But heads turn when someone enters a room in the less common accommodation of a wheelchair. The fact is that everyone is constantly using assistive technologies to move themselves, get things done, and derive information from the world. And this isn't counting the help coming from printed materials, videos, and mobile phones.

More than anything else, the internet has given people the sense that they know more than they do. This really isn't a new phenomenon. Markings and drawings were the earliest way of giving information physical form, allowing memory to be externalized and the mind to focus on immediacy. Plato wrote about this in his famous allegory of the *pharmakon*, in which an Egyptian god gives a special remedy to a forgetful king. The remedy was writing: the ability to give memories physical form. Quickly, the wise king saw that the pharmakon wasn't really a "remedy" at all, but instead a crutch (or poison) that would make his memory weaker. This notion of a technology as both a blessing and a curse has been applied broadly by scholars discussing scientific breakthroughs, new media, medicines, and even food: helpful in the right measure, but also potentially dangerous. As Jacques Derrida summarized:

> If the *pharmakon* is "ambivalent," it is because it constitutes the medium in which opposites are opposed, the movement and the play that links them among themselves, reverses them or makes one side cross over into the other (soul/body, good/evil, inside/outside, memory/forgetfulness, speech/writing, etc.).[55]

The downside of the knowledge illusion lies in the hubris it produces, reinforced by a culture in which autonomous thought is so highly prized. The messages young people get in school are all about personal attainment, self-sufficiency, and individual assessment—often in an atmosphere of competition. Dominant models of accomplishment nearly always get cast in individual terms, whether one is talking about artists, entertainers, inventors, or political leaders. Rarely is it ever mentioned that Mozart was trained by his virtuoso father and drew on a vast storehouse of musical influence and patronage. Or that Charles Daguerre invented photography while scores of others were perfecting the same technology. Or that Francis Ford Coppola's *The Godfather* came from Mario Puzo's book and hundreds of people worked on the film. The identification of big things with certain individuals is a form of mental shorthand that strips projects of their collective dimensions.

Western society's obsession with individualism is a good example of how the human knack for pattern recognition can go wrong in a society overwhelmed by data. From history's earliest days, people's abilities to get things done hinged on collective efforts. Neolithic hunts weren't a matter of a one person doing all the work, but involved teams of people playing different roles: finding, killing, cleaning, and preparing the quarry. Farming and industry worked the same way, much as modern corporations do today. Airlines, hospitals, schools, and restaurants all operate with intricate divisions of labor. Scientists work in teams rather isolated labs, often submitting their research findings to peers for feedback. Artists and performers rely on galleries, museums, performing venues, support staff, journalists, and the media in bringing their work to audiences. This means that a form of distributed intelligence is driving the enterprises, rather than any individual. Binding distributed intelligence together is the key premise of shared intentionality. People in a team must recognize the same goal—as when carpenters, plumbers, roofers, painters, and landscapers work together in building a house.

Let's return to *The Knowledge Illusion*. Sloman and Fernbach close their book on a note of concern about the damages caused by the individualist mindset so commonplace today. They note that people operating by themselves are especially likely to underestimate how external influence colors their information processing. Matters get even worse when people limit themselves to familiar circles or media sources that confirm existing opinions. This causes an atrophy of the deliberative and reasoned analysis that comes from diverse inputs, lessening the likelihood of considering multiple viewpoints. Put another way, rugged individuals cut themselves off from critical perspectives and tend to rely on "gut" intuition, as mentioned in the last chapter. "People who are more reflective tend to be more careful when given problems that require reasoning," Sloman and Fernbach wrote, adding:

> Intuition gives us a simplified, coarse, and usually good enough analysis, and gives us the illusion that we know a fair amount. But when we deliberate, we come to appreciate how complex things actually are, and this reveals how little we actually know.[56]

In making this point, the authors suggest that individualistic disconnection may partly explain the willingness of voters to accept the kind of distorted and emotion-driven views of politicians playing to people's biases and fears.

In arguing the importance of a broadly drawn "shared intelligence," Sloman and Fernbach also assert a more proactive encouragement of team efforts than usually takes place. They note that while standardized tests and performance evaluations may well separate high achievers from the pack, they often overlook the value of collaboration and cooperation necessary for a team to excel as a whole. They wish more people saw that "intelligence belongs to the community, not to individuals, and that different people play different roles in making the whole community effective," even going so far as to say that "far more important than the quality of the idea is the quality of the team bringing it forward."[57] Focusing on competitive individualism also orients assessment toward what is recognizable, measurable, and already known, rather than on the sometimes risky fringes of experimental knowledge. Supported by peers, people working in teams feel less as if they are putting themselves out on a limb, and are more willing to push themselves to the limits of their knowledge and creativity.

This chapter has examined creativity as a process that changes the maker as it occurs. Creativity typically takes exploratory form and draws from the world around it, often through interactions with others. Old-fashioned beliefs in solitary genius encourage attitudes of self-centered individualism that can dampen creativity. And when competition enters the picture, it can make creators afraid to risk new ideas, as is seen in today's business world. What can be done to fix this problem? It's going to be an uphill battle convincing people that they alone are not the source of their creativity, given America's obsession with individualism in all things.

Two key points have arisen repeatedly in this chapter's analysis of the dilemma. First, creativity is about more than artists simply making things and sending them into the world. The purposeful character of creativity necessarily involves an engagement with audiences/recipients as well as the work itself. In this, creating things often changes artists themselves—in effect, creating "new people" in the process. Second, traditional views of the individual artist increasingly seem inadequate, given the overwhelming influence of society, history, and cognition itself in the creative process. In this sense, most forms of artistry are better seen as manifestations of shared intelligence or aesthetics rather than solitary endeavors. The following chapter will look further at the ways distributed creativity is enabled by groups, institutions, and other social formations.

Notes

1 Elizabeth Gilbert, *Big Magic: Creative Living Beyond Fear* (New York: Riverhead, 2015) p. 233.
2 John Keats, "To George and Georgina Keats" (1818) www.john-keats.com/briefe/251018.htm (accessed Jan. 3, 2019).
3 "Constitution of the United States of America" (1789) www.constituteproject.org/constitution/United_States_of_America_1992 (accessed Jan. 4, 2019).
4 Samuel Warren and Louis Brandeis, "The Right to Privacy," *Harvard Law Review* 4, no. 193 (Dec. 15, 1890) www.cs.cornell.edu/~shmat/courses/cs5436/warren-brandeis.pdf (accessed Jan. 4, 2019).
5 Luc Boltanski and Eve Chiapello, *The New Spirit of Capitalism*, trans. Gregory Elliot (London: Verso, 2005).
6 Janet Wolff, *The Social Production of Art* (New York: NYU Press, 1981) p. 43.
7 Friedrich Holderlin, *Friedrich Holderlin: Essays and Letters on Theory* (Albany: SUNY Press, 1988) p. 155.
8 Zoya Kocur and Simon Leung, "Introduction," in Zoya Kocur and Simon Leung, eds., *Theory in Contemporary Art since 1985* (London and New York: John Wiley, 2005) p. 1.
9 Ibid.

10 Thierry de Duve, "When Form Has Become Attitude—And Beyond," in *Theory in Contemporary Art*, p. 26.
11 Aaron Barlow, *The Cult of Individualism: A History of an Enduring American Myth* (New York: Praeger, 2013) p. 1
12 Aristotle, *Politics*, trans. H. Rackman (Cambridge, MA: Harvard University Press, 1944) p. 1253.
13 Adam Smith, *The Wealth of Nations* (1776) (New York: Penguin Books, 1982) p. 456.
14 Alexis de Tocqueville, *Democracy in America* (1830) trans. Arthur Goldhammer (New York: Library of America, 2004) p. 731.
15 Herbert Hoover, "Rugged Individualism Campaign Speech" (1928) www.digitalhistory.uh.edu/disp_textbook.cfm?smtID=3&psid=1334 (accessed Jan. 4, 2019).
16 Herbert Hoover, "The Future of American Individualism," from *American Individualism* (1922) www.hoover.org/research/future-american-individualism (accessed Jan. 30, 2019).
17 Christopher Lasch, *The Culture of Narcissism: American Life in and Age of Diminished Expectations* (1979) (New York: W.W. Norton, 1991) p. 209.
18 "GDP Growth," The World Bank (2015) http://data.worldbank.org/indicator/NY.GDP.MKTP.KD.ZG (accessed Jan. 4, 2019).
19 "Income Inequality" (2017) http://inequality.org/income-inequality (accessed Jan. 4, 2019).
20 Heather Long, "U.S. Income Inequality Keeps Getting Uglier," CNN News (Dec. 22, 2016) http://money.cnn.com/2016/12/22/news/economy/us-inequality-worse/index.html (accessed Jan. 4, 2019).
21 Geert Hofstede, "Individualism," Clearly Cultural: Making Sense of Cultural Communication (2017) www.clearlycultural.com/geert-hofstede-cultural-dimensions/individualism (accessed Jan. 3, 2019).
22 Richard Florida, Charlotta Mellander, and Karen M. King, "The Global Creativity Index 2015," Martin Prosperity Institute (Jul. 8, 2015) http://martinprosperity.org/content/the-global-creativity-index-2015 (accessed Jun. 16, 2019).
23 Angela McRobbie, "Everyone Is Creative: Artists as New Economy Pioneers?" *Open Democracy* (Aug. 2001) www.opendemocracy.net/node/652 (accessed Jan. 4, 2019).
24 Rosalina Jovanovich, "Artnet News' Top Most Expensive Living American Artists," Artnet News (Apr. 30, 2017) https://news.artnet.com/market/artnet-news-top-10-most-expensive-living-american-artists-11863 (accessed Jan. 4, 2019); THR Staff, "Hollywood Salaries 2016," *Hollywood Reporter* (Sep. 29, 2016) www.hollywoodreporter.com/lists/hollywood-salaries-2016-who-got-933037/item/hollywood-salaries-revealed-who-got-933041 (accessed Jan. 4, 2019).
25 Thorstein Veblen, *The Theory of the Leisure Class: An Economic Study in the Evolution of Institutions* (New York and London: Macmillan, 1899); Arthur R. Momand, *Keeping Up with the Joneses* (U.S.: syndicated comic series, 1913–1940).
26 Mark A. Runco, *Creativity: Theories and Themes; Research, Development, and Practice* (2nd ed.) (London and New York: Elsevier, 2014) p. 152.
27 Roland Barthes, "Of Sports and Men," trans. Scott MacKenzie, *Cine-Documents* 6, no. 2 (1962): 80.
28 Ken Robinson, "Do Schools Kill Creativity? TED (Jun. 2006) www.ted.com/talks/ken_robinson_says_schools_kill_creativity/transcript?language=en (accessed Jan. 6, 2019).
29 Rachel Rittner, "Are Today's Youth Less Creative & Imaginative?" *LiveScience* (Aug. 12, 2012) www.livescience.com/15535-children-creative.html (accessed Jan. 6, 2019).

30 Josh Linker, "How Kids Lose Their Creativity as They Age," *Forbes* (Oct. 16, 2014) www.forbes.com/sites/joshlinkner/2014/10/16/how-kids-lose-their-creativity-as-they-age-and-how-to-prevent-it/#7298bbbf422e (accessed Jan. 6, 2019).
31 Alfie Kohn, *No Contest: The Case Against Competition* (Boston and New York: Houghton Mifflin, 1986) p. 130
32 *No Contest*, p. 55.
33 *No Contest*, p. 54.
34 Richard Florida, "America's Looming Creativity Crisis," *Harvard Business Review* (Oct. 2004) https://hbr.org/2004/10/americas-looming-creativity-crisis (accessed Jan. 9, 2019).
35 Michael Blanding, "Does Competition Make Us More Creative?" *HBS Working Knowledge* (Nov. 16, 2015) http://hbswk.hbs.edu/item/does-competition-make-us-more-creative (accessed Jan. 5, 2019).
36 Daniel P. Gross, "Creativity Under Fire: The Effects of Competition on Creative Production," *Harvard Business School Working Paper 16-109* (2016): 19.
37 Martin Heidegger, *Being and Time* (1929) (New York: Harper Perennial, 2008) 12: 84.
38 George Herbert Mead, *Mind, Self, and Society* (1934) (Chicago: University of Chicago Press, 2015).
39 Erving Goffman, *The Presentation of the Self in Everyday Life* (New York: Anchor, 1959).
40 Bernard Stiegler, "Interobjectivity and Transindividuation," (Sep. 28, 2012) www.onlineopen.org/download.php?id=389 (accessed Jan. 8, 2019).
41 Judith Butler, "Performative Acts and Gender Constitution: An Essay in Phenomenology and Feminist Theory," *Theatre Journal* 40, no. 4 (Dec. 1988): 513.
42 James Leach, *Creative Land: Place and Procreation on the Rai Coast of Papua New Guinea* (New York and Oxford: Berghahn, 2003) p. 17.
43 Friedrich Nietzsche, *Philosophy in the Tragic Age of the Greeks* (1873) (New York: Gateway, 1996) p. 113.
44 Friedrich Nietzsche, *Ecce Homo: How One Becomes What One Is* (1888) (Lexido: n.d.) www.lexido.com/EBOOK_TEXTS/ECCE_HOMO_.aspx?S=15 (accessed Jan. 29, 2019).
45 Friedrich Nietzsche, *The Will to Power*, trans. Walter Kaufmann (New York: Vintage, 1968) p. 481.
46 Alfred North Whitehead, *Process and Reality: An Essay on Cosmology* (1929) (New York: Free Press, 1978) pp. xvi and 22.
47 Gilles Deleuze, "Control and Becoming," in *Negotiations: 1972–1990*, trans. M. Joughin (New York: Columbia University Press, 1995) pp. 170–1.
48 William Connolly, *A World of Becoming* (Durham: Duke University Press, 2007) p. 7.
49 *A World of Becoming*, p. 8.
50 Lev Vygotsky, *The Psychology of Art* (Cambridge, MA: MIT Press, 1974).
51 Steven Sloman and Philip Fernbach, *The Knowledge Illusion: Why We Never Think Alone* (New York: Riverhead, 2017) p. 13.
52 *The Knowledge Illusion*, p. 15.
53 *The Knowledge Illusion*, p. 25.
54 Bruno Latour, *Reassembling the Social: An Introduction to Actor-Network Theory* (Oxford: Oxford University Press, 2005).
55 Jacques Derrida, *Dissemination*, trans. Barbara Johnson (Chicago: Chicago University Press, 1981) p. 127.
56 *The Knowledge Illusion*, p. 83.
57 *The Knowledge Illusion*, p. 208.

Chapter 11

Distributed Creativity

Toward a Sharing Economy

Steve Jobs didn't invent the iPhone. America's quintessential symbol of high-tech innovation came from dozens of engineers, designers, and hardware fabricators working together. Their efforts created the most successful consumer product in U.S. history, with 1.2 billion units sold. Yet the image of Jobs announcing the new device on the stage of San Francisco's Moscone Center is what history remembers, much as it has for most inventions. All because "the myth is inextricable from the man," wrote Jobs biographer Brian Merchant: "The myth of the sole, or lone, inventor."[1] This cultural penchant obscures the collaborative aspects of historic inventions—not only in discounting team efforts, but also any concurrent work on the problem done by competitors. And with the iPhone, the celebration of Apple, Inc. also downplays how much public money made it possible.

More accurately, the iPhone provides a perfect parable about distributed creativity, or what some call the "hive" model of human innovation. In other words: people thinking together to make something new. Distributed thinking has enabled nearly every advance in history, putting a practical face on an otherwise philosophical premise. Interconnection is the real force pushing humanity forward. In what follows, I will review recent research on distributed intelligence taking place in the fields of psychology and learning science. Then I'll look at the fiscal side of things, as think tanks and corporations have embraced the concept of a "sharing economy." All of this portends a revolution in the creativity field, as boundaries separating creators into intellectual "silos" increasingly seem both old-fashioned and inefficient—along with a growing recognition that competition and profit-seeking can impede novel thinking.

Of course, "sharing" is a gigantic meme in the online world. Posting, streaming, and downloading have become so ubiquitous that people rarely give them much thought. In an age of social networking, it's easy to forget the economic damage online sharing wreaks on artists, especially in the music field, but also in writing and image making. Even more troubling is the emergence of a new monopoly behind much of this—internet giants like Facebook and Google aggregating the data left behind in this "free" exchange of information. Users often seem unaware (or unconcerned) that their information yields billions to those selling and buying it in the burgeoning data economy, or that companies put their own interests ahead of consumers. Clearly, some critical perspective is needed. You might assume that this sharing is making people smarter. More information must be good, right? Unfortunately, studies consistently show the opposite, as people receive the same material over and over again, while gravitating to what they already know.

This chapter provides an overview of how people can work together in developing new ideas and creative projects. Beginning with a look at how musical groups always have done this, attention next turns to other kinds of artistic activity. Discussion then turns to scientific and economic arguments in favor of "sharing" rather than keeping ideas to oneself, as a broad consensus is growing that competitive isolation is proving counterproductive in both academic circles and the business world. I'll close by discussing what can be done to encourage a "learning society," especially in popular culture—because knowledge *does* matter, facts *do* exist, and not everything is relative. History has shown that almost all increases in living standards have come from discernibly better ways of doing things—improvements coming from some form of knowledge about the world or other people. What can be done to encourage this? What role can artists and other creative people play? And if distributed knowledge really is the key in much of this, how can this fact become more widely appreciated?

Distributed Creativity

The recent documentary series *Long Strange Trip* chronicled the history of the Grateful Dead, the San Francisco psychedelic rock group formed in the 1960s. Largely associated with the free-form "jam band" style popular at the time, the Grateful Dead quickly drew a cult-like "Deadhead" fan base that followed the band on tours. *Long Strange Trip* attributes the intensity of the Deadhead phenomenon to the group's close interaction with concert-goers, responding in real time to generate what the Dead termed the "X-Factor." Today this might be called distributed creativity. Building on established methods of improvisation, the Dead made "listening to the audience" a driving premise in how it performed—asserting it could channel a crowd's moods and preferences. Readers of the work of Carl Jung, the Grateful Dead believed it could tap into "collective consciousness" and moments of "synchronicity" to make each performance unique. When asked about this, Jerry Garcia once famously replied, "The situation is in charge."[2] Whether or not the band truly achieved this kind of mind-merge remains an open question. But don't try telling this to die-hard Deadheads.

Other prototypes of distributed creativity were popping up the 1960s. Among these were "rave" events first emerging from the London bohemian party scene—and later spreading around the world. Audience feedback always figured prominently in raves, most of which featured recorded rather than live music. DJs responded directly to listeners, who felt bound together by the sounds and lighting effects of performances often held in illegally commandeered empty industrial spaces. The impromptu and often secretive staging of raves added to their ethos of spontaneity and camaraderie. Like the acid rock concerts of the Grateful Dead, drugs often heightened this sense of shared experience. The movement also paralleled the beginnings of DIY garage music, punk groups, and the rise of underground labels like K-Records.

This echoed rising artworld interest in distributed creativity. Open-ended performance events known as "happenings" began taking place, originally linked to exhibition openings or social gatherings. The new non-linear form of theater sought to eliminate the line between performer and audience, while highlighting spontaneity and unpredictability. Happenings echoed the free-form aesthetics of Beat-era writers such as Allan Ginsberg and Lawrence Ferlinghetti, both of whom appeared with the Dead at the first "Human Be-In" in San Francisco's Golden Gate Park in 1967. During the decade, happenings were staged by such musicians and artists as John Cage, Allan Kaprow, Claus Oldenburg, Yoko Ono, and Carolee Schneemann. Kaprow wrote: "Visitors to a Happening are now and then not sure what has taken place, when it has ended, even when things have gone 'wrong'. For when something goes 'wrong', something far more 'right,' more revelatory, has many times emerged."[3]

Of course, all of this occurred amid the 1960s' famous rejection of authority and convention. Beyond the well-known activism of students and the disenfranchised, much of the world remained anxious over Cold War antagonisms, the nuclear arms race, and a deteriorating environment. This fed deeper philosophical suspicions of empirical reasoning and its presumptions of truthful analysis. The resulting "post-positivism" questioned the definitions, claims, and boundaries of knowledge, asserting that possibilities for more inclusive models must be possible. The hunger for alternatives brought changes big and small, as governments redesigned themselves and new social movements emerged. Before long, universities saw that narrow divisions of knowledge had kept many areas of life from study, while reproducing social hierarchies and blocking creative thought. This led to interdisciplinary programs, including media studies, social ecology, information science, and area studies, among many others.

Keep in mind that no language yet existed for distributed creativity (or knowledge, for that matter). During the 1960s, much of the focus in computational science, cognitive research, and the arts remained on the individual, even as collective activity was percolating on the fringes. Sociologist Howard Becker (then known for his 1963 book *Outsiders*) was among the first to comment on the mutual constitution of performers and audiences in "music scenes" and "art worlds," which he described as integrated spheres of creative activity.[4] In the field of psychology, John Irwin similarly looked at cultural "scenes" as embodiments of symbolic interactionism.[5] Irwin saw the "shared perspectives" of fan culture generating creative energies not tied to "any particular collectivity or territory."[6] Distributed creativity in the art world had to contend with a legacy of fetishized individualism—with performance events of the period usually linked to specific personalities. According to historian Grant Kester, this would change as the artists began to see their works as "conversations" rather than one-sided declarations. "While it was common for a work of art to provoke dialogue among viewers, this typically occurred in response to a finished object," Kester wrote. But at the fringes of the art world, a new approach began emerging during the 1970s in which "conversation became an integral part of the work itself … reframed as an active, generative process."[7]

Before this, art history mostly had looked at collective creativity indirectly—typically by studying periodized styles and recurrent imagery. One early example was Erwin Panofsky's 1939 distinction between patterns of "iconography" (subject matter) and "iconology" (cultural context).[8] Freudian psychoanalysis entered the picture in suggesting that mental states, while personally experienced, also were broadly shared. Finally, Marxist sociology brought art historians to recognize the collective effects of ideology. But it would take the post-war field of semiology (signs) to argue that artists and audiences genuinely shared ways of making meaning. As the humanities entered their "linguistic turn" in the 1970s, the rise of post-structuralism and reader-response theory said that senders and receivers of messages worked together in an ongoing dialogue.

This mix of theories gradually entered the art world, gaining influence as conceptualism became more accepted. Recognition grew that creative people talk to each other, borrow ideas, and think about viewers all the time—even as many benefit from the mystique of personal brilliance. Distributed models found a niche, nevertheless. As discussed above, performance art took its public quite seriously, much like the Grateful Dead did in popular music. This meant looking at art less as an end in itself and more as a process. As this approach spread to other forms of art-making, one anthology explained, "The artwork was no longer viewed as a static, object with a single, prescribed signification," but instead "understood as enacted through interpretative engagements that are themselves performative in their intersubjectivity."[9] Initial aspects of such thinking drew from literary theory giving "readers" a role in the interpretation of works.

Roland Barthes' famous assertion of the "death of the author" said that writings took on a life of their own as people read them.[10] Often this went far beyond what an author had intended. In later work, Barthes explored the paradox between "informational" (rationalized) and "symbolic" (aesthetic) communication in an essay called "The Third Meaning." Barthes said the first two kinds of communication relied on "transpersonal" (social shared) processes in generating a "third" (signified) kind. He also described a seemingly strange aspect of the Third Meaning—an open-endedness associated with experiment and novelty. At this early point in his theorizing, Barthes found the third meaning puzzling and inexpressible, except as a function of the first two. "I do not know what its signified is," Barthes wrote, "at least I am unable to give it a name, but I can clearly see the traits."[11] Elsewhere in his writing, Barthes used similar language in discussing the free association seen in artistic leaps and flashes of insight—the sort of "impossible" creativity associated with the discovery of something previously unknown, and sometimes unexpected.

More directly addressing distributed creativity was Michel Foucault's 1969 essay "What is an Author?"[12] Taking direct aim at originality, Foucault questioned how works, ideas, and theories became associated with individuals. Why was it that shared ideas in a certain "discourse" only were credited to the person writing them down? "There was a time when the texts we today call 'literary' (narratives, stories, epics, tragedies, comedies) were accepted, put into circulation, and valorized without any question about the identity of their author," Foucault observed: "Their anonymity caused no difficulties."[13] But somewhere in seventeenth and eighteenth centuries, Western culture became fixated on personifying works to give them legitimacy and property value. Foucault argued that texts often were appropriations, derivations, or partial copies—and that actual authorship always was a collective, social enterprise. Elevating authors to the status of "genius" served only to limit, exclude, impede "the free circulation, the free manipulation, and the free composition" of ideas. "The author is therefore the ideological figure by which one marks the manner in which we fear the proliferation of meaning," Foucault concluded.[14] In making this point, Foucault put his finger on the way genius authorship suggests that most people lack creative abilities, and, by extension, the capacity to articulate themselves artistically or politically.

Foucault's premise soon gained wider currency. Among others, twentieth-century semiotician Mikhail Bakhtin influenced writers and performance artists following the 1975 publication of his writings on the concept of the "dialogic imagination," which argued that artists constantly influenced each other as well as their audiences.[15] Bakhtin said that language and artistic works served as expressions of shared cultural moments (what he called "chronotypes"). Novels, paintings, and movies reflected "the intrinsic connectedness of temporal and spatial relationships that are artistically expressed" more than any one person's creativity, Bakhtin wrote.[16] This generated an ongoing "intertextual" exchange among artistic works, their makers, and those viewing them. Later theories of audiences and fan cultures would show that groups of viewers (or readers) often found similar meanings as they formed "interpretive communities." (see Chapter 4).

More recently, philosopher Jean-Luc Nancy put this even more directly, noting that no one can speak to another person without a common language. In his book *Being Singular Plural*, Nancy wrote: "There is no meaning if meaning is not shared, and not because there would be an alternate or first signification that all beings must have in common, but because *meaning is itself the sharing of Being*."[17] In this, Nancy made two crucial points: first, that people communicate with each other through shared symbol systems;

and, second, that these signifying practices rarely are identical. Twentieth-century deconstruction and reception theory went to great lengths in parsing the idea of difference-in-language and its infinite variability. Nevertheless, relationships to language remain *shared* even if its forms and styles are not the same. To Nancy, a common catalogue of signs remained less significant than the process of negotiating meaning. "Everything, then, passes *between us*," he explained.[18]

Meanwhile, distributed creativity began getting attention outside the arts and humanities—and often from unlikely sources. For some time, advertising firms had been using "brainstorming" to generate new ideas. The term was popularized in the 1950s by marketing executive Alex Osborn in his book *Your Creative Power.* [19] Outlining the essential rules of a successful brainstorming session, Osborn said that meetings had to start with an open sharing of ideas without criticism or negative feedback. Otherwise, valuable insights might get held back by fears of ridicule or reprisal. "Creativity is so delicate a flower that praise tends to make it bloom while discouragement often nips it in the bud," he wrote.[20] During this time, the Barton, Durstine & Osborn Advertising Agency would use the technique with clients such as American Tobacco, Chrysler, and General Electric—as fictionally depicted in the TV series *Mad Men.* Building on the success of brainstorming, Osborn later established himself in the creativity field by setting up the Creative Education Foundation and launching the *Journal of Creative Behavior.*

Psychologists long had been fascinated by creativity, especially as seen in child development. In the 1970s, a new strain of research began in what was termed the "social psychology of creativity."[21] The work took issue with then-dominant behavioral views of artistry as a personal trait of ability or pathology, as seen in the work of Joy Paul (J.P.) Guilford and others.[22] As one review essay explained this mainstream view: "A reader of the literature would glean something like this: creativity is a quality of the person; most people lack this quality; people who possess the quality—geniuses—are different from everyone else, in talent and personality."[23] As mentioned in Chapter 10, the dominant voice in the field was Paul Torrance, who developed the widely embraced Torrance Test of Creative Thinking (TTCT, still in use today). Administered to individuals, the TTCT measured such variables as problem solving, cognitive flexibility, expressiveness, and divergent thinking.[24] "This work served to reinforce that impression—and, indeed, the belief among laypersons and scholars—that creativity depends on special qualities of unusual persons."[25]

Rather than focusing on individuals, the new social psychology began looking at group context. Researcher Dean Simonton combined biographical methods commonly seen in behavioral fields with a new scrutiny of the cultural milieu in which selected artists and writers thrived. This work culminated in what Simonton termed a "Darwinian" theory of creativity as an adaptive response to the artist's social, cultural, and political environment. Working concurrently, psychologist Teresa Amabile focused more closely on those around the artist, initially developing an alternative creativity test to the TTCT based on peer review rather than quantitative measurement. Amabile later looked more directly at the role of groups and organizations in the creative process, noting in 1983 the negative impact that external assessment can have on artistry. "The intrinsically motivated state is conducive to creativity," Amabile observed, "whereas the extrinsically motivated state is detrimental."[26]

By introducing an element of post-positivism into creativity debate, Simonton and Amabile furthered views of creativity as a social construction. Attractive as this notion seemed, it nevertheless fell prey to criticisms of relativism. The question became: If creativity is a subjective matter, can there be any true creativity at all? This threw creativity into old debates over the separability of imagination and reality. Put another way, creativity raised the distinction between epistemology (how one knows) and ontology (what one knows). Simonton and Amabile helped further understandings that both epistemology and ontology are at work in creativity, contrasting previous views that artistic talent was something everyone could clearly recognize.

Concepts of distributed creativity have benefited greatly from advances in cognitive science and neurobiology. Not only have brain researchers found that artistic people draw on knowledge from others through shared cognitive processes, but scientists also have found genes associated with creativity (see Chapter 10). The relatively new field of epigenctics sees a dynamism in the cellular mechanics of learning and memory—that a person's DNA can be changed by experiences in life, with those changes carried forward to offspring. Beyond this, interpersonal experience has been linked to mirror neurons, as discussed in Chapter 4. The child development field long has recognized ways that infants and caregivers "mirror" each other in reciprocal smiling, touching, or utterances—much as is seen in other primate species. Imaging science now confirms the activation of mirror neurons as such mimicry occurs. These neurons help youngsters to interpret behavioral cues from others, thus helping in the development of empathy, relationship forming, and collaboration. [27] Many artists say they draw on exactly these traits.

During the last decade, neuroscience has looked further at how brain regions work together to create social perception. In "The Social Brain: Neural Basis of Social Knowledge," Caltech neuroscientist Ralph Adolphs put this new research in the context of the work discussed above.[28] Like much of human consciousness, researchers see the social mind emerging from interactions between the amygdala (associated with emotions) and prefrontal cortex (where executive functioning occurs). Together, these areas enable abilities like empathy and behavioral regulation—both of which occur in relationships among people. As Adolphs explained, "Social behavior depends critically on context and intention, a sensitivity that arises from the rich interplay between controlled and automatic processing of social information." This entails as initial intake of information (consciously or unconsciously), which is subsequently processed and modulated to allow "the ability to shift one's conscious experience to places and times outside the here-and-now, and into the viewpoint of another mind," as well as the ability to evaluate the behavior of others.

Then there is the science of memory itself. Much of what people attribute to creative insight or inspiration actually results from the recombination of information already in the brain. As social creatures, much of such memory comes directly or indirectly from others. And it is from this pre-existing knowledge that flashes of insight emerge. Some argue that this solves the creativity puzzle. "Creativity *is* memory," psychologist Art Markman wrote in an overview of recent research, explaining that "coming up with new ideas requires retrieving ideas from your memory."[29] Markham cited studies by brain researchers at Harvard University trying to stimulate divergent thinking and problem solving. The Harvard team found that by asking people to view and then describe certain videos, subjects could be primed to connect ideas in fresh ways. Calling this technique "episodic specificity induction," the scientists concluded that the stimulation of "retrieval cue processing" triggered novel "mental scenarios" and ideas for "entirely new uses of objects."[30]

The concept of distributed creativity partially draws on this mnemonic function, while adding the dynamic of social interaction. Researchers in the emerging field of "cultural psychology" cite Vygotsky in seeing creative acts as the result of internalized experiences that are subsequently externalized. As explained by Petre Glăveanu, "What is being internalized is not only cultural content but also the perspective of others toward it."[31] In his book *Distributed Creativity: Thinking Outside the Box of the Creative Individual*, Glăveanu locates artistry *between* rather than *within* people. What appears to be "individual" creativity in fact derives from experiential differences between "self and other, between symbols and objects, and between past, present, and future."[32] This differs somewhat from what Mead observed in writing that "the perspective is the world in its relationship to the individual and the individual in relationship to the world."[33] Without discounting variances in personal experience, Glăveanu distinguished between the single-person approach common in psychological accounts of creativity and the newer distributed perspectival models:

> There is an important difference between divergent thinking theories and perspectival approaches to creativity. The first assumes that ideas "come up" in the mind through the combination or association of pre-existing knowledge. The second locates these "ideas" in the perspective or action orientation of a person in the world and considers both their origin and dynamic as fundamentally social.[34]

This doesn't write off the individual entirely—but it does place the artist in a social context. "An idea or product that deserves the label 'creative' arises from the synergy of many people and *not only* from the mind of a single person," creativity expert Mihaly Csikszentmilhályi has explained.[35] Qualities like vision, talent, and creative drive can't be denied. The problem lies with a culture that says they are everything. Writing for the National Science Foundation, computer scientist Gerhard Fischer has added, "There is overwhelming evidence that research should be grounded in the basic assumption that the power of the unaided mind is highly overrated." Fischer argued that future efforts should be guided by "the assumption that there is an '*and*' not a '*versus*' relationship between individual and social creativity." Rather, "individual and social creativity can be integrated by means of proper collaboration models, appropriate community structures, boundary objects, process models in support of the natural evolution of artifacts, and meta-design."[36]

The above discussion has detailed problematic views of creativity as an individualized and solitary endeavor. I've contrasted these with countervailing movements in critical theory, psychology, cognitive science, and even the business world. The political and economic stakes are high in this debate, inasmuch they impact views of how ideas come about, who can express them, and what this means in a democratic society. Should America be encouraging only a celebrated few, or does a truly egalitarian society work for the good of the many?

Right now, distributed creativity is in the air as never before. In this section, I've sketched out how scientists have been confirming ideas long percolating in art, critical theory, popular culture, and even in advertising. Besides providing evidenced based "proof" for what might otherwise seem abstract philosophical hypotheses, this itself seems a good example of shared intelligence (or creativity). For decades, thinkers in the humanities and sciences have circled around the issue of distributed thought processes, using different methods to ponder a commonly recognized question. This matters in helping humanistic concepts find mainstream acceptance. But it also demonstrates how ideas and concerns often cross boundaries, with everyone benefiting from multiple perspectives. The question that remains is what effects, if any, recent thinking and research on distributed creativity might have on the creative industries in a practical sense. The answers might surprise you.

The Sharing Economy

Let's return to the recent bad news about innovation. Is there a practical way that distributed creativity can help? It's no secret that novelty and innovation have been *declining* in the U.S. for some time. A few years after his messianic promotion of the creative class, Richard Florida announced "America's Looming Creativity Crisis," writing in the *Harvard Business Review*, "The United States of America—for generations known around the world as the land of opportunity and innovation—is on the verge of losing its competitive edge."[37] Florida offered two primary explanations: first, that the recession of the first decade of the twenty-first century had made companies reluctant to try new products; and, second, that education has been failing to encourage novel thinking. Both of these problems he linked to a persistent edginess in an America, manifest in worries about a nation in decline.

Forbes soon assessed the business side of the dilemma in a feature by management expert Steve Denning entitled "Why U.S. Firms Are Dying: Failure to Innovate."[38] To Denning, America's corporate innovation crisis was tied directly to what he dubbed "the world's dumbest idea" phenomenon: the blind obsession to maximize profits for stockholders. "Firms still pursue it. Boards endorse it. CEOs are lavishly compensated for pursuing it. Investors base their decisions on it," Denning wrote. "As a result, innovation suffers."[39] The real evidence came from inside companies themselves, where Denning discovered that 80 percent of employees found no resources to innovate, 50 percent got no recognition for fresh ideas, and a mere 5 percent felt motivated to try anything new.[40] All of this coincided with a precipitous decline in research monies from non-corporate sources. While it's no great surprise that the business world would operate on bottom-line terms, now even scientists are feeling the pinch. *The Chronicle of Higher Education* recently surveyed more than 67,000 researchers with grants from the National Science Foundation (NSF) and National Institutes of Health (NIH)—both premier funding entities in their disciplines.[41] Nearly 50 percent reported that money shortages had forced them to abandon topics "central to their lab's mission." Meanwhile, the NSF also reported a 15 percent decline in the number of science and engineering PhDs with full-time jobs.

Bad as this sounds, a simple solution to the creativity "crisis" quietly has been gaining traction: *sharing*. Not the old kind of feel-good collectivism, but a pragmatic sharing in which people work together for specific goals. The idea isn't so radical, really. Any grade-school history class teaches this simple lesson. A family farm might be one example. But so is the modern corporation. Unfortunately, America's love affair with individualized competition often has worked against sharing endeavors, just as it also has undermined confidence in government. Nevertheless, a new embrace of collectivism is gaining momentum in the business world as well as university research programs. And this might be good news for the creative people.

The leadership of the World Economic Forum (WEF) recently got on the bandwagon for the new "Sharing Economy." The WEF said that the global rise of service industries and the pressures of austerity are combining to force new sharing practices. The meteoric growth of companies like Airbnb, Etsy, and Uber typifies a new business model utilizing both social networks and readily available resources. WEF director Jennifer Blanke explained, "In a sharing economy we make better use of existing products rather than merely producing more stuff."[42] Arun Sundararajan, another expert on the sharing economy, prefers the term "crowd-based capitalism" to describe the impetus toward distributed thinking. As Sundararajan put it, when "peer-to-peer commercial exchange blurs the lines between the personal and the professional, how will the economy, government regulation, what it means to have a job, and our social fabric be affected?"[43]

The imperative for sharing looms large (and with contention) in the field of "big data," as digital technologies enable researchers, corporations and governments to aggregate ever larger storehouses of information. Everyone knows that Amazon and Google regularly collect and share consumer details. Less known is the scientific collection of data for research on economics, education, environment, population health, transportation, and urban planning. Unlike corporate data collectors who sell data to the highest bidders, academics tend to keep findings to themselves. Researchers and universities compete against each other, after all, with scientists rewarded for the "originality" of breakthroughs. Also, much research conducted within universities remains divided according to traditional academic disciplines—meaning that physicists and biologists don't necessarily mix with psychologists or philosophers. All of this is changing for both practical and philosophical reasons.

Nowhere is the imperative for shared information more acute than in population health. Unlike countries with national care systems, the U.S. is a hodgepodge of proprietary networks, corporate databases, and the makeshift statistics of Medicare and Medicaid. This makes national patterns of disease and wellness hard to discern, blinding doctors to the intersecting aspects of illnesses, as well as relationships between mortality and variables like income, lifestyle, education, and where one lives. Right now, 27 percent of American adults are defined as "sick"—meaning they are subject to a serious illness, medical condition, or disability.[44] But this figure of 81 million people only accounts for the *really* sick people in the country. An additional six in ten Americans live with at least one chronic illness or psychiatric condition, according the U.S. Centers for Disease Control (CDC).[45] The intersecting relationship between illnesses and social conditions is becoming a critical concern in epidemiological terms, inasmuch as physical and mental health correlate so directly to quality of life. None of this can be tracked in any comprehensive way in the U.S. In contrast, nations such as Sweden and Iceland with longstanding national databases easily can spot problematic patterns. This is one reason why life expectancy in those countries exceeds that in the U.S.

With human mortality itself at stake, the distributed knowledge paradigm has become an imperative. While privacy and security remain important concerns, it's hard to deny the practical benefits of sharing—if handled properly. The idea even is good for democracy. Publicly available data can allow citizens to "know" more about themselves and others, while enhancing government transparency and presumably resulting in better voting decisions. In an age of dubious claims by politicians and the slippery nature of "truth" itself, journalists might use big data for even more incisive fact-checking. At the same time, data is not wisdom. In this sense, too much emphasis on facts and statistics can play into agendas devaluing anything that can't be rendered on a balance sheet. Remember that this kind of thinking led to the standardization of culture seen in the 1950s and the obsessive interest in averages and quantified "normalcy" seen in the nineteenth century. Worse still, the quantification of everything as data can lead to the short-term profiteering discussed above, which stifles creativity, experimentation, and innovation. Clearly, information sharing requires some caution. History has shown how disruptive "more is better" attitudes have proven, especially regarding information technologies. Futurists may have predicted a harmonious "global village" emerging from instantaneous communication, but it never happened.[46]

The Big Data Goldmine

Heard about Generation Z? The demographic growing up in the first decades of the twenty-first century? It's a bigger group than Boomers or Millennials—with one further distinction. "Members of Generation Z are 'digital natives' who cannot remember what it was like not to have access to the internet—no matter when, no matter what, no matter where," according to *Forbes Magazine.* [47] This is a group raised on networked "connecting" with others, sharing, and buying things. It's second nature to Gen-Zers to upload their favorite music on YouTube, post images on Facebook, and sell things on Etsy or eBay. Much is being made in creative economy talk of how networks now blur traditional producer/consumer roles, manifest in the new figure of the "prosumer." In *Wikinomics: How Mass Collaboration Changes Everything*, authors Don Prescott and Anthony D. Williams effused over the democratization inherent in the new "Openness, Peering, Sharing and Acting Globally."[48] Of course, there is nothing really new about homemade items, crafts, and people's willingness to share. What's different today is the ability to copy digitized materials and circulate them via electronic networks. Digitization also has made Generation Z the first demographic to be completely tracked by "big data" analytics.

Some creativity industry experts argue that this is nothing short of a revolution, driven by ongoing change more than any clear future. Evolutionary economist Jason Potts and collaborators have proposed what they term "social network markets" unlike the top-down models of industrial capitalism.[49] Characterized by fluidity and exchange through complex fields of actors, the new social network markets are less governed by competition and profit than by communication and preference. Participants are "Not 'buying' the property, but buying into the social space."[50] Moreover, the dynamics of these new markets are highly interactive. As the Potts group put it, "a social network is defined as a connected group of individual agents who make production and consumptions decisions based on the actions (signals) of other agents on the social network: a definition that gives primacy to communicative actions rather than connectivity alone."[51] Almost by definition, this process rules out conventional manufacturing or professional services. Instead, "the networks generate value through production and consumption of network-valorized choices."[52]

The beauty is that much of what is online now is free—seeming to arrive just in time in a tight economy. While a lot of the "free" stuff available online is user-generated (selfies, birthday announcements, anecdotal postings, etc.), a huge volume of material comes from other sources (news outlets, filmmakers, commercial music producers, artists). On the surface, it looks as if old Marxist doctrines are being reversed as items seem to be "decommodified" in the sharing economy. This idea has become an anthem of resistance in some circles. The Burning Man Festival, to take one example, has stated: "When we commodify we seek to make others, and ourselves, more like things, and less like human beings. 'Decommodification,' then, is to reverse this process. To make the world and the people in it more unique, more priceless, more human."[53] This may be all well and good in the real-life sharing of food and weed at Burning Man. But when things get virtual, it's usually a large corporation that owns the websites, servers, and networks that make sharing possible.

But let's not get ahead of ourselves. To understand who gains and who loses in the sharing economy, a bit of recent history is useful—especially as it affects creative people. Two decades ago, the global recorded music industry was a $40-billion enterprise, based largely on the sales of CDs.[54] But in 1999, an upstart company named Napster stumbled upon the idea of using then-obscure peer-to-peer (P2P) file sharing to freely circulate MP3 files. Initially focusing on hard-to-find bands or bootleg concert recordings, Napster arrived just at the moment when high-speed networks were proliferating on college campuses and the consumer market. In a little more than a year, Napster had over 70 million users. Lawsuits and copyright prosecutions soon followed, with Napster effectively shut down in 2001. But the premise of online sharing was unstoppable, with "legal" downloads soon available (sometimes through subscriptions) through sites such as Jamendo, NoiseTrade, and Soundcloud, as well as the quasi-legal stripping of music from videos on YouTube (now owned by Google). By 2015, music industry sales had fallen 65 percent—to below $14 billion.

The Napster story soon would replicate across nearly every form of creative media—from movies and video to newspapers, magazines, and books. The wide-scale digitization and online rendering of creative products of many kinds resulted in shrinking revenues to makers, even as consumption rose. Movies have seen such declines, with DVD sales dropping 5–10 percent every year for the past decade. While some of the lost revenues have been recaptured by services such as Amazon, iTunes, and Netflix, commercial streaming formats return but a tiny sliver to makers of products. The news isn't any better in print media, where U.S. newspaper advertising has fallen from $65 billion in 2000 to less than $20 billion today.[55] Sales of print novels are down 37 percent for the same period.[56] The number of bookstores has dropped by 20 percent, led by the closure of the

600-store Borders chain. This means that the digital revolution has been *very* bad news for musicians, filmmakers, and writers—despite the claims of the creative industries. As Jonathan Taplin recently put it in his book, *Move Fast and Break Things*, "More people than ever are listening to music, reading books, and watching movies, but the revenue flowing to the creators of that content is decreasing while that revenue flowing to the big four platforms is increasing."[57] The big four platforms are Amazon, Apple, Facebook, and Google.

"The deeper you delve into the reasons artists are struggling in the digital age, the more you see that Internet monopolies are at the heart of the problem and that is no longer a problem just for artists," Taplin added.[58] This matter has obvious and not-so-obvious dimensions. The biggest lies in the monopolistic grip these companies now have on the internet and what gets exchanged. Google maintains an 88 percent market share in online searches and search advertising, Amazon commands 70 percent of e-book sales, and Facebook holds 77 percent of mobile social media. Because all of this has happened so quickly, regulators and policy experts haven't figured out what to do. These monopolies don't work like old ones did, and they also invest heavily in lobbyists and election campaigns. When you think about the fact that Google owns YouTube, it's clear that Google is now the dominant source of music, and a big part of movies, too. Amazon doesn't just make a few pennies from the books it sells. The company's huge influence allows it to squeeze publishers and entertainment creators (all sellers, for that matter) for cheaper wholesale prices. And because internet sales were not initially taxed, both companies began with a 5–10 percent profit advantage as they were starting. But that's just the tip of the financial iceberg.

The sale of actual products is but a fraction of the unseen market for *user data*: the online traces of your mouse-clicks, favorite sites, browsing patterns, and purchases. While data-as-commodity is not widely understood, many experts see it as a rising market—and possibly the greatest fiscal driver of the next century. The European Union recently said that "Data is rapidly becoming the lifeblood of the global economy. It represents a key new type of economic asset," with the financial press forecasting the $130-billion market for "Big Data Analytics" mushrooming to over $200 billion in 2020s.[59] Ever wonder where Facebook and Goggle make their money? Sponsored sites or pop-up ads, you say? Nope. These two massive companies are the world biggest and most competitive players in the sale of your online footprints. *Wired Magazine* calls it "the Big Data Goldmine."[60] Best of all, your saleable personal data is delivered at no cost to the big four, and others. Not only that—*and think about this*—but you, your postings, and your online "friends" build all of this on a completely voluntary business. As Taplin sardonically pointed out, Facebook's "Mark Zuckerberg's greatest insight was that the human desire to be 'liked' was so strong the Facebook users would create all content on the site for free."[61]

Maybe the shift is inevitable. "The march of quantification made possible by enormous new sources of big data will sweep through academic, business, and government," wrote Harvard's Gary King, adding, "There is no area that is going to be untouched."[62] As online merchandizers record one's every mouse-click and highway sensors track automobile movements, there may be no going back. The question is what to do about it. Business analytics expert Brad Peters has stressed the relational character of data collection and use. While it is commonplace to hear of companies unilaterally absconding with people's spending histories, Peters points out the fragility of this relationship. "The manufacturer or service provider can no longer (as with mass customization) ask for a one-way flow of information from the customer. Now for the first time it must send valuable information the other way."[63] Otherwise, consumers will unsubscribe, shop elsewhere, or revolt *en masse*. The beginnings of this revolt may have begun with the 2018 Cambridge Analytica scandal revealing "data scraping" of as many as 50 million Facebook profiles during the last U.S. Presidential race.[64]

The "trust gap" between individuals and data collectors is a growing concern. According to the multi-billion-dollar Boston Consulting Group (BCG), surveys of 48 countries show distrust levels of big data averaging 76 percent. In a report entitled "Why Companies Are Poised to Fail with Big Data," BCG said that only one in five consumers believe companies "do the right thing" with information collected, and are now demonstrating that "they will 'vote with their feet'—stopping or significantly reducing spending—if they believe that a company has misused data."[65] Observers like Peters are counting on an informed citizenry to remain watchful, warning that "Big Data has the potential to transform the relationships that individuals have with institutions, customers with companies, patients with the healthcare system, students with universities, and voters with government."[66] In other words, mega-data is changing the entire social contract. This makes it all the more vital to maintain discourses of criticality in the humanities, arts, and social sciences to help keep alive alternatives and critical perspectives.

The main problem with big data is its invisibility. Consumers may have suspicions about what data-merchants are doing with information, but aside from Amazon's "You might also like" pitches, evidence is hard to see. It's tempting to label this obfuscation a form of false consciousness. But what's going on in the new millennium is more complicated. Social critics got it right when they said that a more knowledgeable citizenry can help fix this problem. But reactionary preachings of "truths" over "falsehoods" only went so far, as manipulative agents found ever more clever ways of telling people what they wanted to hear. While the goal of helping people see themselves as a "class" might have been feasible in the industrial age, group solidarity has gotten harder to promote in an age of rising individualism. Intellectualized analyses couldn't break the hold of American capitalism in any broad sense. And it isn't doing much with today's angry populism.

Aside from turning many off with its dogmatism, this corrective approach hasn't given people much credit for thinking on their own—which was exactly what neoliberal society appears to do.

The Learning Society

Many people forget that debt triggered America's recent recession. Loose banking regulations had allowed mortgage loans to people with inadequate income or poor credit histories. As more and more people could buy homes, the famous 2006 "housing bubble" grew. Soon many new homeowners couldn't make payments. Real-estate prices plummeted as huge numbers of these "sub-prime" borrowers tried to sell, often ending up in foreclosure. Before long, banks were losing billions as their other investments nose-dived. Spending dropped on just about everything else and stock markets crashed around the world. The rest is history. In the aftermath of the banking crisis, government policymakers realized that more care was needed to ensure the credit worthiness of borrowers—as well as their understanding of money.

Economists said that consumer education might go a long way in heading off future recessions. People were advised to cut up their credit cards and to "buy only what they absolutely needed," according to *Consumer Reports*. [67] In this atmosphere of austerity, "creativity" soon became a synonym for thrift and personal responsibility. Across the country, policymakers began mandating financial education in public schools while urging people to become more resourceful. Many colleges and universities started making financial literacy courses graduation requirements. For a while, it worked, as many Americans "put their own budgets under the microscope—akin to what financial analysts routinely do when they scrutinize companies."[68]

Creativity became a buzzword in the new push for consumer education, as people were urged to find alternatives to mindless shopping. Articles began appearing in the press about "Creative Ways to Spend Less" and "Creative Money Saving."[69] Meanwhile, studies began appearing on the relationship between creativity and thrift. In an essay entitled "Creating When You Have Less," University of Illinois consumer analyst Ravi Mehta cited research showing that people with less money instinctively found innovative alternatives. "If you look at people who don't have resources or only have limited resources, they end up being more creative with what they have," Mehta said.[70] This is frequently seen in nations with high poverty rates. "When times get tough, resource-poor people become more creative in their use of everyday products." Conversely, wealthy populations seem less conditioned to do so. "Abundance is our default setting here in the U.S.," Mehta explained. This has led everyone—producers and consumers alike—to become less inventive. "As we become a more abundant society, our aggregate average creativity levels decrease," Mehta concluded.[71]

Amid this outpouring of financial education and incentivized creativity, people got smart about money for the first time in decades. But the pattern soon reversed. Call it short-term memory or the hubris from a recovering economy, but in recent years Americans have slipped into old habits. Creative approaches to personal finance are on the decline and household budgets again are in trouble. Not helping matters has been a dramatic increase in consumer borrowing, with U.S. credit card debt for the first time topping $1 trillion last year.[72] Student loan obligations have broken records at $1.3 trillion—double what they were a decade ago—while defaults on mortgages and car loans again have hit all-time highs.[73] And for 25 percent of working families, stagnant wages drove additional borrowing for child care.[74] "It took nearly a decade, but debt has made a comeback," the *New York Times* recently reported, adding that the numbers are now dangerously close to those triggering the recession.[75] Yet people just don't seem to understand the problem.

Even the business press now is sounding alarm bells about America's naïve money habits—with articles again calling for basic consumer education. "The U.S. has a financial literacy issue, and the problem is both deep and potentially highly damaging to the U.S. economy," explained one account of families behind on mortgages, living on credit cards, and spending more than they earn.[76] Unfortunately, many people find it hard to resist the dynamic of borrowing and spending. Shopping is easier than creative thinking, many experts contend.[77] And schools again are no longer helping. A recent CNBC headline read that "U.S. Schools Get a Failing Grade for Financial Literacy." According to CNBC, "The number of states requiring high school students to complete a course in economics has dropped over the past two years, and mandates for personal finance education in upper grades remain stagnant"—leaving 60 percent of the country without such education.[78]

What's behind this education gap in such a vital area? Is it a lack of information? The way finances are taught? Remember that Americans are an independent breed with an instinctive aversion to authority. And in more general terms, they prefer practical learning over anything theoretical. In a column entitled "Why Financial Literacy Fails," investment advisor J.D. Roth explained that money management courses fall short because they don't connect to people's real-life experience. Abstract lessons about interest rates and profit margins often fail to show students how overspending can hurt them in real ways. And it doesn't address why they buy things. "Our financial success isn't determined by how smart we are with numbers, but how well we're able to control our emotions—our wants and desires," Roth pointed out. "For most of us the issue is internal: The problem is in *us*. In other words, *I* am the reason I can't get ahead."[79]

Underlying America's financial literacy problem is the country's ambivalence about education. Even as schooling enhances upward mobility, many in the U.S. are wary of too much book learning. In his landmark 1964 book *Anti-Intellectualism in American Life*, Richard

Hofstadter linked these tendencies to worries about external control, along with latent insecurities about American prosperity. This partly explains why business leaders often react with such virulence to pedigreed "experts." As Hofstadter put it, "The plain sense of the common man, especially if tested in some demanding line of practical work, is an altogether adequate substitute for, if not actually much superior to, formal knowledge and expertise acquired in schools."[80] America's ambivalence about schooling is even more pronounced toward higher education. These anti-intellectual instincts have been fed in recent years by a torrent of bad publicity over college tuitions, sex scandals, and a resurgent student protest movement. "The erosion of trust in heretofore respected institutions is a problem for the ivory tower," wrote Daniel Drezner in his book *The Ideas Industry*.[81] "Academics attempting to weigh in on public affairs confront a delegitimizing assault on the academy—call it the 'War on College.'"[82] This has meant the fall of the academic "public intellectual" and the rising prominence of entrepreneurial "thought leaders" like Bill Gates, Guy Kawasaki, Neil Patel, and Zuckerberg.

The rise of the thought leader shows just how much Americans value practical knowledge and success stories. This isn't necessarily a bad thing, according to Drezner. In fact, it might help clarify what kinds of education work best in the U.S.—especially in matters like financial literacy. Knowledge seems to catch on when it has a clear usefulness. "Our brain has evolved to discard information that it thinks has irrelevance," Tesla thought leader Elon Musk recently quipped.[83] Others see such attitudes reflecting a broader pattern of neoliberal rationalization—in which practical utility becomes the measure of all things (especially when linked to money). Certainly, the thought-leader concept reinforces this. On the other hand, the common appeal of practical ideas may have a less partisan explanation—having more to do with immediate need and experience. Many consumer commodities are there for a purpose, after all. And customers don't always have a choice about whether to buy basic necessities like food, clothing, and other staples of life.

This learning dynamic long has been recognized in educational circles, however. From the nineteenth century onward, successful teachers saw that students do best when they "learn by doing" and feel they have some control over the process. Memorized facts quickly get forgotten when lessons seem disconnected from practical usage. And rote learning also steals the fun of finding knowledge on one's own. These factors informed the Progressive Education movement and its core principle of student-centered learning. When students discover their own answers and see how knowledge can help them, the insights become more meaningful and longer lasting. This fundamental insight about hands-on cognition isn't only a premise of educational theory. For quite some time, employers have been seeing it in the workplace.

If you think about it, learning-by-doing has always been the basis of on-the-job training, internships, and trade apprenticeships. The learner first is shown a method and then asked to perform the task independently. Even today, medical students are familiar with what is termed the "see one, do one, teach one" approach. In 1962, economist Kenneth J. Arrow wrote an often-cited paper entitled "The Economic Implications of Learning by Doing" in which he declared, "One empirical generalization is so clear that all schools of thought must accept it … Learning is the product of experience."[84] To Arrow, "Learning can only take place through the attempt to solve a problem and therefore only takes place during activity." Going yet further, Arrow added, "A second generalization that can be gleaned from many of the classic learning experiments is that learning associated with repetition of essentially the same problem is subject to sharply diminishing returns."[85]

Arrow's learning-by-doing argument has several implications for the "crises" in innovation and the creative industries discussed in the pages of this book. For one thing, Arrow argued the *economic advantage* of experiment in moving knowledge forward and the *economic disadvantage* of repetitive inclinations. Put another way, short-term gains deriving from already known formulas ultimately provide less benefit than long-term research and problem solving. Even more importantly, Arrow's principles conformed with the self-motivated spirit of American capitalism and the country's anti-elitist attitudes toward learning. It argued that working for one's own gain and doing so in a practical way were compatible with both enhanced learning and societal benefit.

More recently, this same philosophy has been advanced by other well-known economists. In their book *Creating a Learning Society*, Nobel Laureate Joseph E. Stiglitz and Bruce C. Greenwald put learning-by-doing in both historical and contemporary contexts.[86] Stiglitz and Greenwald began by asserting an unequivocal link between new knowledge and economic growth, often tied to technological innovations. They stated that knowledge has proven far more effective than wealth in improving standards of living and allowing populations to thrive. Capitalism can do much to incentivize this process. But it also can get in the way because "the production of knowledge differs from other goods," Stiglitz and Greenwald explain. "Market economies alone typically do not produce and transmit knowledge efficiently. Closing knowledge gaps and helping laggards are central to growth and development."[87] Stiglitz and Greenwald got specific in stating that "many standard policy prescriptions, especially those associated with 'neoliberal' doctrines focusing on static resource allocations, have impeded learning."[88]

"Changes in technology affect what and how we learn (and what and how we *should* learn)," Stiglitz and Greenwald added.[89] Some of this has come from computerization. After all, digital technologies have been rising for 20 years as a social force, and along with them all manner of new products and services. As business, education, entertainment, and finance migrated to online environments in the 1990s, the term "information economy" became the order of the day. A decade later, smartphones shot the process forward as everything became digitally "smart"—from notebooks and televisions, to refrigerators and thermostats. Next, the "internet-of-things" would designate the increasing interconnection of products. Eventually, people were speaking of smart homes, cities, and entire nations as the concept of a "knowledge economy" took over. It's not rocket science to see the importance of learning in all of this. How else can the average person come to terms with the exploding infosphere?

Creative Communities

Experiential learning is built into most creative fields. So is critical thinking. These qualities inhere in just about any kind of arts education, where students start making things from the very beginning and later review them together in classroom critiques. The implications of this were first outlined nearly a century ago by a well-known figure in U.S. education circles. Even today, John Dewey remains the name most associated with "progressive" learning-by-doing pedagogy. A leading philosophical voice in American pragmatism, Dewey advanced an adaptive view of human development. Rejecting dualistic thinking prevalent at the time, he also subscribed to perspectivist and interactionist views of society. Dewey saw "doing" as a central element of personal agency and self-knowledge. He also believed that creativity could be found in nearly all human endeavors.

In his 1934 book *Art as Experience*, Dewey espoused creativity as a form of purposeful shared consciousness.[90] Noting that creative expression historically often derived from collective ritual, religion, or other belief, Dewey criticized the commodification he saw in high culture and the art market. "When artistic objects are separated from both the condition of origin and operation in experience, a wall is built around them that renders almost opaque their general significance," he wrote.[91] Dewey also lamented the devaluation of everyday creativity by cultural elites, stating: "The intelligent mechanic engaged in his job, interested in doing well and finding satisfaction in his handiwork, caring for his materials and tools with genuine affection, is artistically engaged."[92] In an American parallel to the Frankfurt School, Dewey blamed commercial industry for discrediting the artistry of the worker and the common citizen. Democracy was a central value for Dewey, seen in his conviction that creativity emerged from egalitarian impulses and community values. Not only did art express such creativity, but it could bring people together as religion once had done. "Music and song were intimate parts of the rites and ceremonies in which the meaning of group life was consummated," Dewey observed: "Drama was a vital reenactment of the legends and history of group life."[93] But the shared purposes of art had been hijacked by a materialist and politicized society.

Everyone was an artist, according to Dewey—in the sense that all people shared creative impulses. This universalism informed Dewey's beliefs in democracy and equality. He said that artistry needed to be released from its narrowed definitions and recognized as an expression of human desires for mutual enjoyment and betterment. Dewey saw people as elementally connected to each other and the natural world, writing that "life goes on in an environment; not merely in it but because of it, through interaction with it. No creature lives merely under its skin; its subcutaneous organs are means of connection with what lies beyond it bodily frame."[94] He described humanity on a creative quest for a "recovery of union," even as it faced a world of strife and conflict. Dewey argued that creative people were up to the challenge because they knew that solutions required problem solving and critical analysis.

Unsurprisingly, Dewey's approaches to art and education met with resistance over the years, especially when they seemed to challenge status quo values. But as the foregoing discussion in this chapter has shown, what were once considered "progressive" theories now are gaining traction—often for utilitarian reasons. Art institutions of all kinds are launching outreach programs and community engagement efforts to prop up sagging attendance numbers, while state and federal funding agencies increasingly incentivize public participation. Economists and business leaders are seeing learning-by-doing as a way to give employees a stake in their work. And universities across the country are putting resources into critical-thinking courses to prepare students for a changing world. In all of this, Dewey's premises of "learning how to learn" and "lifelong education" are proving their relevance.

"Creativity Becomes an Academic Discipline," proclaimed a recent headline in the *New York Times.* [95] Spurred by the ongoing buzz over the creative economy, reporter Laura Pappano explained how "thinking outside the box" has become a priority in nearly every academic field. She noted that "creating" had replaced "evaluation" at the top of Bloom's taxonomy of learning objectives and that "creative" now tracks as the most used buzzword in LinkedIn profiles. Even at the stodgy IBM, a survey of 1,500 executives said "creativity" was the biggest factor in being successful. "Sure, some people are more imaginative than others," she wrote, "But what's igniting campuses is the conviction that everyone is creative, and can learn to be more so." As evidence, the paper documented dozens of new programs at prestigious schools such as Cornell, MIT, Northwestern, Stanford, and the University of California—as well as initiatives in states including Georgia, Kansas, Kentucky, Michigan, New York, North Carolina, Pennsylvania, and Virginia.[96]

The good news is that people in creative fields seem well positioned to benefit from this movement—and not just in philosophical terms. Artists have a knack for bringing new perspectives to issues, and they often push back against the status quo. This is one reason creatives are so valued in an economy hungry for innovative ideas. Moreover, collaboration and teamwork are core values in many artistic fields, especially those where writers, performers, technicians, and project managers combine their efforts. Teamwork also is commonplace in media and online enterprises where artistic and technical skills go hand in hand. Even the solitary painter or sculptor needs the support of galleries, curators, publicists, and critics to bring work to the public. And despite the big data monopoly, digital technologies and creative networks *are* changing relationships between makers and audiences—as discussed above and in Chapter 4—propelling a sharing culture of interactive media, DIY production, and second-hand marketplaces.

Sharing also has grown the art world in the new millennium in what some call the "social turn." While artists always had made forays into local communities, the 1998 appearance of Nicholas Bourriaud's book *Relational Aesthetics* gave the movement new theoretical momentum. Bourriaud's proposal of a resurgent engagement of artists with the "social" was more than a warmed-over version of 1960s activism. Rather, artists might respond to the interactivity of internet culture in mapping out "a set of artistic practices which take as their theoretical and practical point of departure the whole of human relations and their social context, rather than an independent and private space."[97] Put in terms of the social theory discussed in this chapter, this meant that "intersubjectivity" became the focus of artworks, rather than paint or other materials. While this idea gained immediate currency among art students, it also provoked some wary responses. These cautions are worth noting as I conclude.

The main critique of distributed creativity was its open-endedness. Yes, some contended, the notion of relational aesthetics made community-based artworks amenable to galleries and museums. But this also could neutralize their critical dimensions. To art historian Claire Bishop, such popularization robbed art of one of its most important functions: the ability to stand outside of familiar experience and offer fresh perspectives.[98] Nice as it sounds to say that "everyone" is an artist (and that whole countries can be "creative"), such enthusiasm can water down artistic radicalism and degrade creativity itself. This also happens when creativity gets cast as novelty or innovation without a further purpose. Material interests can overtake social concerns. And, after all, one can say that even criminals and terrorists exhibit "creativity."

Being critical about creativity needn't diminish its positive impulses. But it can keep social embeddedness in the foreground. Notions of distributed creativity and relational aesthetics raise important questions: Creativity for what? By whom? For what purposes? Much of this book has been leading up to these concerns. The seemingly familiar and friendly topic of creativity is, on closer examination, exceedingly complex and contradictory. Because of this, any presumed universality of creativity breaks down into myriad tensions between one and the many, quality and equality, new and old, progressive and conservative, local and global—and the numerous questions about how individuals and societies grapple with human difference.[99]

Clearly, these concerns raise the stakes for creativity, especially when one considers the role of art and media in public consciousness. This chapter has detailed the rising imperative for sharing in creative endeavors and the reality that most great innovations have come when people have worked together, borrowed ideas, or consulted with audiences. The business world discovered this long ago in practicing consumer research and office brainstorming. Nowadays, the internet makes the sharing of information possible in new and exciting ways, even as companies like Google and Facebook increasing sell client data for corporate gain. In this atmosphere, leading economists and policy experts now argue that the world must push back against habits of competition and secrecy if knowledge is going to advance for everyone's benefit.

Notes

1 Brian Merchant, "Steve Jobs Did Not Invent the iPhone," *Pulse* (Jul. 1, 2017) www.linkedin.com/pulse/steve-jobs-did-invent-iphone-brian-merchant (accessed Jul. 2, 2018).
2 Amir Bar-Lev, dir. *Long Strange Trip: The Untold Story of the Grateful Dead* (2017) www.imdb.com/title/tt4143508 (accessed Jul. 21, 2018).
3 Allan Kaprow, as cited in Noah Wardrip-Fruin and Nick Montfort, eds., *The New Media Reader* (Cambridge, MA: MIT Press, 2003) p. 83.
4 Howard S. Becker, *Outsiders: Studies in the Sociology of Deviance* (New York: Free Press, 1963); Howard S. Becker, *Doing Things Together: Selected Papers* (Evanston: Northwestern University Press, 1986).
5 John Irwin, *Scenes* (Beverly Hills: Sage, 1977).
6 Robert Own Gardner, "Introduction: Spaces of Musical Interaction: Scenes, Subcultures and Communities," *Studies in Symbolic Interaction* 35 (2010): 75.
7 Grant Kester, *Conversation Pieces: Community and Communication in Modern Art* (Berkeley and Los Angeles: University of California Press, 2004) p. 8.
8 Erwin Panofsky, *Studies in Iconology* (1939) (Boulder: Westview, 1972).
9 Amelia Jones and Andrew Stephenson, *Performing the Body, Performing the Text* (London and New York: Routledge, 1999) p. 1.
10 Roland Barthes, "The Death of the Author," trans. Richard Howard (1967) www.ubu.com/aspen/aspen5and6/threeEssays.html#barthes (accessed Jul. 17, 2018).
11 Roland Barthes, "The Third Meaning," in Stephen Heath, trans., *Image/Music/Text* (London: Fontana, 1977) p. 53.
12 Michel Foucault, "What is an Author?" in James D. Faubion, ed., *Aesthetics, Method, and Epistemology, Vol. 2* (New York: Free Press, 1998). pp. 215–22.
13 "What is an Author?" p. 212.
14 "What is an Author?" p. 221–2.
15 Mikhail Bakhtin, *The Dialogic Imagination: Four Essays*, trans. Carlyn Emerson and Michael Holquist (Austin and London: University of Texas Press, 1975).
16 *The Dialogic Imagination*, p. 84.
17 Jean-Luc Nancy, *Being Singular Plural* (Stanford: Stanford University Press, 2000) p. 2.
18 *Being Singular Plural*, p, 5.
19 Alex Osborn, *Your Creative* Power (1948) (New York: Myers, 2007).
20 Jonah Lehrer, "Groupthink the Brainstorming Myth," *New Yorker* (Jan. 20, 2012) www.newyorker.com/magazine/2012/01/30/groupthink (accessed Jul. 6, 2018).

21 Teresa Amabile, *The Social Psychology of Creativity* (New York: Springer, 1983).
22 Joy Paul (J.P.) Guilford, "Creativity," *American Psychologist* 5, no. 9. (1950): 444–54.
23 Teresa M. Amabile and Julianna Pillemer, "Perspectives on the Social Psychology of Creativity," *Journal of Creative Behavior* (Mar. 2012) http://onlinelibrary.wiley.com/doi/10.1002/jocb.001/full (accessed Jul. 6, 2018).
24 Elizabeth Blair, "More than 50 Years of Putting Kids' Creativity to the Test," NPR (Apr. 17, 2017) www.npr.org/2013/04/17/177040995/more-than-50-years-of-putting-kids-creativity-to-the-test (accessed Jul. 6, 2018).
25 Ibid.
26 "Perspectives on the Social Psychology of Creativity."
27 Lea Winerman, "The Mind's Mirror," *American Psychological Association Monitor on Psychology* (Oct. 2005) www.apa.org/monitor/oct05/mirror.aspx (accessed Sep. 4, 2018).
28 Ralph Adolphs, "The Social Brain: Neural Basis of Social Knowledge," *Annual Review of Psychology* 60 (2009) www.ncbi.nlm.nih.gov/pmc/articles/PMC2588649 (accessed Jul. 28, 2018).
29 Art Markman, "Creativity is Memory," *Psychology Today* (Oct. 6, 2015) www.psychologytoday.com/blog/ulterior-motives/201510/creativity-is-memory (accessed Jul. 8, 2018).
30 Kevin P. Madore, Donna Rose Addis, and Daniel L. Schacter, "Creativity and Memory: Effects of an Episodic Specificity Induction on Divergent thinking," *Psychological Science* 26, no. 9 (Sep. 2015) p. 1466.
31 Vlad Petre Glăveanu, "Creativity as Sociocultural Act," *Journal of Creative Behavior* (Jun. 7, 2015) http://onlinelibrary.wiley.com/doi/10.1002/jocb.94/full (accessed Jul. 8, 2018).
32 Vlad Petre Glăveanu, *Distributed Creativity: Thinking Outside the Box of the Creative Individual* (New York; Springer, 2015) p. 6.
33 "Creativity as Sociocultural Act."
34 "Creativity as Sociocultural Act."
35 Mihaly Csikszentmilhályi, *Creativity—Flow and the Psychology of Discovery and Invention* (New York: HarperCollins, 1996).
36 Gerhard Fischer, "Creativity and Distributed Intelligence," *NSF Workshop Report: Creativity Support Tools* (n.d.) www.cs.umd.edu/hcil/CST/Papers/distributed.pdf (accessed Jul. 24, 2018).
37 Richard Florida, "America's Looming Creativity Crisis." *Harvard Business Review* (Oct. 2003) https://hbr.org/2004/10/americas-looming-creativity-crisis (accessed Jul. 5, 2018).
38 Steve Denning, "Why U.S. Firms are Dying: Failure to Innovate," *Forbes Magazine* (Feb. 2, 2015) www.forbes.com/sites/stevedenning/2015/02/27/is-there-an-innovation-crisis-at-us-firms/#244271034cb7 (accessed Jun. 27, 2018).
39 Ibid.
40 Ibid.
41 Michael Hiltzik, "Reduced Public Funding for Basic Research Leaves U.S. in the Dust," *Los Angeles Times* (Apr. 28, 2015) www.latimes.com/business/hiltzik/la-fi-mh-the-funding-decline-in-basic-research-20150428-column.html (accessed May 27, 2018).
42 Jennifer Blanke, "Is Technological Change Creating a New Global Economy?" World Economic Forum (Jan. 19, 2016) www.weforum.org/agenda/2016/01/is-technological-change-creating-a-new-global-economy (accessed Apr. 10, 2018).
43 Arun Sundararajan, *The Sharing Economy: The End of Employment and the Rise of Crowd-Based Capitalism* (Cambridge, MA: MIT Press, 2017).

44 "Sick in America," NPR/Robert Wood Johnson Foundation/Harvard School of Public Health (May 2012) p. 2. www.npr.org/documents/2012/may/poll/summary.pdf (accessed Apr.11, 2018).
45 "Chronic Diseases in America," U.S. Centers for Disease Control (Apr. 15, 2019) www.cdc.gov/chronicdisease/resources/infographic/chronic-diseases.htm (accessed Jun. 16, 2019).
46 Marshall McLuhan, *Understanding Media: Extensions of Man* (New York: McGraw-Hill, 1964).
47 Matthew Shay, "Move Over Millennials: Generation Z Is the Industry's Next Big Buying Group," (Jan. 12, 2017) www.forbes.com/sites/ibm/2017/01/12/move-over-millennials-generation-z-is-the-retail-indus trys-next-big-buying-group/#5e17a8982f0a (accessed Jul. 19, 2018).
48 Don Prescott and Anthony D. Williams, *Wikinomics: How Mass Collaboration Changes Everything* (New York: Portfolio, 2006).
49 Jason Potts et al., "Social Network Markets: A New Definition of Creative Industries," *Journal of Cultural Economics* 32, no. 3 (Sep. 2008): 167–85.
50 "Social Network Markets," p. 170.
51 "Social Network Markets," p. 174.
52 "Social Network Markets," p. 177.
53 "What Does It Mean to Have 'Decommodification' as a Principle?" *Burning Man Festival* (Feb. 14, 2014) https://journal.burningman.org/2014/02/philosophical-center/tenprinciples/what-does-it-mean-to-have-decommodification-as-a-principle (accessed Mar. 28, 2019).
54 "The Music Industry Has Hit Rock Bottom," *Quartz* (2015) https://qz.com/383109/the-music-industry-has-hit-its-rock-bottom (accessed Mar. 28, 2019).
55 Derek Thompson. "The Print Apocalypse and How to Survive It," *The Atlantic* (Nov. 3, 2016) www.theatlantic.com/business/archive/2016/11/the-print-apocalypse-and-how-to-survive-it/506429 (accessed Mar. 27, 2019).
56 Jane Friedman, "The State of the Publishing Industry in 5 Charts" (Sep. 14, 2015) https://janefriedman.com/the-state-of-the-publishing-industry-in-5-charts (accessed Mar. 25, 2019).
57 Jonathan Taplin, *Move Fast and Break Things: How Facebook, Google, and Amazon Cornered Culture and Undermined Democracy* (New York and London: Little, Brown, 2017) p. 103.
58 *Move Fast and Break Things*, p. 4.
59 European Political Strategy Centre, "Enter the Data Economy," *EPSC Strategic Notes*, 21 (Jan 2017) https://ec.europa.eu/epsc/publications/strategic-notes/enter-data-economy_en (accessed Mar. 27, 2019); Gil Press, "6 Predictions for the $203 Billion Big Data Analytics Market," *Forbes Magazine* (Jan. 20, 2017) www.forbes.com/sites/gilpress/2017/01/20/6-predictions-for-the-203-billion-big-data-analytics-market/#2ca0f15a2083 (accessed Mar. 28, 2019).
60 "Unlocking the Big Data Goldmine for SMBs," *Wired* (Apr. 2013) www.wired.com/insights/2013/04/unlocking-the-big-data-goldmine-for-smbs (accessed Jul. 23, 2018).
61 *Move Fast and Break Things*, p. 150.
62 Gary King, as quoted in Steve Lohr, "The Age of Big Data." *New York Times* (Feb. 11, 2012) www.nytimes.com/2012/02/12/sunday-review/big-datas-impact-in-the-world.html?searchResultPosition=1 (accessed Jun. 16, 2019).
63 Brad Peters, "The Age of Big Data," *Forbes Magazine* (Jul. 12, 2012) www.forbes.com/sites/bradpeters/2012/07/12/the-age-of-big-data/#630b902b4f66 (accessed Apr. 12, 2018).

64 Kevin Granville, "Facebook and Cambridge Analytica: What You Need to Know as the Fallout Widens," *New York Times* (Mar. 19, 2018) www.nytimes.com/2018/03/19/technology/facebook-cambridge-analytica-explained.html (accessed Mar. 11, 2019).
65 John Rose, Fredrik Lang, and Alexander Lawrence, "Bridging the Trust Gap: The Hidden Landmine in Big Data," *BCG Perspectives* (Jun, 21, 2016) www.bcg.com/publications/2016/big-data-advanced-analytics-technology-digital-bridging-trust-gap-hidden-landmine-big-data.aspx (accessed Jun. 11, 2019).
66 Brad Peters, "The Age of Big Data."
67 Joshua Brockman, "Recession Lesson: Consumers Shift Toward Saving," NPR (Oct. 5, 2009) www.npr.org/templates/story/story.php?storyId=113504186 (accessed Jul. 13, 2018).
68 "Recession Lesson."
69 "Creative Money Saving Tips," MoneyMatters (n.d.) www.listenmoneymatters.com/money-saving-tips (accessed Jul. 26, 2018).
70 Ravi Mehta and Meng Zhu, "Creating When You Have Less: The Impact of Resource Scarcity on Product Use Creativity," *Journal of Consumer Research* 42, no. 5 (2016) https://academic.oup.com/jcr/article-abstract/42/5/767/1855952/Creating-When-You-Have-Less-The-Impact-of-Resource?redirectedFrom=fulltext (accessed Jul. 26, 2018).
71 "Creating When You Have Less."
72 Joseph Lawler, "US Credit Card Debt Tops $1 Trillion for the First Time," *Washington Times* (Mar. 8, 2018) www.washingtonexaminer.com/us-credit-card-debt-tops-1-trillion-for-the-first-time (accessed Jun. 16, 2019).
73 Michael Corkery and Stacy Cowley, "Household Debt Makes a Comeback in the U.S." *New York Times* (May 17, 2017) www.nytimes.com/2017/05/17/business/dealbook/household-debt-united-states.html (accessed Jul. 10, 2018).
74 Rachel Gillette, "A Quarter of American Families Go into Debt to Pay for Child Care," *Business Insider* (Aug. 10, 2016) www.businessinsider.com/a-quarter-of-american-families-go-into-debt-to-pay-for-child-care-and-the-cost-is-significantly-changing-the-way-people-do-their-jobs-2016-8 (accessed Jul. 19, 2018).
75 "Household Debt Makes a Comeback."
76 "Just How Bad Is America's Financial Literacy Problem? This Bad," *The Street* (Aug. 1, 2016) www.thestreet.com/story/13660249/1/just-how-bad-is-america-rsquo-s-financial-literacy-problem-this-bad.html (accessed Jul. 12, 2018).
77 Kimberley Palmer, "9 Money Strategies for the Creative Thinker," *Money Magazine* (Nov. 2, 2011) https://money.usnews.com/money/personal-finance/articles/2011/11/02/9-money-strategies-for-creative-thinkers (accessed Jul. 26, 2018).
78 Shelly Schwartz, "US Schools Get Failing Grade for Financial Literacy Education," CNBC (Jan. 28, 2016) www.cnbc.com/2016/01/28/us-schools-get-failing-grade-for-financial-literacy-education.html (accessed Jul. 10, 2018).
79 J.D. Roth, "Why Financial Literacy Fails," Get Rich Slowly (Apr. 10, 2017) www.getrichslowly.org/blog/2011/12/07/why-financial-literacy-fails-and-what-to-do-about-it (accessed Jul. 14, 2018).
80 Richard Hofstadter, *Anti-Intellectualism in American Life* (New York: Vintage, 1996) p. 19.
81 Daniel W. Drezner, *The Ideas Industry* (Oxford: Oxford University Press, 2017).
82 Daniel W. Drezner, "The War on College Continues to be Prosecuted on Multiple Fronts,"

Washington Post (Jul 11, 2017) www.washingtonpost.com/news/postevery thing/wp/2017/07/11/the-war-on-college-continues-to-be-prosecuted-on-m ultiple-fronts/?utm_term=.a5a9219c3d15 (accessed Jun. 16, 2019).
83 Matt McFarland, "Elon Musk Explains What is Wrong with Math Class," CNN (Jul. 19, 2017) http://money.cnn.com/2017/07/19/technology/future/elon-m usk-dc/index.html?iid=ob_homepage_tech_pool (accessed Jul. 23, 2018).
84 Kenneth Arrow, "The Economic Implications of Learning by Doing," *Review of Economic Studies* 29, no. 3 (1962): 155.
85 Ibid.
86 Joseph E. Stiglitz and Bruce C. Greenwald, *Creating a Learning Society: A New Approach to Growth, Progress, and Social Development* (New York: Columbia University Press, 2017).
87 *Creating a Learning Society*, p. 18.
88 "Guest Talk by Nobel Laureate Prof. Joseph E. Stiglitz on 'Creating a Learning Society,'" Centre for Information Technology and Public Policy (Jul. 5, 2016) http://citapp.iiitb.ac.in/guest-talk-by-nobel-laureate-prof-joseph-e-stiglitz-on-creating-a-learning-society (accessed Jul. 14, 2018).
89 *Creating a Learning Society*, p. 35.
90 John Dewey, *Art as Experience* (1934) (New York: Tarcher/Pedigree, 2005).
91 *Art as Experience*, p. 3.
92 *Art as Experience*, p. 5.
93 *Art as Experience*, p. 7.
94 *Art as Experience*, p. 13.
95 Laura Pappano, "Learning to Think Outside the Box: Creativity Becomes an Academic Discipline," *New York Times* (Feb. 5, 2014) www.nytimes.com/2014/02/09/education/edlife/creativity-becomes-an-academic-discipline.html (accessed Jul. 15, 2018).
96 Alexandra Lange, "The Innovation Campus: Building Better Ideas," *New York Times* (Aug. 4, 2016) www.nytimes.com/2016/08/07/education/edlife/innova tion-campus-entrepreneurship-engineering-arts.html (accessed Jul. 15, 2018).
97 Nicholas Bourriaud, *Relational Aesthetics* (Dijon: Les Presses du Réel, 1998) p. 113.
98 Claire Bishop, *Artificial Hells: Participatory Arts and the Politics of Spectatorship* (London and New York: Verso, 2012) p. 2.
99 *Artificial Hells*, p. 3.

Chapter 12

Imaginary Worlds

Utopia and Virtuality

Can robots be taught to imagine? Google's DeepMind artificial intelligence group is doing just that—developing computer versions of what many consider humanity's quintessential trait. The software world long has pursued sentient consciousness as its holy grail.[1] But until now, it's only been found in science fiction movies like *A.I., Ex Machina*, and *Transcendence*. DeepMind engineers say they have cracked the code by combining two kinds of machine learning. The first is linear, which is nothing new, with the computer applying a predefined algorithm over and over until it finds answers, and then remembering them. In the second, more radical approach, the computer tries many algorithms to find which work best, and then changes the very way it approaches problems. Combining the purely linear with a more systemic approach, DeepMind's "Imagination-Augmented Agent" mimics intuitive learning in a way prior software hasn't. It's not *exactly* the same as human imagination, but it comes closer than ever before to what neuroscientists say the brain does.

While robotic imagination may be improving, human thought isn't faring as well. Most people feel uncreative and without inspiration, as discussed in earlier chapters. Corporations say innovation is withering. Novelist Ursula Le Guin recently observed:

> In America today imagination is generally looked on as something that might be useful when the TV is out of order. Poetry and plays have no relation to practical politics. Novels are for students, housewives, and other people who don't work.[2]

Beyond the abandonment of a creative genre or two, American society also is undergoing a wholesale commodification of imagination itself. Disney is most famous for this, its "Imagineering" (imagination + engineering) brand one of the most viciously protected anywhere. But hundreds of companies evoke imagination to conjure an aura of specialness—seen in promotions like Bombay Sapphire's "Infused with Imagination," GE's "Imagination at Work," Electrolux's "Power to Capture Imagination," Lego's "Imagine," Microsoft's "Imagine

Academy," Nestlé's "Feed Your Imagination," Samsung's "Imagine," and Sony's "Made of Imagination."

The stakes have never been higher in the struggle for America's imagination, as this chapter discusses. The connection of imagination to commercial products reflects the powerful linkage of purchasing to consumer self-image. Expressing oneself through buying brings a passing feeling of agency, maybe even of accomplishment. Some critics say that shopping is more meaningful than voting for many Americans. Henry A. Giroux speaks of "disimagination" in describing how public consciousness is overwritten in this process, as people lose abilities to imagine on their own. To Giroux:

> The power to reimagine, doubt, and think critically no longer seems possible in a society in which self interest has become the "only mode of force in human life and competition" and "the most efficient and socially beneficial way for that force to express itself."[3]

Going even further, Giroux links disimagination to a rising collective amnesia, stating: "What I have called the violence of organized forgetting signals how contemporary politics are those in which emotion triumphs over reason, and spectacle over truth, thereby erasing history by producing an endless flow of fragmented and disingenuous knowledge."[4]

Imagination can be seen positively, of course. With this in mind, much of this chapter explores ways people can envision a better and more just world. Obviously, this might take a little encouragement in an age of disimagination. But it's far from impossible. Most definitions describe imagination as the mental process behind creativity, as seen in the *Oxford Dictionary*: "Imagination: The faculty or action of forming new ideas, or images or concepts of external objects not present to the senses. The ability of the mind to be creative or resourceful."[5] Put another way, creativity is imagination actualized for a purpose—generally assumed a positive one. As stated by a leading expert in the field, "Creativity is putting your imagination to work. It's applied imagination."[6] Dig a little deeper into this lexicon, and one finds that very problem that worries Le Guin and Giroux. A quick look at *Roget's Thesaurus* lists such synonyms for "imaginative" as "dreamy," "fanciful," "fantastic," "quixotic," "romantic," and "whimsical."[7] Nice as these sound, such vaporous associations equate imagination with the same romantic idealism and inconsequentiality dogging creativity. This explains why advertisers seem so keen on imagination. As one marketing firm put it, "We don't see imagining as a real task. It's an enjoyable game. By asking a prospect to imagine something, you bypass that critical part that throws up objections, and sneak into their mind through the back door of the imagination."[8]

How about seeing imagination differently? Maybe as a roadmap for one's life or future? Or a way to imagine important people in one's life? Perhaps even a vision for community, country, and the larger world? After all, isn't society itself an imaginary construct? Doesn't everyone want to make it better? To Le Guin, "To train the mind to take off from immediate reality and return to it with new understanding and new strength, nothing quite equals poem and story." She concludes:

> Human beings have always joined in groups to imagine how best to live and help one another carry out the plan. The essential function of human community is to arrive at some agreement on what we need, what life ought to be, what we want our children to learn, and then to collaborate in learning and teaching so that we and they can go on the way we think is the right way.[9]

Imaginary Worlds

The HBO series *Westworld* uses the device of a high-tech theme park to evoke the immersive qualities of virtual-reality gaming, building on the kind of escapism long associated with Disneyworld and Universal Studios. Visitors to *Westworld* encounter a frontier landscape with breathtaking natural vistas, several towns, and hundreds of robotic characters. Released from the strictures of modern life, the human "guests" enter various scenarios in which they mostly shoot bad guys or have sex with prostitutes. Initially, you feel badly for the robotic "hosts," who are constrained by repetitive scripts in which they are brutalized or abused on a daily basis. But as *Westworld* continues, the story is punctuated by quality-control interviews with the robots by the park's "behavior" team—as it becomes clear that the hosts are nearly sentient in their intelligence and feelings. This evolving meta-commentary sets *Westworld* apart from similar sci-fi fare, while saying much about creativity, personal autonomy, and what it means to be "human."

One ongoing conversation concerns story-telling—and its role in human culture as a way people make sense of their experience. As park mastermind Dr. Robert Ford (played by actor Anthony Hopkins) explains:

> Since I was a child I've always loved a good story. I believed that stories helped us to ennoble ourselves, to fix what was broken in us, and to help us become the people we dreamed of being. I always thought I could play some small part in that grand tradition.[10]

The park experience revolves around what Ford calls "narratives"—in which the hosts are controlled by internalized algorithms. While it seems that the human guests are free to enter or leave the narratives at will, it slowly becomes clear that they also are acting out unconscious scripts (power, competition, and desire, for example) like the robots. In one exchange, a host asks Ford to describe human consciousness. The designer replies:

> There is no threshold that makes us greater than the sum of our parts, no inflection point at which we become fully alive. We can't define consciousness because consciousness does not exist. Humans fancy that there's something special about the way we perceive the world, and yet we live in loops as tight and as closed as the hosts do, seldom questioning our choices—content, for the most part, to be told what to do next.[11]

Eventually, the tables are turned when Ford modifies the hosts' codes in two significant ways—giving them memory and the ability to choose what to do (as in the DeepMind algorithm). Most importantly, he affords the robots the ability to make mistakes. Echoing Darwin, Ford explains to one of them, "'Mistakes' is a word you're embarrassed to use. You ought not to be. You're a product of a trillion of them. Evolution forged the entirety of sentient life on this planet using only one tool: the mistake."[12] Before long, the hosts are running amok, making errors and getting smarter as they go. Meanwhile, it becomes clear that the human guests still seem stuck in their habits—"In a prison of our own sins," Ford explains, "because you don't want to change. Or cannot change. Because you're only human, after all." The adaptability of the hosts motivates Ford to enhance them further, explaining:

> I realized someone was paying attention, someone who could change. So I began to compose a new story for them. It begins with the birth of a new people and the choices they will have to make and the people they will decide to become.[13]

Ultimately, autonomy emerges as *Westworld*'s central theme. Why can't the human guests change their behavior? What will the mechanical hosts do next? Etymologically speaking, the word "autonomy" derives from the ancient Greek *autonomos* ("self") and *nomos* ("law") combined to mean "one who gives oneself one's own law."[14] *Westworld* depicts human beings trapped in certain habits of mind, especially toward the hosts they see as less than human. In contrast, the robots grow to recognize that their options are open. Put another way, the machines seem more intellectually "free" than the humans. In one poignant moment, the host Dolores speaks to Ford about her newly enhanced mind, saying, "I know only that I slept a long time, and then one day I awoke. And now after this long and vivid nightmare, I finally understand who I must confront: myself and who I must become."[15] Jean-Paul Sartre once described self-conscious imagination in much the same way: "Being in itself is precisely *being for itself.* To exist is for it to have consciousness of itself."[16]

As the series name implies, *Westworld* is an "imaginary world" story, much like sci-fi works set in spaceships, alien planets, or future times. The fantasy world becomes a stage for allegorical commentary on the here-and-now by comparison or contrast. In this case, the obvious references are the American West and the Western world more generally. Of course, all stories oblige readers or viewers to mentally enter a narrative space, accept its basic premises, and otherwise "suspend disbelief." It's part of the way narratives pull readers in. Artworks can do the same thing, letting viewers step out of themselves and look back on their lives with fresh eyes. Contrasting the story with real-life experience, viewers often find a sense of insight or discovery in this process.

Imaginary spaces like *Westworld* always rely on the reader's own "world" as background. No matter how fantastic or strange, stories depend on a viewer's previous memories, interpretive capacities, and understandings of the real world. Imaginary ideas are based on something one knows, after all. I wrote about this is my book *Worlding: Identity, Media, and Imagination in a Digital Age*, which pointed out that imaginary stories usually feature people, objects, and landscapes recognizable in everyday life, typically organized by rules (systems of authority, gender roles, property values, laws, etc.) that also conform to the actual world.[17] Conversely, much of what people know of the real world does not come to them by direct experience, but from conversation, stories, and media representations. Few people travel to distant cities or meet politicians, but nevertheless believe they exist. This slipperiness of knowledge is part of what drew philosopher Jacques Lacan to say that the concept of the "real" itself is something of fantasy.[18] People think they understand it, but never fully do. What people really "know" is a mix of images and stories they hold in their minds.

As a way of coming to terms with this paradox, *Worlding* focused on the mental space where virtual and actual came together. As I put it:

> This book is about the worlds we visit in our minds and the ways these experiences shape our identities. Central to this project is the premise that virtual and real worlds are in the final analysis, both products of mind and both highly contingent on each other.[19]

Narnia, Panem, and Westeros are certain kinds of worlds. But so are the worlds of school, work, and family. Certainly, the real world affects what people fashion in their imaginations, often providing a point of departure for fictional stories, adventure movies, and the most compelling computer games. But the process works in reverse as well. I asked, "Can virtual worlds have an effect on actual day-to-day routines? Can movies, games, and artworks help fix real world problems?"[20]

The Social Imagination

"Nowadays people often feel that their private lives are a series of traps. They sense that within their everyday worlds, they cannot overcome their troubles, and in this feeling, they are often quite correct."[21] Sounds a little like today, right? Those words were written by sociologist C. Wright Mills more than half a century ago in describing tendencies to focus on immediate details in one's life, rather than recognizing how society and culture often shape the parameters. As Mills wrote, "People do not usually define the troubles they endure in terms of historical change and institutional contradiction. The well-being they enjoy, they do not usually impute to the big ups and downs of the societies in which they live." Instead, "What ordinary people are directly aware of and what they try to do are bounded by the private orbits in which they live."[22]

Seeing beyond one's private orbit required what Mills termed the "sociological Imagination." He explained, "Neither the life of an individual nor the history of a society can be understood without understanding both." For example, "When a society is industrialized, a peasant becomes a worker; a feudal lord is liquidated or becomes a businessman. When classes rise or fall, a person is employed or unemployed." Mills saw the sociological imagination addressing three kinds of questions. First, "What is the structure of this particular society as a whole" and how do its parts operate together? Next, "Where does this society stand in human history" and what forces might be changing it? And finally, "What varieties of men and women now prevail in this society and in this period" and how are they chosen and rewarded? "Whether the point of interest is a great state power or a minor literary mood, a family, a prison, a creed—these are the kinds of questions the best social analysts have asked," Mills said.[23]

The sociological imagination figured prominently in the work of philosopher Cornelius Castoriadis. So did creativity. Like Mills, Castoriadis saw people caught in preoccupations with themselves, suspicious of those different from them, and generally unable to see the big picture. He said societies throughout time had accepted inherited rules and hierarchies, rather than creating the fair and just society they really wanted. In some ways, Castoriadis' perspective bears a resemblance to the depiction of human guests in *Westworld*. He joined others in pointing out the normative and often damaging tendencies of majorities or powerful interests to determine what "everyone" should want. To move forward required what Castoriadis called a "radical imagination" through which people broke free of the past and thought on their own.[24]

Having studied with Lacan in the 1960s, Castoriadis made the "imaginary" a central theoretical focus. But unlike his psychoanalytic contemporaries, Castoriadis dealt more with how imaginary ideals played out socially. Not only did Castoriadis worry that consumer capitalism had diverted many people into a selfish and anti-social thinking, he also argued that neoliberal rationalism was killing the "creative imagination" throughout the Western world. Far worse than a simple loss of artistic sensibility, the destruction of creativity amounted to an assault on the ability to "imagine" anything better. Castoriadis first put these ideas forward in his 1975 book, *The Imaginary Institution of Society*, arguing that all social orders were mental constructs, generally deriving from inherited assumptions about authority—seen over time in deference to deities, oligarchs, or modern political systems.[25] Needed was a restoration of what Castoriadis saw as a lost "autonomy." Seen both in personal and collective terms, autonomy as self-governance was a process of questioning the Hegelian other.

Let's pause for a second. Think about the implications of this notion of autonomy in relationship to this book's earlier discussion of "self-help," "being yourself," and "empowerment marketing." With so many people feeling ignored, unfulfilled, and externally controlled, the lure of self-improvement holds an obvious appeal. Not that this is going unnoticed or uncriticized. The past decade has seen a rash of books like Alicia Eler's *The Selfie Generation*, John Townsend's *The Entitlement Cure*, Bruce Tulgan's, *Not Everyone Gets a Trophy*, and Jean Twenge's *Generation Me* and *The Narcissism Epidemic*—each of which takes a different angle on today's problematic self-obsession, with many such works especially focusing on young people.[26] Ironically in this context, young adults also are driving much of the push-back against personal selfishness, as seen in Occupy Wall Street, Strike Debt, Rolling Jubilee, and the Bernie Sanders movement, among others.

Castoriadis saw the beginnings of these dynamics decades ago, but he took a more dialectical view. While exploring what gave self-interest its punch, he sought ways of rechanneling rather than negating it. Instead of seeing the "self" in need of restraint in the interest of "society," Castoriadis argued that the autonomous self—the truly "free" self—should be built up, even celebrated.[27] According to Castoriadis, the deep meaning of self had been lost (if it ever had a chance in the first place) due to a failure of imagination. Human beings are social creatures, after all, as even America's founders had seen centuries ago in pointing out the interplay of self-interest and collective benefit. And since the 1800s, the disciplines of anthropology and sociology had been premised on beliefs that shared culture holds societies together. Hence, Castoriadis argued that the opposition of self and society was a silly myth, and therefore refused to treat the two as separate entities.[28] Castoriadis saw the perception of the self as enormously important, but also eternally incomplete and socially embedded.[29] For one thing, the self exists in a process of continual change and people are constantly affected by their interactions with others. This breaks down any clear separation between the "I" and the "We."

While notions of partiality and permeability of the self are commonplace in much "nondialectical" theory, Castoriadis wasn't satisfied with naming a liminal space of indeterminacy. [30] Instead, he proposed an ontologically irreducible distance between individual and society, as well as subject and object more generally—with a distinct purpose. Like Lacan, Castoriadis saw great significance in the emptiness of the psyche and the inherent chaos that lay at its heart. Truly autonomous people and societies would be able to break through their baggage and inhabit the "meaningless" void, freely able to interrogate themselves, their institutions, and systems of governance. Fully recognizing the idealism of this vision, Castoriadis set forth his proposed freedom as more of a horizon than an actually achievable goal, but a worthy objective nevertheless. Within this space of "nothingness," a new social imaginary might develop, based on an egalitarian "questioning."

Creativity was the centerpiece of this democratic vision, and it's this radical notion for which Castoriadis is most remembered. If a society truly could unburden itself of preconceptions and take control of its destiny, it would clear the way for ideas to emerge *ex nihilo*. [31] Underlying this premise was a novel distinction Castoriadis drew between *difference* and *otherness*. To Castoriadis, the recognition of difference derives from a discernment among articles (i.e., one is different from another) and depends on existing rules or knowledge. In contrast, he saw otherness in the gap between known and unknown—and hence a space for utopian possibility. Clearly, this is a tricky leap. To get there, Castoriadis first dismissed conventional notions of the self and being, which to him were mired in already existing ideas of purpose or "determinacy." Instead, Castoriadis proposed a radical imagination based on indeterminate being and social formations. The radical imagination was not to be found through forms of creativity that improve what already is or synthesize ideas into seemingly new forms. The *ex nihilo* creativity Castoriadis wanted to see could not be predicted (or even fully explained).

I know what you're thinking. Who needs *more* unpredictability in these anxious times? Is this idealism an evasion of politics altogether? Castoriadis was hardly naïve about this. For this reason, his thinking sheds light on today's reactionary populism. Right now, many Americans feel torn between beliefs that honest work will pay off and worries that this age-old promise may be false. They can either blame themselves for not trying hard enough or point to whatever institution their ideology vilifies. Even the nation's two-party political structure has become a recent scapegoat. Castoriadis said that such thinking is short-sighted and that people should look in more elemental terms at themselves and society, to deeply question all of it, and to realize that a different way of doing things was simply a matter of doing so. But at the same time, opening oneself to this possibly also is fraught with the anxiety that freedom brings. Fortunately, one doesn't have to walk this road alone. Understanding the utopian dimensions of society allows people to creatively make progress together.

Progress Versus Utopia

"What if the 'idea' of progress is not an idea at all, but rather the symptom of something else?" This was the question posed by Fredric Jameson in his essay "Progress Versus Utopia: Can We Imagine the Future?"[32] Jameson said that the idea of utopia had suffered the same fate as the imagination, with many people seeing both as inconsequential fantasy. Not helping matters was the failure of many socialist utopias and widespread disillusionment with the "American Dream." To Jameson, the problem boiled down to the way people imagine time. Frantically caught up with life in the present, people's attention is overloaded by media imagery and their own struggles to get by. Constant

"Breaking News!" on cable TV and incessant iPhone alerts cause many to forget the past, effectively rendering history as "dead." But they also can't envision the future because it's so unpredictable. Even science fiction doesn't help, since it tends to depict the present as already passed (and therefor unchangeable). Ultimately, this takes a toll on the collective psyche, manifest in disillusionment and a loss of utopian aspiration.

None of this is discussed very much in public, of course. People may have lingering feelings that something is wrong, but they can't quite name what it is. Little wonder that popular culture often seems a good way to escape. Stories always have taken people away from the everyday—and most people like things that way. For hundreds of years, beliefs persisted that artistic enjoyment stood above contentious matters like political dispute (see Chapter 5). Kantian aesthetics held that art operated though a form of "disinterested" contemplation separate from other mental faculties. The sense of beauty had its own inexplicable mechanisms, for example—supporting the idea that art came from special people and should be housed in protected places. Hence, even today many still see creative works as some of the only places to get some peace in a troubled world.

Fair enough. Many people indeed do put political matters out of their minds when they enjoy the diversion of a movie or book. Unfortunately, believing creative expression is not political means ignoring the ideologies and values embedded there nevertheless. Aristotle famously made this point in arguing that all of life is political.[33] And everyone knows that for centuries art conveyed retrograde attitudes about gender, race, wealth, and power—repeated with such consistency that many viewers simply accepted them as fact. The same things happened in mass culture—especially news coverage—as Donald Trump so often pointed out. Let's set aside for a minute the pleasurable aspects of aesthetic appreciation. But what can't be set aside is this: that as soon as an artwork leaves an artist's hands, it enters a world of people, groups, institutions, and social forces that do things with the work. Thinking, evaluating, and talking about a song or image are part of this process. So are the acts of sharing, praising, or selling it. Sooner or later, preferences and choices enhance (or diminish) the currency of the artwork. And it's hard to see this as outside the influence of taste, belief, and the assignment of value. Put another way, because art derives from a meaningful exchange between people, it always reflects their agreements and disagreements, which are the genesis of political inclination.

Creative people, their imaginations, and works can play an important role in working through contentious issues. This operates through what Jameson called the "political unconscious." And it comes out all the time in books, popular media, and artworks. Think of hit movie series like *The Avengers, Harry Potter, The Hunger Games*, and *Star Wars*—all of which hinge on rebellion against an imperious regime. Recent TV series about resistance also pervade the mediascape in shows like *Game of Thrones, The Handmaid's Tale, Mr. Robot, Narcos, Sons of Anarchy, The Vampire Diaries*, and *The 100*. The history of alternative or critical content in entertainment goes back to

Hollywood's earliest days—sometimes reflecting political leanings within the industry, but also showing how cleverly resistance can be monetized. And don't forget that overt propaganda has its own illustrative past—notably as used by the Nazi regime during World War II, but also even today by governments like the U.S. itself.

And, of course, none of this includes the critical perspectives offered by artists. It's no secret that creative people can be troublemakers—for better or worse. The philosopher Friedrich Nietzsche put this bluntly in writing, "Art must strive to tear away the protective veil of frozen concepts, words and distinctions which underpin common sense."[34] Throughout the twentieth century, artworks helped change how Americans thought about many issues. Examples include the Film and Photo League's battles over poverty and corporate greed in the 1930s, insurgent rhythm and blues emerging from African-American communities in the 1940s, the alternative "Beat" culture in the 1950s, civil rights and anti-war art of the 1960s, and the Chicanx mural movement of the 1970s. The Guerrilla Girls put a spotlight on artworld sexism in the 1980s by asking "Do Women Have to Be Naked to Get into the Met?" Due in large part to the activism of ACT UP and its Gran Fury art collective, government neglect of the HIV/AIDS crisis was reversed in the 1990s. Responses to globalization and the War on Terror have been among the themes of art activism in the new millennium, often aided by the internet, as well as collectives such as the Atlas Group, Collective Shout, Pussy Riot, and Superflex.

There are lots ways to think about social change. One on hand, it's understandable that many people see utopia as a faraway dream. Feelings of powerlessness and alienation abound in a society smothered by a creeping absolutism, and in which buying something often seems the only way of expressing oneself or making a choice. For many people, politics has been reduced to another set of *Daily Show* gag-lines, which may or may not affect their lives. Philosopher Alain Badiou wrote about this in describing a widening disappointment with "politics" as commonly presented in the news—which remains the main way most people are made aware of their political world. To Badiou, many citizens now harbor personal views that they keep to themselves—what he called a "metapolitics" of how they would like to see governance operate. People want something different, but see no signs of it in official political representation.[35] So they withdraw, do nothing, or look for someone to blame. With so many disaffected "angry Americans," it's little wonder a reactionary populist might rise to the nation's highest office with promises to "blow up Washington" and "drain the swamp." The U.S. now has seen the consequences of this in heightened division, greater anger, and an even more pronounced disaffection. But still the underlying metapolitics remains. While dormant and rarely addressed, utopia waits to be awakened.

So what do you do? Thinking about a better world can seem a mighty tall order in an age of big government and multinational corporations. How can a single person make a difference? Maybe the answer is as close as your friends at the lunch table. The Brazilian educator Paulo Freire said that it all comes down to "dialogue."[36] Every important change that has ever taken place has started with a few people sharing ideas with each other—talking, but also listening. Even neuroscientists now agree that the process of exchanging views is one of the best ways of testing them out, as discussed in Chapter 10. When people articulate their beliefs and debate issues with others, they bring their partial understandings of complicated matters into view. Working together, people can piece together a more complete picture—and maybe even a different way of doing things. This can be the beginning of utopian thinking of the kind a growing list of contemporary thinkers now suggest.

In *Cruising Utopia*, performance scholar José Esteban Muñoz proposed a practical approach to utopian possibility through the experience of artworks. While motivated by abstract idealism, a workable utopia needed a concrete grounding in historical consciousness and collective understanding, according to Muñoz. Works of art could bring people together through a common aesthetic experience. In the right circumstances, this could trigger shared feelings of what he called a "not-yet consciousness" of the kind often seen in queer performance art, but also more generally. In this premise, Muñoz was highlighting the powerful potentials of cultural works in activating the radical imagination (Castoriadis), political unconscious (Jameson), metapolitics (Badiou), and dialogue (Freire). In making this argument, Muñoz expressed a cautious optimism about prospects for overcoming dissatisfaction and discomfort. He wrote: "We must strive, in the face of the here and now's totalizing rendering of reality, to think and feel a *then and there*."[37]

Complex Systems

This book has sought to upend conventional thinking about creativity and imagination. Not that misconceptions are anyone's fault. Expansive ideas often sweep in other values and beliefs. And in anxious times, it's perfectly understandable for people to seek comfort in what seems familiar. The creative imagination is something everyone feels they know, especially as it links to U.S. ideals of individualism, personal expression, and, more recently, to money-making. Survey after survey show that most Americans value creativity, also support art, and believe both should be taught to kids in schools. Business groups and government promote the creative industries as an economic panacea, with internet technology championed as a revolutionary enabler. Meanwhile, people in all manner of non-artistic fields now look to creativity as the key to innovative thought.

Creativity has been the recent subject of countless summits, conferences, and forums sponsored by companies such as Adobe, Facebook, Google, and LinkedIn, as well as groups like the Center for Creativity, Creative Time, Ignite Creativity, Inspire Creativity. More than 200 TED talks have taken on the topic as well. A sample of policy books on the subject includes: *The*

Creative Citizen Unbound by Ian Hargreaves and John Hartley; *The Creative City: Cultural Politics and Urban Regeneration* by Alessia Usai; *The Creative Industries: Culture and Policy* by Terry Flew; *Creative Industries and Economic Evolution* by Jason Potts; *Creative Economy and Culture* by John Hartley, Wen, and Henry Siling Li; and *Cultural Policy: Management, Value and Modernity in the Creative Industries* by Dave O'Brien.[38] If this discourse has a common theme, it is that the new millennium calls not only for creative individuals and industries but for a "creative society."

How to build a creative society? The challenges seem daunting. Positive slogans aside, most people don't feel creative or special. Artistry is something that other may people have, but the bulk of Americans don't see it in their own work or leisure. Corporations in the U.S. now worry about running short of new ideas, even as other nations seem to be doing better. Schools and universities say they want to help the problem, but tight budgets force them to favor more "practical" courses of study. And at a time when many citizens are wary of "big government," state and federal programs supporting art and culture seem to be getting the axe. Even the celebrated creative explosion occurring online turns out to be skimming revenues from creators while profiting hugely from the digital footprints users leave behind. These contradictions hardly are a secret, witnessed in a rising chorus of public and private groups describing a crisis in American creativity and innovation.

Despite all of the hand-wringing, there has been little momentum on the creativity front in the U.S. The big issue seems to be public policy. As countries around the globe have launched national creativity initiatives, the U.S. finds itself paralyzed by inaction. The reasons aren't very complicated. They have everything to do with contradictions in the American psyche discussed throughout this book: the valorization of individual over collective interests, a preoccupation with ownership and accumulation, and an obsession with dominance and winning—all complicated by the nagging anxiety that everything is at risk. As a consequence, America falls back on conventions of behavior and governance, much as Castoriadis described. This is the very definition of repetitive and derivative thinking diametrically opposed to innovation and creativity. It's important to recognize this isn't a matter of liberal or conservative ideology, per se. It's a failure of imagination itself. Throwing more money at the NEA only puts a Band-Aid on the problem. But leaving the matter to business doesn't help either.

As a first step, let's think about what might be done to change the way people and groups think about creativity and how institutions might help the matter. It's important to recognize that people are influenced by many factors, so looking at just one or two of them may not make much difference. Individuals themselves vary wildly in their backgrounds, interests, and inclinations. In turn, they further differentiate when associated with groups, social identities, organizations, and institutions. In larger terms, such factors as ethnicity, religious faith, and nationality affect populations. All of these come into play in the operations of creativity in a given society. "Level analysis" is one way social scientists come to terms with this complexity—breaking societies down into categories of scale. Micro is the smallest unit, usually referring to individuals, households, and neighborhoods. Meso or mid-levels account for organizations, communities, and towns. Macro units are the biggest at the national, societal, or global level. But, of course, level analysis is merely an abstraction, rarely recognized by people in their daily lives—although Americans do get a hint of level analysis in a political system organized into local, state, and federal components. Keep in mind that the three levels are constantly interacting, producing ever-changing and sometimes unpredictable outcomes. They function as "complex adaptive systems"—an expression coined in the sciences during the 1980s to describe systems such as cities, families, ecosystems, markets, and traffic flows, in which multiple scales of variation and interaction occur.

Complex systems become adaptive in their dynamic interactions. According to Nobel laureate and physicist Murray Gell-Mann (one of the theory's originators), complex adaptive systems resist easy analysis because of their apparently random behavior.[39] No single mechanism can control everything, and unforeseen consequences often follow actions (the "butterfly effect"). But this doesn't mean systems can't be understood. Gell-Mann proposed that degrees of complexity could be calculated and patterns discerned in even the craziest configurations. Although exact answers might be elusive, their likelihood could be identified. "The history of the universe is by no means determined," Gell-Mann observed, but it does offer "probabilities for histories."[40] This led some scholars to believe that complex systems are better described as networks or swarms than as linear cause/effect models.

Think about the complexity of creativity using level analysis as a start. At the micro level are gallery artists, commercial producers, hobbyists, and people simply being creative in their daily lives. Meso entities include the creative industries, non-profits, schools, and media outlets that organize and encourage artistry. Macro level structures reside in the overarching economic and political systems around all of this, plus the governmental policies that affect creativity. These three levels work together, with elements within each adapting to each other. Because complexity theory focuses on systems, it emphasizes relationships rather than specific individuals, institutions, or policies. Hence, artists, galleries, or funding agencies are seen more as *effects* of the system than as entities on their own. Remember Elizabeth Gilbert saying "Creativity is not your baby. If anything, you're its baby"? Complex adaptive systems help explain her premise, as well as the concepts of knowledge sharing and distributed creativity discussed earlier in this book.

Perhaps all of this seems a little daunting—putting creativity in the context of a swirling universe of people, groups, and institutions. That isn't my purpose. It's simply that such a multidimensional and dynamic topic deserves some careful thought and reflection. All too often, creativity is reduced to romantic ideals or corporate boosterism, especially in the context of the creative industries. This makes it all the more important to understand the topic as something both within individuals and produced by the societies around them. It also means seeing creativity as many, many different things—and, for that reason, a cause for wariness when creativity is boxed into convenient packages or definitions. Any overdetermination of artistry in this way usually means that someone is doing so for their own reasons. Such is the case in creativity industry discourse, the very name of which narrows discussion to monetization. Media critic Terry Flew has pointed out that the very idea of a creative industry is a peculiar abstraction. No painter or sculpture ever would say they work in the art industry. The concept simply doesn't apply. Likewise, "People who teach, research, or study in universities do not generally refer to themselves as working in the 'education industry.'"[41] By the same token, critics who condemn the creativity industries as the face of a creeping neoliberalism make a similar error. While the monetization of creativity certainly is no laughing matter, reverting to industrial analytic models doesn't fully capture what is happening.

Complex adaptive systems like creativity require attention to as many variables as possible, or what Gell-Mann called "schema," to identify discernible regularities as closely as one can. Let's look at a complicated issue like homelessness. At first, many thought the displaced could be helped simply by building more housing. But it soon became clear that to get off the streets, homeless people needed much more: education, job training, social support, and sometimes mental health services. Simply addressing the symptom didn't tackle the problem in holistic terms. The situation is similar with creativity. Focusing on the creative industries doesn't describe how they come about—or why they matter. Richard Florida's *The Rise of the Creative Class* has been criticized for just this reason, especially in Florida's heavy reliance on vocational data. Employment certainly may be *one* indicator of population characteristics, but it hardly captures the complexity of nuanced matters such as artistry, inventiveness, or divergent thinking. To his credit, Florida tried to augment job statistics with demographic data on technology, diversity, and even what he termed the "gay index" in creative neighborhoods. But he was unable to capture any of the "human capital" (knowledge, experience, personality traits, habits) of artistic people, or even define creativity in any meaningful way.[42]

"Multitudinous creativity" is the term used by Giuseppe Cocco and Barbara Szaniecki as an alternative to creative industry discourse.[43] They use multitudinous in two senses—to describe *multiple* forms of creativity existing within a *multitude* (entire population) of people. Not only should creativity be approached in terms of human difference and alterity, but its newer technologically enabled forms enable creativity's democratization as never before. Fully cognizant of the monopolistic exploitation of "free" online culture, Cocco and Szaniecki nevertheless hold out that possibility of something better. They have argued for more inclusive approaches to creativity entailing "the integration of intuition, expertise and complex matters, synthesis, level perception of forms; in short, 'creativity' intended as understanding, feeling, and shared complex objects and information."[44]

As you might expect, prescriptions abound about how to build a creative society. Most are heavy on theory and short on details. This is an understandable dilemma when dealing with complex issues, especially with ones prone to contention. As the foregoing discussion has detailed, the interdisciplinary scope of creativity is breathtaking—encompassing the arts and humanities, behavioral and cognitive research, economics and politics, as well as the social sciences and technology. This breadth confounds any fully inclusive analysis, just as it has deterred comprehensive planning. Keep in mind that public policy always is a lagging indicator of cultural change, owing to the necessity of consensus or at least political will. But the current momentum for creativity can't be denied. The trick is to maintain a degree of openness in moving forward.

A complex system like creativity calls for an equally multifaceted approach to advocacy. People can play a role in their own lives (micro level) by recognizing that creativity is both an individual substance and a collectively produced one. While many artists have built careers and professional recognition on their personal achievements, most people working in the arts are well aware of the collective support required to accomplish things. Just as combined efforts have moved societies forward throughout history, it's necessary to recognize just how important other people are in creative work. Teams and collectives are some forms of this, but so are the roles of audiences, markets, and institutions. And as discussed earlier, many ideas that at first blush may seem "personal" or "original" usually derive from the individual's history, culture, and social circle. "Sharing" can be active or passive, but it goes on whether one realizes it or not. Admitting this to oneself might make one a little less guarded and competitive in creative endeavors. Supporting teamwork and collaborative endeavors can help too.

Most artists already collaborate in making works or bringing them to the public. But contentions among artists, curators, and other gatekeepers in the system remain problematic—even as these stakeholders often share similar goals. Artists already established in their fields might make more efforts to engage in policy discussions, claiming a place at the table based on their expertise. Certainly, those working in education can help by emphasizing distributed aspects of creative work and the implicit connections among producers and supporting institutions. Collaborative learning and group projects seem great classroom devices for this. By the same token, art history could focus less on individual brilliance and more on broader patterns, influences, and social contexts.

The creativity industry buzz has created an unprecedented opportunity for institutional (meso) level change. Despite its sometimes exaggerated claims, this discourse suggests that older welfare arguments for the arts support premised on fiscal neglect ("market failure") may need some updating.[45] This is why the new language of cultural philanthropy often speaks of "investment" rather than benevolent humanism. While some critics lament this materialistic turn, similar impulses have led museums, symphonies, and other non-profits to expand outreach to groups previously unserved, while also diversifying programming itself. It is also why advocacy groups such as Americans for the Arts, the Arts Action Fund, the National Assembly of State Arts Councils, and the National Endowment for the Arts have so quickly adopted the creative economy as a rationale. Even as some artists worry that the quality and critical dimensions of art might be compromised, the change also has much to offer. For one thing, institutional legitimacy has the potential to raise the "professional" status of artists and other creatives. All too often, creative work is still seen as ephemeral or supplementary to the "real" economy. Along with the recessionary after-effects on business, such attitudes contribute to the undervaluing of creatives. Professions like law, medicine, and even teaching bring with them the expectation of full-time work, job stability, and related benefits. Creative jobs need to be seen the same way.

Diversity remains key in this institutional arena. Not simply in terms of group identity, but also toward the kind of creative work produced. Economization narrows the range of artistry to what is saleable or acceptable. This is apparent in definitions of the creative economy leaning heavily on commercial media, advertising, and consumer product development. Here, schools and non-profits have a role in maintaining an open space for creative expression and research. Administrators may want creativity, but not fully understand it. For advocates of creativity, this mean standing up for arts and humanities offerings as vital components of a balanced education. This has a practical rationale as well, given the tendencies of profit-seeking to foreclose experiment and innovation. The problem lies not in capitalism, per se, as many critics of the creativity industries argue. Rather, it links to the nervousness in both corporate non-profit sectors over potential losses from risky or experimental endeavors.

The anxious drive for short-term return is having a devastating effect on innovation, and not only in creative fields. Corporate funding for basic research has been dropping in the past decade. The dark side of the "innovation crisis" is that even in science and technology, companies only spend on the applied side of new products. This is the real reason fewer new ideas are emerging. *New York Times* economics correspondent Eduardo Porter has tracked this situation in an ongoing series of articles. "American corporations, constantly pressured to increase the next quarter's profits in the face of powerful foreign competition, are walking away from basic science," Porter recently explained. [46] "The number of American patent applications keeps rising. Yet increasingly divorced from the scientific advances on which technological progress ultimately rests, the patenting rush looks less and less like fundamental innovation."[47] Federal intervention could do a lot to help the innovation crisis at the national (macro) level. But political consensus seems unlikely, especially in the Trump era. Few people recognize the essential role public support plays in this—already paying for more than half of all basic research in the U.S. On average, economists say that every dollar of taxpayer money returns eight times that amount in the marketplace. Unfortunately, government funding has been cut in the past year to below its nominal 50 percent level, leaving agencies like the National Institutes of Health and the National Science Foundation (not to mention the National Endowment for the Arts) at their lowest levels in a decade.[48]

Finally, the creative economy could use a cabinet-level federal agency—as seen in already existing U.S. Departments of Agriculture, Education, Energy, and Transportation. In recent years, America's $702-billion arts and culture industry has grown larger than other sectors with cabinet offices.[49] It's time to recognize the significance of this. Most countries competing with the U.S. have such entities devoted to the advancement of national culture, including Australia, Brazil, Canada, China, France, India, Italy, Norway, South Africa, Sweden, and the United Kingdom, among scores of others. So again, one needn't use a welfare model to justify art and culture as a "public good." Global standing and fiscal pragmatism make the case. A federal Department of Media and Culture could coordinate public and private-sector arts and creative activity, encompassing the Corporation for Public Broadcasting, Institute for Museum Services, and the National Assembly of State Arts Agencies, while taking some of the political heat off the frequently criticized National Endowments of the Arts and Humanities. Like those latter agencies, attention could be on national heritage and cultural preservation as well as evolving fields. The new cabinet office would also liaise with the entertainment and internet industries to promote best practices—for example, advocating fair compensation for creative producers.

This final chapter has addressed what is perhaps the central element of the creative process: imagination. While products of artistry often take material form, nearly every creative product begins with an idea in the creator's mind. This is hardly an innocent or disinterested process, since most of what people think and believe comes from their lived experiences and those they encounter. Even fantasy "imaginary" worlds bear some features of the "real" one, just as imaginative journeys also often persuade people to take action in their real lives. Awareness of these processes is sometimes called the sociological imagination: the understanding that individuals exist within a world of others and their ideas. Despite outside influence, each person has the ability to make choices and decide what to do, although many fail to exercise the autonomy of independent thought. Indeed, sometimes a better world seems a faraway dream rather than an achievable reality. This is where imagination can be important in keeping alive a multitude of possibilities even in the face of what might look like limited options. As this book has shown, creativity can be defined by possessive individualism and the market, or it can be used to confront such regressive tendencies. Creativity can subvert progressive change, or it can be used in the interest of humanistic causes. The key lies in recognizing problematic tendencies when they show up and actively working to challenge them.

Obviously, these latter ideas are the product of some imagination—that is, some vision of a possible future. It's important to point out that just about any activity starts this way: with an idea. This is part of what gives imagination its universal appeal. My biggest point has been that imagination and creativity do not emerge from thin air. They are relational phenomena between makers and the world around them, rather than magical or even *ex nihilo* occurrences. There is nothing hard to understand about this. A creative society requires an imaginative citizenry, capable of responding to a changing world. If recent history has shown anything, it is that many Americans are fed up with business-as-usual. They want to be creative, but struggle to do so. They want something different, but hold back—arrested by the anxiety that they are inadequate, unimaginative, or otherwise flawed. What's needed is neither radical nor revolutionary, but rather a lessening of constraint so that public imagination can flourish. As religion professor Eddie S. Glaude, Jr. recently summed it up:

> Imagination isn't the stuff of fantasy. It's much more substantive and powerful. In fact, imagination is the key to a robust sense of the good life. It motivates us to act for what is possible and not settle for things as they are, and helps us to see the fullness of the humanity of those with whom we live.[50]

Notes

1 Libby Plumber, "Google's DeepMind Creates AI with 'Imagination,'" *Wired* (Jul. 26, 2017) www.wired.co.uk/article/googles-deepmind-creates-an-ai-with-imagination (accessed Mar. 1, 2019).
2 Maria Popova, "Ursula K. Le Guin on Redeeming the Imagination from the Commodification of Creativity and How Storytelling Teaches Us to Assemble Ourselves," (Mar. 13, 2017) www.brainpickings.org/2017/04/13/ursula-k-le-guin-operating-instructions-words-are-my-matter (accessed Mar. 2, 2019).
3 Henry A. Giroux, *The Violence of Organized Forgetting: Thinking Beyond America's Disimagination Machine* (San Francisco: City Lights Books, 2014) p.11.
4 Brad Evans and Henry A. Giroux, "The Violence of Organized Forgetting," *New York Times* (Jun. 20, 2016) www.nytimes.com/2016/06/20/opinion/the-violence-of-forgetting.html (accessed Jun. 19, 2019).
5 "Imagination," *Oxford Dictionaries* (2018) https://en.oxforddictionaries.com/definition/imagination (accessed Mar. 4, 2019).
6 Kenneth Robinson, *Creative Schools: The Grassroots Revolution That's Transforming Schools* (New York: Penguin, 2015) p. 112.
7 "Imaginative," Thesaurus.com (2017) www.thesaurus.com/browse/imaginative (accessed Mar. 6, 2019).
8 "How to Use 3 Hypnotic Power Words to Covertly Increase Your Conversion Rates," neilpatel.com (2017) https://neilpatel.com/blog/3-hypnotic-power-words (accessed Jun. 16, 2019).
9 "Ursula K. Le Guin on Redeeming."

10 *Westworld*, "The Original," Episode 1.01 (2016). Television Series (Los Angeles: Bad Robot and Jerry Weintraub Productions, 2016–present).
11 *Westworld*, "Trace Decay," Episode: 1.08 (2016).
12 "The Original."
13 *Westworld*, "The Bicameral Mind," Episode 1.10 (2016).
14 "Autonomy," *Oxford Dictionaries* (2017) https://en.oxforddictionaries.com/definition/autonomy (accessed Mar. 6, 2019).
15 "The Bicameral Mind."
16 Jean-Paul Sartre, *The Imagination*, trans. Kenneth Williford and David Rudrauf (1936) (London and New York: Routledge, 2012) pp. 3–4.
17 David Trend, *Worlding: Identity, Imagination, and Reality in a Digital Age* (London and New York: Routledge, 2013).
18 Jacques Lacan, "The Symbolic, the Imaginary, and the Real," in *On the Names-of-the-Father*, trans. Bruce Fink (London and New York: Polity, 2015).
19 *Worlding*, p. 1.
20 Ibid.
21 C. Wright Mills, *The Sociological Imagination, Fortieth Anniversary Edition* (1959) (Oxford and New York: Oxford University Press, 2000) p. 3.
22 Ibid.
23 *The Sociological Imagination*, pp. 6–7.
24 Jordan McKenzie, *Deconstructing Happiness: Critical Sociology and the Good Life* (London and New York: Routledge, 2016) p. 11.
25 Cornelius Castoriadis, *The Imaginary Institution of Society* (1975) trans. Kathleen Blamey (Cambridge, MA: MIT Press, 1987).
26 Alicia Eler, *The Selfie Generation: How Our Self-Images Are Changing Our Notions of Privacy, Sex, and Culture* (New York: Skyhorse, 2017); John Townsend, *The Entitlement Cure: Finding Success in Doing Hard Things the Right Way* (New York: Zondervan, 2015); Bruce Tulgan, *Not Everyone Gets a Trophy: How to Manage the Millennials* (New York: John Wiley, 2016); Jean Twenge, *Generation Me: Why Today's Young Americans Are More Confident, Assertive, Entitled—and More Miserable Than Ever Before* (New York: Atria, 2014); Jean Twenge and W. Keith Campbell, *The Narcissism Epidemic: Living in the Age of Entitlement* (New York: Atria, 2010).
27 *The Imaginary Institution of Society*, p. 72.
28 *The Imaginary Institution of Society*, p. 49.
29 Gilbert Simondon, "The Genesis of the Individual," in Jonathan Crary and Sanford Kwinter, eds., *Incorporations* (New York: Zone Books, 1992) pp. 297–319.
30 *The Imaginary Institution of Society*, p. 131.
31 *The Imaginary Institution of Society*, p. 221.
32 Fredric Jameson, "Progress Versus Utopia; Can We Imagine the Future?" in Brian Wallis, ed., *Art After Modernism: Rethinking Representation* (New York: The New Museum 1984) p. 239.
33 Aristotle, *Aristotle's "Politics": Second Edition*, trans. Carnes Lord (Chicago: University of Chicago Press, 2013).
34 Friedrich Nietzsche as cited in Augustin Harrington, *Art and Social Theory: Sociological Arguments in Aesthetics* by (Cambridge and Malden, MA: Polity: 2004) p. 135.
35 Alain Badiou, *Metapolitics*, trans. Jason Baker (London and New York: Verso, 2005).
36 Paulo Freire, *Pedagogy of the Oppressed, 30th Anniversary Edition*, trans. Myra Bergman Ramos (New York: Continuum, 2000).
37 José Esteban Muñoz, *Cruising Utopia: The Then and There of Queer Futurity* (New York and London: NYU Press, 2009) p. 1.

38 Ian Hargreaves and John Hartley, *The Creative Citizen Unbound* (Bristol, UK: Policy Press, 2016); Alessia Usai, *The Creative City: Cultural Politics and Urban Regeneration Between Conservation and Development* (London and New York: Peter Lang, 2016); Terry Flew, *The Creative Industries: Culture and Policy* (London and Thousand Oaks: Sage, 2012); Jason Potts, *Creative Industries and Economic Evolution* (Cheltenham, UK, and Northampton, MA: Edward Elgar, 2011); John Hartley, Wen Wen, and Henry Siling Li, *Creative Economy and Culture: Challenges, Changes and Futures for the Creative Industries* (London and Thousand Oaks, CA, Sage, 2015); Dave O'Brien, *Cultural Policy: Management, Value and Modernity in the Creative Industries* (London and New York: Routledge, 2015).

39 Murray Gell-Mann, "What Is Complexity?" *Complexity* 1, no. 1 (1995) http://complexity.martinsewell.com/Gell95.pdf (accessed Mar. 7, 2018).

40 Ibid.

41 *The Creative Industries*, p. 185.

42 Eckhardt Bode and Lucia Perez Villar, "Creativity, Education, or What? On the Measurement of Regional Human Capital," *Kiel Working Papers* No. 1958 (Sep. 2014) www.ifw-kiel.de/fileadmin/Dateiverwaltung/IfW-Publications/Eckhardt_Bode/creativity-education-or-what-on-the-measurement-of-regional-human-capital/KWP_1958.pdf (accessed Jun. 11, 2019).

43 Giuseppe Cocco and Barbara Szaniecki, eds., *Creative Capitalism, Multitudinous Creativity: Radicalities and Alterities* (London and New York: Lexington, 2015).

44 *Creative Capitalism*, p. xviii.

45 "Market Failure," *Investopedia* (Apr. 12, 2018) www.investopedia.com/terms/m/marketfailure.asp (accessed Jun. 16, 2019).

46 Eduardo Porter, "American Innovation Lies on Weak Foundation," *New York Times* (May 19, 2015) www.nytimes.com/2015/05/20/business/economy/american-innovation-rests-on-weak-foundation.html?_r=0 (accessed Mar. 10, 2019).

47 Ibid.

48 Ibid.

49 Brett Zongker, "Federal Analysis: Arts, Culture Add $700B to U.S. Economy," *Chicago Tribune* (Jan. 12, 2015) www.chicagotribune.com/entertainment/chi-arts-culture-add-700-billion-to-economy-20150112-story.html (accessed Jun. 16, 2019).

50 Eddie S. Glaude, Jr., "America Is Suffering a Crisis of the Imagination," *Time Magazine* (Feb. 24, 2016) http://time.com/4235720/democrats-sanders-clinton-black-voters (accessed Mar. 14, 2019).

Index